INSECTS AND PESTICIDES

INSECTS AND PESTICIDES

Edited by

A. B. Saxena

ANMOL PUBLICATIONS PVT LTD

NEW DELHI - 110002 (INDIA)

ANMOL PUBLICATIONS PVT. LTD.
4374/4B, Ansari Road, Daryaganj
New Delhi - 110 002
Ph.: 23261597, 23278000
Visit us at: www.anmolpublications.com

Insects and Pesticides

ISBN 81-7488-352-5

PRINTED IN INDIA

Published by J.L. Kumar for Anmol Publications Pvt. Ltd., New Delhi - 110 002 and Printed at Mehra Offset Press, Delhi.

Preface

This fascinating title has been carefully organised to meet the long-felt needs of increasingly large number of those who deals with different aspects of Entomology. It provides a balanced and integrated treatment of the entire field of entomology. The title is intelligible to the educated laymen but it deals with some complex ideas. It is an adequate text for all requirements in this area for most university students. This ground work with its many new and unfamiliar names is, at any time, dry, uninteresting and difficult, especially as the beginner does not at first see the necessity for it. We have endeavoured to make it interesting by introducing as much comparative morphology and phylogeny, as possible in the limited space at our disposal.

The primary aim throughout has been clarity so as to make the text selective and meaningful. We hope also that general Entomologists and students of Agricultural Entomology will find in this book valuable and recent informations about the subject and also allied branch of the subject. The book provides a nucleus around which the teacher can plan a successful course without much bearing on complementary and supplementary readings. The theme of the book is novel and for the first time the various chapters in different volumes have been organised reviewing the available literature. All the topics have been illustrated with line diagrams, graphs and tables wherever necessary.

We hope that the book will definitely prove to be a boon to teachers, students and research scholars.

Author

Contents

1

How to know about Pesticides

Everything including the earth, is composed of chemical elements. These include such familiar elements as oxygen, nitrogen, iron, sulfur, and hydrogen. The smallest part of unit of an element is the atom. An element is pure in that all of its *atoms* are alike. It is not possible to chemically change an element. A *compound* is a combination of two or more different elements. Water, salt, and DDT are examples of compounds. When different atoms combine or join, they form a molecule, the smallest part or unit of a compound. And because a compound is a combination of different elements, chemical changes can be made in it by changing the combinations of elements. Elements have a certain way of combin- ing with other elements, and only certain elements will combine. Conversely, certain elements will not combine. Combinations of two or more elements bound together by chemical bonds are *chemical compounds.* The most familiar compound is *water,* a mixture of hydrogen and oxygen, bound together by chemical bonds, which everyone recognizes as H_2O. Proper chemical terminology notes that H_2O is composed of two atoms of hydrogen bonded to one atoms of oxygen to form one molecule of water.

To aid in the writing of chemistry, a chemical shorthand has been developed by chemists. Each elements was given a *symbol*, which makes both the writing and the setting of type for printed matter simpler and easier to read. These symbols are frequently the first letter along or the first letter plus another letter in the element's name.

TABLE 1.1

Chemical Elements from Which Pesticides are Made.

Element	*Symbol*	*Symbol derivation*
Arsenic	As	
Boron	B	
Bromine	Br	
Cadmium	Cd	
Carbon	C	
Chlorine	Cl	
Copper	Cu	Cuprum
Fluorine	F	
Hydrogen	H	
Iron	Fe	Ferrum
Lead	Pb	Plumbum
Magnesium	Mg	
Manganese	Mn	
Mercury	Hg	Hydrargyrum
Nitrogen	N	
Oxygen	O	
Phosphorus	P	
Sodium	Na	Natrium
Sulfur	S	
Tin	Sn	Stannum
Zinc	Zn	

Some of the elements or atomic names and the chemical symbols and illustrating molecues are given in the table. The table contains only those elements used in the making or synthesis of *pesticides.* Indeed, only 21 of the more than 105 chemical elements are used to pesticides. This list can be further reduced by considering those elements used most frequently in pesticides: carbon, hydrogen, oxygen, nitrogen, phosphorus, chlorine, and sulfur. Some, however, may include the metallic and semimetallic elements, iron, copper, mercury, zinc, arsenic and others. And because only about one-half the elements appearing in the table make up the bulk of the most frequently used pesticides, it becomes a simple matter to learn their symbols. For practical generalizations, almost all pesticides are *organic* compounds, that is they contain carbon in their molecules.

Only a few contain no carbon and are thus identified as *inorganic* compounds.

CHEMICAL FORMULAS

A *chemical formula* is the printed description of one molecule of a chemical compound. There are several types of chemical formulas the must be distinguished from each other, since all are used in this book. The *molecular formula* uses the symbols of the elements to indicate the number and kind of atoms in a molecule of the compound, for example, H_2O for water and C_6H_6 for benzene. The structural *formula* is written out, using symbols to indicate the way in which the atoms are located relative to one another in the molecule. The structural formulas for water and benzene can be presented as shown in the margin.

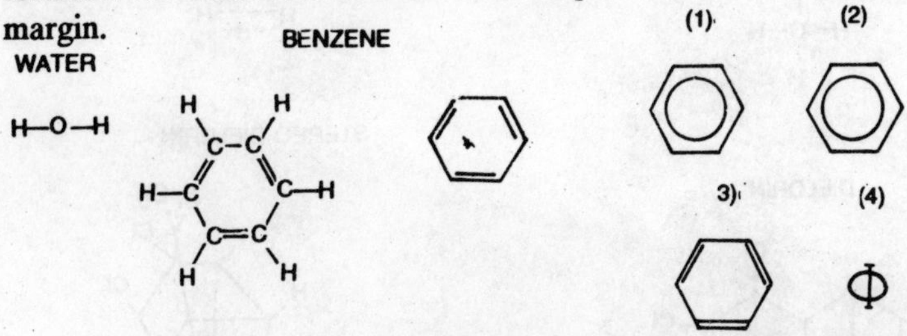

This six-carbon ring, with its six hydrogens as presented, is benzene and, for ease of representation, is usually indicated as a hexagon with double bonds. When other groups have replaced one or more of the hydrogens, the ring is referred to as the *phenyl radical* rather than as the *benzene radical*. For example, turn to the section on DDT. The old chemical name of DDT contains the words di-phenyl, which refers to the two benzene rings with other groups attached.The *benzene* or *phenyl* ring is found quite frequently in pesticide structures, and, because it is an awkward structure for printers to set from type, simpler ways of presentation have been devised and are commonly used in pesticide literature. Two of these have already been shown. The four remaining designs are : (1) the hexagon containing a broken circle or oval; (2) similar but with an unbroken circle or oval;

(3) the hexagon with the three double bonds extending to its sides; and, finally, (4) the greek letter *phi*, ϕ.

The *structural formula* is a two dimensional representation of a three-dimensional object. The actual shapes of organic molecules are not shown by structural formulas, but one can become familiar with these structures without knowing their shape and spatial design. *Stereo* or *three-dimensional formulas* are printed so that the reader can visualize the depth and spatial conformation of certain molecules. The structural and stereo formulas for methane and dieldrin are illustrated.

METNANE

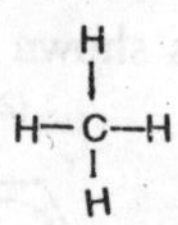

STREO METHANE

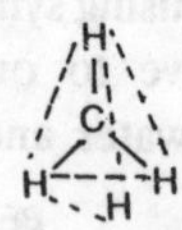

DIELDRIN

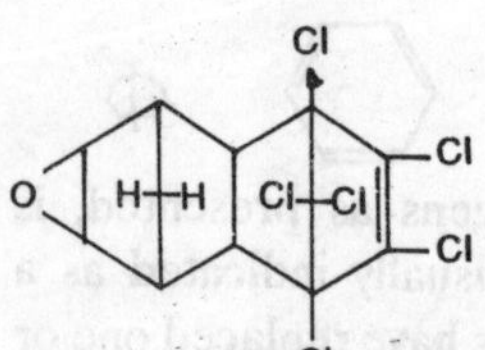

STEREO DIELDRIN

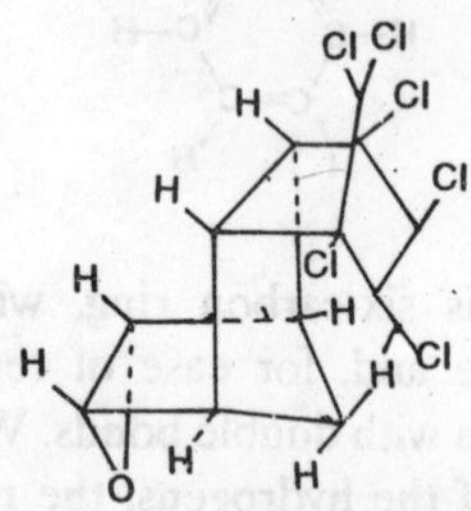

Generic Terms, Trade Names and Chemical Names

The term *pesticide* is an all-inclusive but nondescript word meaning "killer of pests." The various generic words ending in-*cide* (from the Latin -cida, "killer") are classes of pesticides, such as *insecticides* and *herbicides*. Table 1-2 lists the various pesticides and other classes of chemical compounds not commonly considered pesticides.

The term *biocide* is frequently used by laypersons to connote the adverse effects of pesticide and is often used as a synonym for pesticide. But biocide is not a scientific word and has no exact meaning. It is usually used to describe a chemical substance that kills any living plant or animal, and its tone is sensational and emotional rather than scientific. In contrast, the terms in Table 1.2 are precise and offer useful distinctions between various types of chemical substances.

Pesticides are legally classed as *economic poisons* in most state and federal laws and are defined as "any substance used for controlling, preventing, destroying, repelling, or mitigating any pest." Thus, pesticides include groups chemicals that do not actually kill pests (bottom part of Table 1.2). However, because they fit practically as well as legally under this umbrella term, they are included.

TABLE 1.2

A List of Pesticide Classes their Use and Derivation

Pesticide class	*Function*	*Root-word derivation*[a]
Acaricide	Kills mites	*Gr. akari*, "mite or tick"
Algicide	Kills algae	*L. alga*, "seaweed"
Avicide	Kills or repels birds	*L. avis*, "bird"
Bactericide	Kills bacteria	*L. bacterium; Gr. baktron*, "a staff"
Fungicide	Kills fungi	*L. fungus; Gr. spongos*, "mushroom"
Herbicide	Kills weeds	*L. herba*, "an annual plant"
Insecticide	Kills insects	*L. insectum*, cut or divided segments"
Larvicide	Kills larvae (usually mosquito)	*L. lar*, "mask or evil spirit"
Miticide	Kills mites	Synonymous with *Acaricide*
Molluscicide	Kills snails and slugs (may includes oysters clams mussels)	*L. molluscus*, "soft-or thin-thread"
Nematicidekills nematodes*nematoda Gr nema "thread"*		
Ovicide	Destroys eggs	*L. ovum*, "egg"
Pediculicide	Kills lice (head, body, crab)	*L. pedis*, "louse"
Piscicide	Kill fish	*L. pedis*, louse"
Predicide	Kills predators (coyotes usually)	*L. piscis*, "a fish"
L.praeda "prey"		
Rodenficide	Kills rodents	*L. rodere*, "to gnaw"
Silvicide	Kills trees and brush	*L. silva*, "forest"
Slimicide	Kills slimes	Anglo-Saxon *slim*
Termiticide	Kills termites	*L. termes*, "wood-boring worm"
Chemicals classed as pesticides not bearing the-cide suffix		
Attractants	Attract insects	
Chemosterilants	Sterilize insects or pest vertebrates (birds. rodents)	

Defoliants	**Remove leaves**
Desiccants	**Speed drying of plants**
Disinfectants	**Destroy or inactivate harmful microorganisms**
Growth regulators	**Stimulate or retard growth of plants or insects**
Pheromones	**Attract insects, or vertebrates**
Repellents	**Repel insects mites and ticks, or pest vertebrates (dogs, rabbits, deer, birds)**

[a]*Gr.* indicates Greek origin; *L.* indicates Latin origin.

Toxicity is another term frequently used in reference to pesticides and other chemicals. Toxicity is given as the LD_{50}, expressed as milligrams (mg) of toxicant per kilogram (kg) of body weight, the dose that kill 50 per cent of the test animals, usually the laboratory rat.

In this book, each pesticide is identified in five ways. For example, let us look at diazinon, a commonly known household insecticide.

At the top is the name, diazinon (1), the *common name* for the compound. Common names are selected officially by the appropriate professional scientific society and approved by the American National Standards Institute (formerly United States of America Standards Institute) and the International Organization for Standardization. Common names of insecticides are selected by the Entomological Society of America; herbicides by the Weed Science Society of America; and fungicides by the American Phyto patho- logical Society. The *proprietary name* (2), *trade name,* or *brand name* for the pesticide is given by the manufacturer or by the formulator. It is not uncommon to find several brand or trademark names given to a particular pesticide in various formulations by their formulators. To illustrate, diazinon is known as Dianon®, Diazide®, Diazol®, Knox Out® FM, Neocidal®, and Sarolex®, among others, in the United States and abroad. Common names are assigned to avoid the confusion resulting from the use of several trade names, as just illustrated. The *structural formula* (3), as mentioned earlier, is the printed picture of the pesti- cide molecule. The long *chemical name* (4) beneath the structural formula is just that, the chemical name. It is usually presented accord- ing to the principles of nomenclature used in *Chemical Abstracts*, a scientific abstracting journal that is generally, accepted as the world's standard for chemical names. And finally, we sometimes give the *molecular* or *empirical formula* (5) that indicates

(1)
DIAZINON

(2)
(spectracide)

(3)

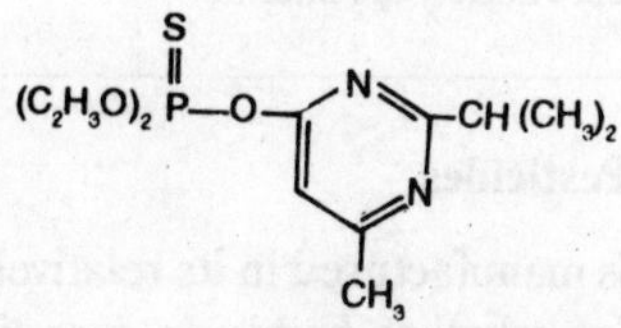

(4)
O,Odiethyl O-(2-isopropyl-4 methyl -6-primidinyl) phosphorothioates

(5)
$C_{12}H_{21}N_2O_3PS$

the various numbers of atoms for comparative purposes. The molecular formula is illustrated only when the structural formula is not known, which is rare.

Because pesticides are used in all countries and must be identified by some readily identifiable name, several organizations of scientific significance in the United States and abroad are involved in their naming, especially their common names.

TABLE 1.3

Abbreviations Used in Connection with the Naming of Pesticides

ACC	American Chemical Society
ANSI	American National Standards Institute
AOAC	Association of Official Analytical Chemists
APS	American Phytopathological Society
BCPC	British Crop Protection Council
BIOS	British Intelligence Objectives Subcommittee
BP	British Patent
BPC	British Pharmacopoeia Commission
BSI	British Standards Institution
CIPAC	Collaborative International Pesticides Analytical Committee
ESA	Entomological Society of America

FAO	Food and Agricultural Organization (of the United Nations)
ISO	International Standardization Organization
IUPAC	International Union of Pure and Applied Chemistry
JMAF	Japanese Ministry for Agriculture and Forestry
US Pharm	United States Pharmacopoeia
WHO	World Health Organization (of the United Nations)
WSSA	Weed Science Society of America

The Formulations of Pesticides

After a pesticide is manufactured in its relatively pure form – the *technical grade material*, whether *herbicide, insecticide, fungicide,* or other classification – the next step is *formulation.* It is processed into a usable form for direct application or for dilution followed by application. The formulation is the final physical condition in which the pesticide is sold for use. The technical grade material may be formulated by its basic manufacturer or sold to a formulator. The formulated pesticide may be sold under the formulator's brand name, or it may be custom-formulated for another firm. As mentioned earlier, pesticides are sold in more than 35,000 formulations. *Formulation* is the processing of a pesticidal compound by any method that will improve its properties of storage, handling, application, effectiveness, or safely. The term formulation is usually reserved for commercial preparation prior to actual use and does not include the final dilution in application equipment. The real test for a pesticide is acceptance by the user. And, to be accepted for use by the homeowner, the grower, or commercial appli- cator, a pesticide must be effective, safe easy to apply, and relatively economic although householders commonly pay 10 to 30 times the price that growers pay for a given weight of a particular pesticide. Price depends to a great extent on the formulation; for instance, the most expensive form of insecticide is the pressurized aerosol.

For satisfactory storage, for effective application, for safely to the applicator and the environment, for ease of application with readily available equipment, and for economy pesticides are formulated into many usable forms. These goals are not always simply accomplished, due to the chemical and physical characteristics of the technical grade perticide. For example, some materials in their "raw" or technical condition are liquids, others solids; some are stable to air and sunlight whereas others are not; some volatile, others not; some are water soluble, some oil soluble, and others may be insoluble in either water or oil. These characteristics pose problems to the formulator, since the final formulated product must meet the standards of acceptability by the user.

More than 98 per cent of all pesticides used in the United States in 1982 are manufactured in the formulations appearing in the simplified classification presented in Table 3-1. Familiarity with the more important formulations is essential to the well-informed user. Let us examine the major formulations used in agriculture, in the home and garden, as well as those employed in structural and commercial pest control.

SPRAYS

Emulsifiable Concentrates

Traditionally, pesticides have been applied as water sprays, water suspensions, oil sprays, dusts and granules. Spray formulations are prepared for insecticides, herbicides, miticides, fungicides, algicides, growth regulator, defoliants and desiccants. Consequently, more than 75 per cent of all pesticides are applied as sprays. The bulk of these are currently applied as water emulsions made from emulsifiable concentrates, sometimes, abbreviated as EC. Emulsifiable concentrates synonymous with emulsible concen- trates, are concentrated oil solutions of the technical grade material with enough emulsifier added to make the concentrate mix readily with water for spraying. The emulsifier is a detergent like material that makes possible the suspension of microscopically small oil droplets in water to form an emulsion. When an emulsifiable concentrate is added to water, the emulsi- fier causes the oil to disperse immediately and uniformly throughout the water, if agitated, giving it an opaque or milky appearance. This oil-in-water suspension is a normal emulsion. There are a few rare formulations of invert emulsions, which are water-in-oil suspensions, and are opaque in the concentrated form, resembling salad dressing or face cream. Invert emulsions are employed almost exclusively as herbicide formulations. The thickened sprays result in every little drift and can be applied in sensitive situations. Emulsifiable concentrates, should remain suspended without further agitation for several days after dilution with water. A pesticide concentrate that has been held over from last year can be easily tested for its emulsifiable quality by adding 1 ounce to 1 quart of water and allowing the mixture to stand after shaking. The material should remain uniformly suspended for at least 24 hours with no precipitate. If a precipitate does form, the same condition may occur in the spray tank, resulting in clogged nozzeles and uneven application. In the home, this can be remedied by adding 2 tablespoons of liquid dishwashing detergent to each pint of concen- trate and mixing thoroughly. In an agricultural or other circumstance, where several gallons of costly pesticide are involved, additional emulsifier can be obtained from the formulator. This should be added to the concentrate at the rate of 0.2 to 0.5 pound for each gallon outdated material. The bulk of pesticides available to the homeowner are formulated as emulsifiable concentrates and generally have a shelf life of about 3 years. Every gallon emulsifiable concentrate contains from 4 to 7 pounds of petroleum solvent, usually one of the more expensive aromatic solvents such as xylene. Within the past three years petroleum solvents have increased in price 300 per cent or more, increasing significantly the costs of formulation of consequently the cost of the formulated product to the buyer. These increases are not likely to stabilize within the foreseeable future.

TABLE 1.4

Common Formulations of Pesticides

1. Sprays (insecticides, herbicides, fungicides)
   a. Emulsifiable concentrates (also emulsible concentrates)
   b. Water-miscible liquids
   c. Wettable powders
   d. Water-soluble, e.g., prepackaged, tank drop-ins
   e. Oil solutions, e.g., barn and corral ready-to-use sprays
   f. Soluble pellets for water-hose attachments
   g. Flowable or sprayable suspensions
   h. Flowable microencapsulated suspensions
   i. Ultralow-volume concentrates (agricultural and forestry use only)
   j. Fogging concentrates e.g., public health mosquito and fly abatement foggers
2. Dusts (insecticides fungicides)
   a. Undiluted toxic agent
   b. Toxic agent with active diluent, e.g., sulfer diatomaceous earth

   c. Toxic agents with inert diluent, e.g., home garden insecticide-fungicide combination in pyrophyllite carrier
   d. Aerosol "dust," e.g., silica aerogel or broic acid in aerosol form
3. Aerosols (insecticides disinfectants, or "germicides")
   a. Push-button
   b. Total release
4. Granulars (insecticides herbicides, algicides)
   a. Inert carrier impregnated with pesticide
   b. Soluble granules, e.g., swimming pool chlorine
5. Fumigants (insecticides nematicides)
   a. Stored products and space treatment, e.g., liquids, gases, moth crystals
   b. Soil treatment liquids that vaporize
6. Impregnates (insecticides, fungicides, herbicides)
   a. Polymeric materials containing a volatile insecticide, e.g., No-Pest Strips ®, pet collars
   b. Polymeric materials containing nonvolatile insecticides, e.g., pet collars, adhesive tapes, pet tags, livestock eartages
   c. Shelf papers, strips cords containing a volatile or contact insecticide
   d. Mothproofing agents for woolens
   e. Wood preservatives
   f. Wax bars (herbicides)
   g. Insecticides soaps for pets
7. Fertilizer combinations herbicides, insecticides or fungicides
8. Baits
   a. Insecticides
   b. Molluscicides
   c. Rodenticides
   d. Avicides
9. Slow-release insecticides
   a. Microencapsulated materials for agriculture, mosquito abatement, and household
   b. Paint-on lacquers for pest control operators and homeowners
   c. Adhesive tapes for pest control operators and homeowners
   d. Resin strip containing volatile, organophosphate fumigant, e.g., No-Pest Strips
10. Insect repellents
   a. Aerosols
   b. Rub-ons (liquids, lotions, paper wipes and "sticks")
   c. Vapor-producing candles, torch fuels, smoldering wicks
11. Insect attractants
   a. Food, e.g., Japanese bettle traps
   b. Sex lures, e.g., pheromones for agricultural and forest pests

12. Animal systemics (insecticides, parasiticides)
    a. Oral (premeasured capsules or liquids)
    b. Dermal (pour-on or sprays)
    c. Feed additive

The most recent innovation in emulsifiable concentrates is the Transparent Emulsion Concentrates (TEC). These are the mixture of the pesticides and the emulsifier with little or no hydrocarbon solvent. In these formulations the emulsifier or surfactant serves as the solvent, replacing the aromatic solvents, yet yielding a product that can be readily diluted with water.

Water-Miscible Liquids

Water-miscible liquids are those mixable in water. The technical grade material may be initially water miscible, or it may be alcohol miscible and formulated with an alcohol to become miscible. These formulations resemble the emulsifiable concentrates in viscosity and colour, but do not become milky when diluteds with water. Few home and garden pesticides are sold as water miscibles, since few of the pesticides that are safe for home use have these physical characteristics. Water-miscible liquid are labeled as Water-soluble Concentrate (WSC), Liquid (L), Concentrate (SC), or Solution (S).

Wettable Powders

Wettable powders, (WP), are essentially concentrated dusts containing a wetting agent to facilitate the mixing of the powders with water before spraying. The technical material is added to the inert diluent, in this case a finely ground talc or clay, in addition to a wetting agent or surfactant and mixed thoroughly in a ball mill. Without the wetting agent, the powder would float when added to water, and the two would be almost impossible to mix. Because wettable powders usually contain from 50 to 75 per cent clay or talc, they sink rather quickly to the bottom of spray tanks unless the spray mix is agitated constantly. Many of the insecticides sold for garden use are in the form of wettable powders because there is very little chance that this formulations will burn foliage, even at high concentrations. In contrast, the original carrier in emulsifiable concentrates is usually an aromatic solvent, which in relatively moderate concentrations can cause foliage burning at temperatures above 32.5°C.

Water-Soluble Powders

In *water-soluble powders* (SP) the technical grade material is a finely ground water-soluble solid are may contain a small amount of wetting agent to assist its solution in water. It is simply added to the spray tank, where it dissolves immediately. Unlike the wettable powders and flowables, these formulations do not require constant agitation; they are true solutions and form no precipitate. Because of their sometimes dusty quality, soluble powders may be packaged in convenient, water-soluble to be dropped into the spray tank.

Oil Solutions

In their commonest form, oil solutions are the ready-to-use household and garden insecticide sprays sold in an array of bottles, cans, and plastic containers, all usually equipped with a handy spray atomizer. Not to be confused with aerosols, these sprays are intended to be used directly or pests or places they frequent. Oil solutions may be used as roadside weed sprays, for marshes and standing pools to control mosquito larvae, in fogging machines for mosquito and fly abatement programmes, or for household insect sprays purchased in supermarkets. Commercially they may be sold as oil concentrates of the pesticide to be diluted with kerosene or diesel fuel before application or in the dilute, ready-to-use form. In either case, the compound is dissolved in oil and is applied as as oil spray; it contains no dust diluent, emulsifier, or wetting agent.

Soluble Pellets

Despite their seeming convenience and ease of handling with a water-hose, soluble pellets are not very effective. They are sold in kits, including the water-hose attachment, fertilizer, fungicide, insecticide, and even a car-wax pellet. The actual amount of active ingredients is very small, and even distribution with a watering hose is difficult.

Flowable or Sprayable Suspensions

There are several types of Flowable (F) or Sprayable (S) Suspensions. Physically they all are thick and creamy, and they vary in appearance from tan to white. The original flowable was an ingenious solution to a formulation problem. Some pesticides are soluble in neither oil nor water, but are soluble in one of the exotic solvents, which makes the formulation quiet expensive and may price it out of the marketing competition. To handle the problem, the technical materials is wet-milled with a clay diluent and water, leaving the pesticide-diluent mixture finely ground but wet. This "pudding" mixes well with water and can be sprayed but has the same tank-settling characteristic as the wettable powders. An example of a second kind of innovative flowable is the blending of the finely ground insecticide carbaryl with molasses. This formulation (Sevimol®) reduces to some extent the pesticide's drift off target during aerial application, increases its adherence to foliage and thus reduces removal by rain, and increase the mortality of moths that are attracted to feed on the molasses.

A third flowable is made by mixing an emulsifiable concentrate, containing a very high percentage of a water-stable toxicant and a thickener, with two to four volumes of water. The result is a thick, concentrated normal emulsion, which is then diluted for use with the appropriate volume of water just before using. This formulation responds to the need for an emulsifiable concentrate made with the least solvent possible to avoid foliage burn (phytotoxicity) in citrus groves. These formulations are occasionally identified as spray concentrates. A fourth innovation is the flowable microencapsulated formu- lation. The insecticide is incorporated by a special process in small, permeable spheres of a polymer or plastic, 15 to $50\mu (1\mu = 10^{-6}m)$ in diameter. These spheres are then mixed with wetting agents, thickeners, and water to give the desired concentration of insecticide in the flowable, usually 2 pounds per gallon. (These are discussed in more detail in the section on slow-release insecticides.)

A fifth modification of flowables, is an improved water-or rain-resistant formulation containing latex as a binding and sticking agent, which extends residual activity to 7-10 days. Additionally, it poses a reduced hazard to honeybees because it adheres to the foliage rather than to foraging bees. Particle size of XLR is reported to range from 5 to 10 in diameter, compared to an average particle size of 20 for the 80 per cent wettable powder.

The most recent flowable is a newly patented flowable aqueous pesticide. It is made by wet-milling a water-soluble pesticide, wetting agents, thickeners, anticaking and antifoaming agents, and a freeze-point depressant in water, resulting in a thick, water-dispersible concentrate :

Ultralow-Volume Concentrates

Ultralow-Volume Concentrates (ULV) are available only for commercial use in the control of public health, agricultural, and forest pests. They are usually the technical product in its original liquid form or, if solid, the original product dissolved in a minimum of solvent. They are usually applied without further dilution, by special aerial or ground spray equipment that limits the volume from .6 liter (l) to a maximum of 4.7 per hectare (ha), as an extremely fine spray. The ULA formulations are used where good results can be obtained while economizing through the elimination of the normally high spray volumes, varying from 11 to 38 L/ha. This technique has proved extremely useful where insect control is desired over vast areas. A recent innovation in ULV formulations is the aerial application of a pyrethroid insecticide at .11 to .23 kg Active Ingredient (AI) in 2.4 L of semirefined cottonseed or soybean oil per hectare. The only crop for which this is currently registered by the EPA is cotton.

Fogging Concentrates

Fogging concentrates are the formulations sold strictly for public health use in the control of nuisance or disease vectors, such as flies and mosquitoes, and to pest control operators. Fogging machines generate droplets whose diameters are usually less than 10 μm but greater than 1 μm. They are of two types. The thermal fogging device utilizes a flash heating of the oil solvent to produce a visible vapour or smoke. The ambient fogger atomizes a tiny jet of liquid in a venturi tube through which passes ultrahigh-velocity air stream. The materials used in fogging machines depend on the type of fogger. Thermal devices use oil only, whereas ambient generators use water, emulsions, or oils. In dwellings and food establishments the oil most commonly used is a deodorized kerosene or deobase.

DUSTS

Historically, dusts have been the simplest formulations of pesticides to manufacture and the easiest to apply. Examples of the undiluted toxic agent are sulfur dust, used on ornamentals, and one of the older household roach dusts, sodium fluoride. An example of the toxic agent with active diluent is any of the garden insecticides having sulfur dust as its carrier or diluent. A toxic agent with an inert diluent is the most common type of dust formulation in use today, both in the home garden and in agriculture. Insecticide-fungicide combinations are applied in this manner, the carrier being an inert clay, such as pyrophyllite. In this instance, particles small enough to pass through a 60-mesh screen are considered dusts. *Mesh* is a unit that refers to the number of grids are inch through which they will pass. The last type, the aerosol dust, is a finely ground silica or boric acid in a liquefied gas propellant that can be directed into crevices of homes and commercial structures for insect control.

Despite their ease in handling,formulation, and application, dusts are in the least effective and, ultimately, the least economical of the pesticide formulations. The reason is that dusts have a very poor rate of deposit on foliage, unless it is wet from dewor rain. In agriculture, for instance, an aerial application of a standard dust formulation of pesticide will result in 10 per cent to 40 per cent of the material reaching the crop. The reminder drifts upward and downwind. Psychologically, dusts are annoying to the nongrower who sees great clouds of dust resulting from a aerial application, in contrast to the grower who believes he or she is receiving a through application for the very same reason. The same statement may be relevant to the avid user of dusts in his garden and the abstaining neighbour ! Under similar circumstances, an aerial or garden hand sprayer application of a water emulsion spray will deposit 60 to 99 per cent of the pesticide on target.

AEROSOLS

In the aerosol culture we have raised: bug bombs, hair sprays, underarm deodorants, home deodorizers, oven cleaners, mace, shaving cream, lubricants, mouth fresheners, furniture polish, car wax, electrostatic inhibitors, adhesives, starch, fabric finishers, pruning sealer, spot removers, engine cleaner, motor starting fluids, water repellents, window sprays, repellents, paints garbage can and shower-tub disinfectants, and supremely, the anti-itch remedies -- the foot and groin sprays. In 1981, 2.20 billion aerosols were packaged in the United State. Of these, 173 million, or 8 per cent, were insecticides. The use of insecticide aerosols has increased steadily over the years, an average of 4 per cent annually from 1971 to 1980. The production of all kinds of aerosols peaked at 2.90 billion units in 1973, the year the fluorocarbon-ozone controversy began. With purchases dropping steadily, producers scrambled to replace fluorocarbon propellants with less ecologically suspect substi- tutes, such as carbon dioxide. They also mounted a publicity campaign to convince consumers of the safety and utility of pressurized products. Ozone is by now a reasonable dead issue, so far as aerosols are concerned. To produce an aerosol, the active ingredients must be soluble in the volatile, petroleum solvent in its pressurized condition. The pressure is provided by a propellant gas. When the petroleum solvent is atomized, it evaporates rapidly, leaving the microsized droplets of toxicant suspended in air. Aerosols are effective only against resident flying and crawling insects and provide little or no residual effect. More recently the total-release serosol has been designed to discharge, its entire contents in a single application. Once the nozzle is depressed, it locks into place, permitting the container's total emission while the occupants leave and remain away for a few hours. These products are available for homeowners as well as commercial pest control operators.

Caution :

Aerosols commonly produce droplets well below 10 μm in diameter. Such droplets are respirable, that is, they are absorbed by alveolar tissue in the lungs rather than impinging on tne bronchioles, as do larger droplets. Consequently, aerosols of all varieties should be handled with discretion, and the user should inhale as little as possible.

GRANULAR PESTICIDES

The granular pesticides are small pellets formed from various inert clays and sprayed with a solution of the toxicant to give the desired content. After the solvent has evaporated, the granules and packaged for use. Granular materials range in size from 20 to 80 mesh. Only insecticides and a few herbicides are formulated as granules. They range from 2 to 25 per cent active ingredient and are used almost exclusively in agriculture, although systemic insecticides as granules can be purchased for lawn and ornamentals. Granular materials may be applied at virtually any time of day, since they can be applied aerially in winds up to 20 mph without problems of drift, an impossible task with sprays or dusts. They also lend themselves to soil application in the drill at planting time to protect the roots from insects or to introduce a systemic to the root for transport to above- ground part in lawns and ornamentals. Microgranules, mesh size ranging from 40 to 80 mesh, have been used in the United States on an experimental basis, somewhat unsuccessfully on cotton and field crops. They are used routinely in the Orient, particularly in Japan.

FUMIGANTS

Fumigants are a rather loosely defined group of chemical substances. The plastic insecticide-impregnated strips and pet collars of the same materials are really a slow-release formulation that permits the insecticide to work its way slowly to the surface and volatilize. Moth crystals and moth balls – paradichlorobenzene and naphthalence, respectively – are crystalline solids that evaporate slowly at room temperatures, exerting both a repellent as well as an insecticidal effect. Soil fumigants are used in horticultural nurseries, greenhouses, and on high-value cropland to control nema- todes, insect larvae, and adults and sometimes to control diseases and weed seeds. Depending on the fumigant, the treated soils may require covering with plastic sheets for several days to retain volatile chemical, allowing it to exert its maximum effect.

IMPREGNATING MATERIALS

Impregnating materials include treatment of woolens for moth-proofing and timbers against wood-destroying organisms. For several years, woolens and occasionally leather garments have been moth-proofed in the final stage of dry cleaning (using chlorinated solvents). The last solvent rinse contains an ultralow concentration of the chlorinated, biodegradable insecticide ethylan, which has long residual qualities against moths and leather-eating beetle larvae. Reilroad ties, telephone and light poles, fence posts, and other wooden objects that have close with or are actually buried in the ground soon begin to deteriorate as a result of attacks from fungal decay microorganisms and insects, particularly termites, unless treated with fungicides and insecticides. Such treatment extends the wooden object's functional life to some 40 or 60 years. The insecticides of choice for wood exposed to potential termite damage are dieldrin and chlordane.

Impregnated Shelf Papers

Impregnated shelf papers, strips, and cords containing insecticides have practically disappeared from the market. Thoroughly effective against stored products insect pests, they usually contain one of the chlorinated insecticides to give long residual activity. Because these insecticides, along with most others, cannot be used where food and food utensils are stored according to regulations established by the Environmental Protection Agency, the use of impregnated materials has declined substantially in the past decade.

Impregnated Wax Bars

Impregnatged wax bars contain a herbicide that is selective against broad-leaf plants. When dragged over grass lawns in a uniform pattern, enough rubs off on weeds to eliminate them, leaving the grass unaffected. This type of application is very selective, represents a spot application that is not disruptive to the environment, and is the type that should be strongly encouraged.

FERTILIZER COMBINATIONS

Fertilizer combinations are formulations fairly familiar to the urbanite who has purchased a lawn or turf fertilizer that contains a herbicide for carbgrass control, insecticides for grubs and sod webworms, or a fungicide for numerous lawn diseases. Fertilizer-insecticide mixture have been made available to growers, particularly in the corn belt, by special order with the fertilizer distributor. The fertilizer and insecticide can then be applied to the soil during planting in a single, economical operation.

Baits

Baits can be purchased or formulated at home. Those that are purchased contain low levels of the toxicant incorporated into materials that are relished by the target pests. Spot application – the placing of the bait in selected places accessible only to the *target* species – permits the use of very small quantities of oftentimes highly toxic materials in a totally safe manner, with no environmental disruption.

Slow-Release Insecticides

Slow-release insecticides are relatively new and only a few are available to the homeowner. Dichlorvos, a volatile organophosphate insecticide, was incorporated into penels of polychlorovinly resin, which permitted the insecticide to volatilize at a slower rate, killing flying and some crawling insects in the vicinity. The most recent slow-release formulation appeared in 1979, a microencapsulated concentrate of diazinon for use in agriculture and for household pests. The principle of this form of slow-release involves the incorporation of the insecticide in a permeable covering, microcapsules or tiny spheres, with dismeters ranging from 15 to 50 μm that permits its release at a reduced but effective rate. The insecticide escapes through the sphere wall over an extended period, thus preserving its effectiveness much longer, usually two to four times longer, than if formulated as an emulsifiable concentrate.

A paint-on slow-release formulation containing chlorpyrifos, Killmaster II®, is a recent innovation in structural pest control. The insecticides is dissolved in a volatile petroleum solvent containing a unique combination of dissolved plastic and lacquers in small quantities. Following its paint-on application as a spot treatment in homes, restaurants, and food-handling establishments, the solvent quickly evaporates, leaving the insecticides incorporated in a *thin*

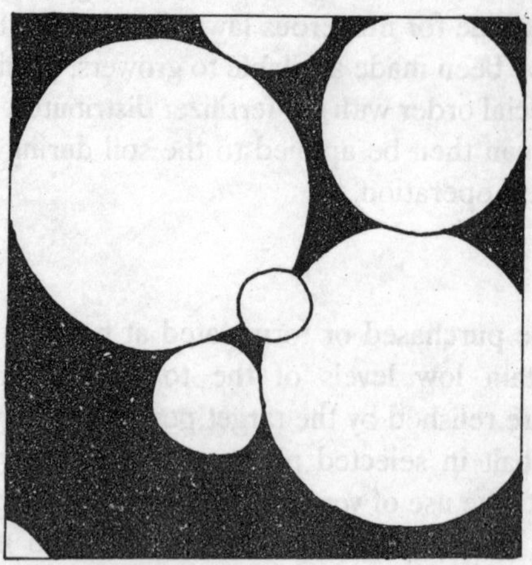

Fig. 1.1. KNOX OUT 2FM

transparent film. Over time the insecticide *"blooms"* or escapes to the surface at a constant rate, presenting at all times a freshly exposed surface to crawling insects. Insecticides adhesive tapes work as contact insecticides against crawling insects. The insecticide is laminated into the multilayered polymeric strips. The adhesive back is exposed by removing a protec- tive film, and the tape is attached beneath counters, under shelves, and in other protected places. Adhesive and point-on slow-release formu- lations are now available to homeowners under various brand names.

TABLE 1.5

Slow-release Insecticide Formulations

Product, active ingredient, and manufacturer	*Physical form*	*Use*
No-Pest Strips® dichlorvos (20%) Texize, Division, Morton-Norwich	Polychlorovinyl resin strip	Fumigant for flying insects
Killmaster II® chloropyrifos (2%) Velsicol Chemical Co.	Ready-to-use solvent containing lacquers and plastics	Most crawling household pests
Positive Control® Pest Aid "180"®, chlorpyrifos (1%) Positive Formulators, Inc.	Ready-to-use solvent	Most crawling household containing lacquers and plastics
Penncap M® methyl parathion (22%) Pennwalt Crop.	Cross-linked nylon polymer, microencapsulated concentrate (30-50 μm diam.)a	Many agricultural pests air or growth
Penncap E® ethyl parathion (22%) Pennwalt Crop. diam)	Cross-linked nylon polymer, microencapsulated concentrate (30-50 μm	Several agricultural pests air or growth
Knox Out 2FM® diazinon (23%) Pennwalt Crop. diam)	Cross-linked nylon polymer, microencapsulated concentrate (30-50 μm (professional	Many agricultural pests and most crawling household pests for house
Sectrol #90 Concentrate pyrethrins + synergists (1.1% + 5.9%) 3M, Industrial Tape Div.	Polyurea microencapsulated concentrate (15-20 μm diam)	Most crawling and flying household insectsspray or air

Insectape® 10% diazinon, chlorpyrifos, or propoxur Herculite Products, Inc	Ready-to-use, adhesive, laminated, multilayered, polymeric strips	Most crawling household pests except professional
Roach-Tape propoxur (4%) Herculite Products, Inc	Ready-to-use adhesive, laminated, multilayered, polymeric strips	Most crawling household pests hand (home)
Altosid Briquet 7.9% methoprepe Zoecon Corp.	Ready-to-use briquets (68 g) of slowly disintegrating medium	Mosquito growth regulator to prevent adult mosquito emergence

2

Resistance to Chemicals

Resistance of pests to the chemicals has been expressed in insects, mites, bacterial and fungal plant pathogens, domestic rodents, weeds, and several disease pathogens of humans and domestic animals. Most *Staphylococcus* or staph infections are resistant to *penicillin;* houseflies have developed resistance to *DDT, malathion,* and other insecticides; cockroaches have become resistant to *chlordane,* and in several areas, to *diazinon;* some strains of the venereal disease gonorrhea cannot be controlled with any but the very latest antibiotics; some recently developed fungicides with specific modes of action fail to control certain plant diseases; and certain weeds, once readily controlled with a particular herbicide, now go virtually unscathed. Certainly other pests are even now developing resistance to various pesticides. The concept of resistance to chemicals has been understood for approximately 35 years. It is a simple mechanism that can be attri- buted to the singular mode of action of a particular pesticide, which disrupts only one genetically controlled process in the metabolism of the pest organism. The result is that resistant populations appear suddenly, either by selection of resistant individuals in a population or by a mutation, which appears to be the less common of the two means. Generally, the more specific the site and mode of action, the greater the likelihood that a pest population will develop resistance to that chemical. When a pesticide is lethal and specific in its mode of action, it must be expected to favour the survival of resistant strains of the target population.

Resistance to Insecticides

The susceptible insects are killed, leaving behind only those that genetically resistant to the *toxicant.* Resistant individual make up an increasingly large part of the pest population and pass their resistance on to the next and future generation. Resistance in insect pests simply represents the result of applying a stringent biological selection mechanism, the insecticide, over several generations. The greater the number of generations exposed to an insecticide, the greater the potential for a population to develop resistance as a consequence of this intensive selection mechanism. Resistance to chemical by insects is a very serious problem. In 1944, only insect species were known to be resistant to insecticides. Today's estimates place this number well beyond 250, more than half of which are agricultural pests.

As more insecticides become ineffective aganist certain species, the problems of insect pest control increase proportionally. Populations of several insect pests now have such a high proportion of individuals resistant to all known insecticides that substitute materials are no longer available and insecticides are not recommended as control measures. In these instance chemical control is no longer the method of choice. Nor is resistance to insecticides confined to pests of agriculture. Public health problems resulting from resistance in insects that transmit or vector disease may become far more important than our agricultural difficulties. One of the less familiar phases of insecticide resistance is cross-resistance. Insect populations that have developed resistance to one *organochlorine* insecticide are frequently resistant to another organochlorine, even though they have not been previously exposed. In many instance, insect populations that became resistant to the organochlorines quickly became resistant to the organophosphate insecticides. And because the modes of action of the organo- phosphates and carbamates are similar, populations that became resistant to the former soon became resistant to one or more of the latter.

The ineffectiveness of an insecticides does not indicate with certainty that the insect is resistant. Effectiveness can also be reduced by the destruction of natural controls, for example, of parasites and predators that normally hold the pest species in check. Only a very small fraction of the total number of insects are considered pests. At normal population densities. Most insects pose no threat to cultivated crops, and many are important to the health and stability of the enviornment because they control other potentially damaging species. Most of the insecticides in use today have broad-specturm effects, that is, they are lethal to a wide range of insects, including beneficial as well as pest species. If the natural enemies of a pest species are destroyed, a resurgence in the pest population may occur, resulting in increased damage to the crop presumably being protected. The use of broad-spectrum insecticides, insects that were once controlled naturally by their enemies and predators sometimes increase in such numbers that they become pests. This is the result of killing the insect's natural enemies. Thus an insect species of secondary importance can become a pest if insecticides are applied to control a primary pest. Resistance in indeed a problem in many inscect pest species, and, unfortunately, the exploration and discovery of new insecti- cides have not kept pace with the loss of once-effective insecticides due to resistance or cancelation for reasons of human and enviorn- mentral safety. Thus, the long-range answers to insect resistance will have to be found in other control methods or in the combination of methods known as *integrated pest management*.

WEED TOLERANCE AND RESISTANCE TO HERBICIDES

Some weeds are remarkably naturally tolerant to the toxic action of herbicides. Such tolerance is based largely on morphological, physilogical, and genetic plant characteristics. In constrast, resistance implies a genetic selection over a period of time and several plant generations, in response to a selective herbicide, resulting in a truly resistant weed biotype. Tolerant plants survive chemical applications and flourish in the absence of competition from susceptible species. The rotation of herbicides is useful in these cases if selective chemicals are available. However, most herbicides in common use today are broad spectrum.

The problem of resistance to herbicides by weeds has not yet plagued the weed scientist as seriously as it has the ento- mologist and the rodent control specialist. However, plant patho- logists are already quite familiar with the problem. For more than 20 years, they have seen the antibiotic fungicides fail to control bacterial plant diseases, because antibiotic- resistant strains of plant pathogenic becteria quickly become prominent in the population of those organisms treated with antibiotics.

TABLE 2.1.

Succeptibility of some common annual weeds to several selective herbicides at usual selective rate[a]

	Alachlor	*Atrazine*	*Chloramben*	*Chlorpropham*	*2,4-D*	*Diuron*	*EPTC*	*Linuron*	*Trifluralin*
Barnyardgrass	S	I	S	I	T	S	S	S	S
Chickweed		S		S	I	S	S	S	S
Cocklebur	T	S	I	T	S	I		S	T
Crabgrasses	S	T	S	I	T	S	S	S	S
Fall panicum	S	I	I	I	T	S	S	I	S
Foxtails	S	S	S	I	T	S	S	S	S
Jimsonweed	I	S	I	I	S			S	T
Johnsongrass seedlings	I	T	I	T	T	I	S	I	S
Knotweed	T	S		S	I	S	S	I	S
Lambsquarters	I	S	S	T	S	S	S	S	S
Morningglory annual	T	S	T	I	S	I	S	S	I
Mustards	I	S	I		S	S	T	S	T
Nightshades	I	S	S		I		I	S	T
Nutsedges	I	I	T	T	I		I	T	I
Pigweeds	S	S	S	I	S	S	I	S	S
Purslane	S	S	S	S	I	S	S	S	S
Ragweed	I	S	S	T	S	S	I	S	I
Smartweed	I	S	S	S	I			S	I
Velvetleaf	T	S	I	T	I	I	I	S	T
Wild cucumber	T	S	I	T	I			S	I

[a]S – susceptible : I – intermediate : T – tolerant.

TABLE 2.2

Examples of Annual Weeds Tolerant to Specific Herbicides.

Herbicides	*Tolerant annual weeds*
Atrazine (AAerx®)[a]	*Pancicum* spp.
Benefin (Balan®)	Grouondcherry, mustard, prickly lettuce, spiny sowthistle, sunflower

Bensulide (Prefar®)	Pigweed, groundcherry, horse purslance, mustard, sowthistle, prickly lettuce
Bromacil (Hyvar_X®)	Spurge, horse purslane, plantain
Cycloate 9Ro-neet®)	Mustard, little mallow, shepherdspurse, legumes
Difenzoquat (Avenge®)	Canary grass
Diuron (Karmex®)[a]	Legumes, little mallow, mustard
EPTC (Eptam®)[a]	Malva, Russian thistle
Paraquat (Ortho Paraquat CL®)	Mexican sprangletop, plantain
Phenmedipham (Betanal®)	Pgiweed, barnyardgrass, knotweed
Prometryn (Caparol®)[a]	Grasses
Prometryn (Chem-Hoe®)[a]	Summer grasses, most broadleaf weeds
Simazine (Princep®)	*Panicum* spp.
Trifluralin (Treflan®)	Groundcherry, mustard, legumes, sunflower
2.4-D (Various)	Most grasses, little mallow, knotweed

[a]Other trade names may be available.

Several instances of weed resistance have been observed in the past few years, and with increasing frequency. The most significant is the triazine-resistant biotypes of at least 10 weed species that have been reported since 1970. These are in the plant families Compositae, Chenopodiaceae and Amaranthaceae. The species that have received the greatest amount of attention are the common groundsel, common lambsquarters, and redroot pigweed *(Amaranthus retroflexus L.).* (Hensley 1981). These occur in the corn belt where corn has followed corn year after year without crop rotation, and *triazines* have been the herbicides of choice, almost to the exclusion of all others. Investigations into the mode of resistance in these resistant species reveal a different mechanism from that found in species tolerant to the triazines, such as corn. Corn's tolerance to the *triazine* herbicides is attributed is its *ability* to metabolize the active triazine herbicide to an inactive form.

Another instance is that of Johnsongrass' resistance to dalapon, Johnsongrass is not a native of the area and was brought in as a forage grass; numerous biotypes were introduced, which shortly became perennial weeds. Apparently, resistance to dalapon was observed after five or six years' usage, when the susceptible biotypes had been eliminated, leaving only the resistant strains. Dalapon, once a very effective herbicide against this perennial is no longer useful.

Why do herbicide-resistant strains of plant pests become so quickly the predominating strains of the population? If a potent herbicide acts specifically, that is, if it kills by blocking a single enzymatically catalyzed metabolic reaction, it exerts a selection pressure on the population of the pest that permits survival of only those strains that are resistant to the poison. The likelihood that such strains exist is high; their resistance is derived from their capacity to bypass the metabolic "roadblock" imposed by the pesticide and continue their metabolic activites using an alternative pathway. Because a single metabolic reaction is governed by a single gene, a one gene difference between an individual strain and its wild sister strains could make one strain resistant to a herbicide that has a highly specific mode of toxic

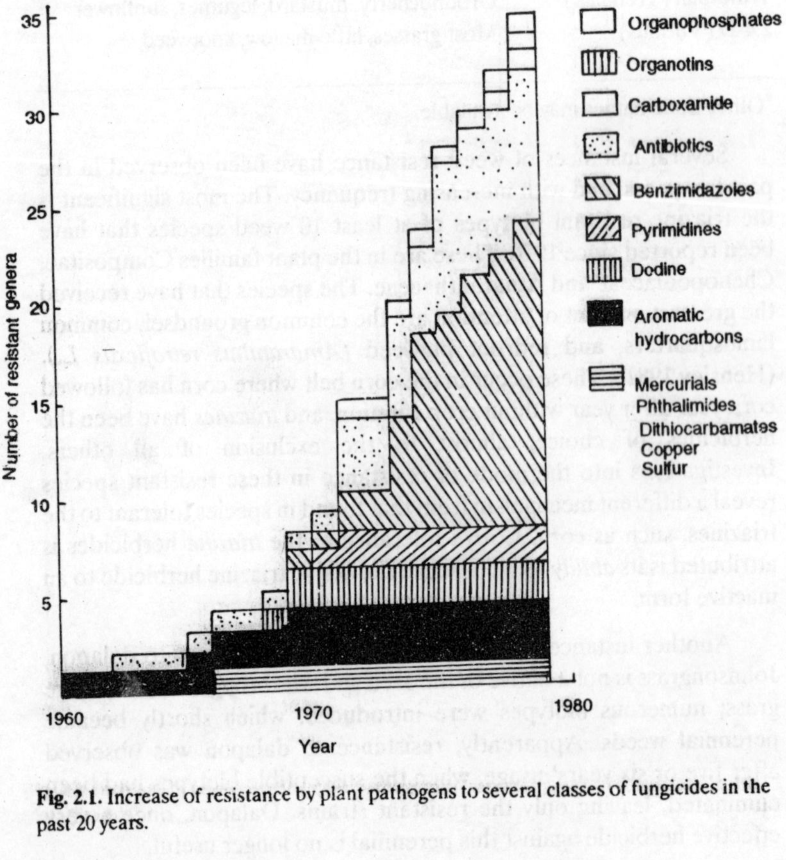

Fig. 2.1. Increase of resistance by plant pathogens to several classes of fungicides in the past 20 years.

action.The potency of a herbicide is also important. A weak herbicide that is effective only in large amountd will be ineffective against a population of pest species that consists of strains that tolerate the poison. Development of herbicide-resistant of plant pests is not an effect unique therapeutic chemicals. As long as the pesticide is potent and specific in its mode of action, it must be expected to favour the survival of resistant strains of the target population.

Disease Resistance to Fungicides

Resistance to chemicals other than the heavy metals occurs commonly in fungal, and on rare occasions in bacterial and plant disease pathogens. Several growing seasons after a new fungicide appears, it becomes noticeably less effective against a particular disease. As our fungicides become more specific for selected diseases, we can expect that pathogens to become resistant. Again, such fungicides exert a selective pressure on the population, and resistant populations appear suddenly, either by selection of resistant individuals or by a single- gene mutation. Generally, the more specific the site and mode of fungicidal action, the greater the likelihood for a pathogen to develop a tolerance to that chemical.

Resistance can be alleviated or prevented by early use of spray programmes designed to preclude long-term exposure of the pathogen to a single disease control agent. Once resistance is a significant part of a pathogen population, the only choice is to use disease control agents to which there is no cross-resistance. When a fungicide with a propensity for resistance is to be used, the programme must be planned to prevent resistance; for example by (1) using mixtures of fungicides (2) rotating fungicides, (3) limiting the use of fungicides, or (4) developing integrated control strategies.

3

Weed Killers

In the recent years the chemical weed killers have largely replaced mechanical methods of weed control in countries where intensive and highly mechanized agriculture is *practiced* these chemical weed killers are called the herbicides. Herbicides provide a more effective and economical means of weed control than cultivation, hoeing, and hand pulling. Together with fertilizers, other pesticides, and improved plant varieties, they have made an important contribution to the increased yields we now have and serve to fight against *combat* rising costs and shortages of agricultural labour. The heavy use of herbicides is confined to North America, Western Europe, Japan and Australia. In the absence of herbicides, it would have been impossible to mechanize fully the production of cotton, sugar beets, grains, potatoes and corn.

Herbicides are also used on a large scale away from the farm, in areas such as industrial sites, roadsides, ditch banks, irrigation canals, fence lines, recreational areas, railroad embankments, and power lines. Herbicides remove unwanted plants that might cause damage, present fire hazards, or impede work crews. They also decrease costs of labour for mowing.

Herbicides are known as selective when they are used to kill weeds without harming the crow and as *nonselective* when they are used to kill all *vegetation*. Both selective and nonselective materials can be applied to weed foliage or to soil containing weed seeds and seedlings, depending on the made of action. The term true selectivity refers to the capability of a herbicide, when used at the proper dosage and time, to be active only against certain species of plants but not against others. But selectivity can also be achieved by placement, as when a nonselective herbicide is applied in such a way that it reaches the weeds but not the crop.

Herbicide classification would be a simple matter if only the selective and nonselective categories were present. However, there are multiple-classification schemes that may depend upon the on selectivity, contact versus translocation, timing, area covered, and chemical classification.

Herbicides are also classified as contact or translocated. Contact herbicides kill the plant parts to which the chemical is contact and are most effective against annuals, those weeds that germinate from seeds and grow to maturity each year. Complete coverage is essential in weed control with contact materials. Translocated herbicides are absorbed either by roots or above-ground parts of plants and then circulate within the plant system to distant tissues. Translocated herbicides may be effective against all weed types; however, their greatest advantage is in the control of established perennials, those weeds that continue their growth from year to year. Homogenous appli- cation is needed for the translocation materials, whereas complete coverage is not required.

Herbicides can also be classified on the basis of their timings application with regard to the stage of crop or weed growth. Timing depends on many factors, including the chemical classification of the material and its persistence, the crop and its resistance to the herbicide, weed species, cultural practices, climate, and soil type and condition. The three categories of timing are preplanting, pre-emergence, and post- emergence.

Preplanting applications for control of annual weeds are made to an area before the crop is planted, within a few days or weeks of planting. Pre-emergence applications are completed before the appearance of the crop or weeds, depending on definition, after planting. Postemergence applications are made after the crop or weed emerges from the soil.

Herbicide application based on the area covered includes four categories: band, broadcast, spot treatments, and directed spraying. A band application treats a continuous strip, as long or in a crop row. Broadcast applications cover the entire area, including the crop. Spot treatments applied to small areas of weeds. Directed sprays are applied to selected weeds or to the soil to prevent contact with the crop.

In this chapter, we emphasize the chemical classification of herbicides. The two major classifications are inorganic and organic herbicides.

Inorganic Herbicides

The chemicals used first in weed control were inorganic compounds. Brine and a mixture of salt and ashes were used to sterilize the land even in the biblical times, by the Romans. In 1896, copper sulfate was used selectively to kill weeds in grain fields. From about 1906 until 1960. sodium arsenite solutions were the standard herbicides of major use. Arsenic trioxide has been used at rates of 450 to 900 kg/ha for soil sterilization, whereas 1 or 2 kg of selected organic herbicides would easily produce the same result. As with the arsenical insecticides, the trivalent arsenicals were nonspecific inhibi- tors or those enzymes containing sulfhydryl groups. They also uncouple oxidative phosphorylation.

Ammonium sulfamate ($NH_4SO_3NH_2$), another salt, was introduced in 1942 for brush control. Other salts have been used over the years. These include ammonium thiocyanate, ammonium nitrate, ammonium sulfate, iron sulfate, and copper sulfate, each used heavily as a foliar spay. Their mode of action are desiccation and plasmolysis (shrinkage of cell protoplasm away from its wall due to movement of water out of its large central vacuole).

The borate herbicides were another family of inorganics, for example, sodium tetraborate ($Na_2B_4O_7-5H_2O$), sodium metaborate ($Na_2B_2O_4-4H_2O$), and amorphous sodium borate ($Na_2B_8O_{13}-4H_2$). The amount of boron or boric acid determined their effectiveness. Borates are absorbed by plant roots, are translocated to above-ground parts and are nonselective, persistent herbicides. Boron accumulates in the reproductive structures of plants, but its mechanism of toxicity is vague. The borates are still used to give a semipermanent form of sterility to areas where no vegetation of any sort is required.

A nonselective herbicide used on a major scale for the last 40 years is sodium chlorate ($Na\ ClO_3$). It acts as a soil sterilant at rates of 225 kg ha, but it can be used as a foliar spray at 5.7 kg/ha as a defoliant of cotton. Caution must be taken with sprays of sodium chlorate to be certain the formulation contains flame retardants, usually sodium metaborate. Sulfuric acid has also been used as a foliar herbicide, but its use is restricted considerably by its corrosiveness to metal spray rigs. Their modes of action are those described for the miscellaneous salts, namely, desiccation and plasmolysis.

Several of the inorganic herbicides are used even today in weed and brush control, but are slowly being replaced by organic materials. Although organic herbicides are not superior to inorganic ones, the EPA has placed intensive restrictions on some inorganic herbicides because of their persistence in soils. The inorganics should not be used around the home except by experts aiming to remove all vegetation from an area.

ORGANIC HERBICIDES

Petroleum Oils

First used organic herbicides were the petroleum oils, which are a complex mixture of long-chain hydrocarbons containing traces of nitrogen-and sulfur-linked compounds. These mixtures include alkanes, alkenes, and often alicyclics and aromatics, formed by distillation and refinement of crude oils. The petroleum oils are important contact herbicides for all vegetation. Many homeowners today make use of old crankcase oil, gasoline, kerosene, and diesel oil for spot treatments. This is not an ecological choice in view of the drive to save energy and avoid further release of hydrocarbons in our already overburdened atmosphere. Although only temporary in effect, they are fast in action and very safe to use around the home. The petroleum oils exert their deadly by penetrating and disrupting plasma membranes. Other herbicidal oils used in agriculture include naphtha, stoddard solvent, fuel oils, and mineral spirits. Petroleum oils are used mostly for nonselective, but occasionally selective, weed control.

Organic Arsenicals

Extensively used as agricultural herbicides are the organic arsenicals, namely the arsinic and arsonic acid derivatives. Cacodylic acid (dimethylarsinic acid) and its sodium salt are the only derivatives

ARSONIC ACID

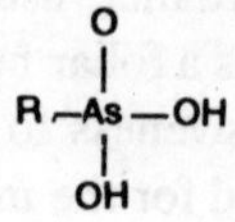

ARSINIC ACID

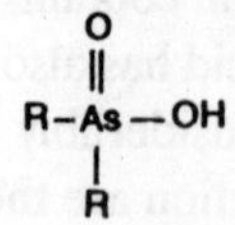

CACODYLIC ACID

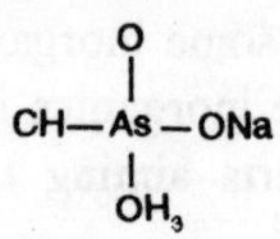

sodium salt

CACODYLIC ACID

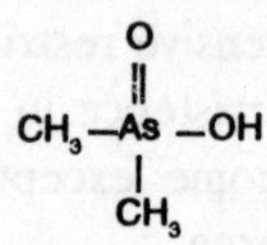

hydroxyclimemethylarsine oxide

MSMA

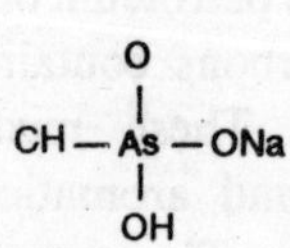

monosodium methanearsonate

DSMA

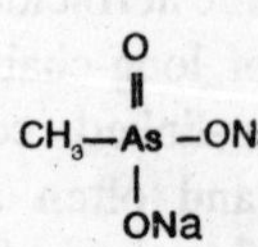

disodium methanearsonate

of arsinic acid. Disodium methanearsonate (DSMA) and monosodium methanearsonate (MSMA) are salts arsonic acid.

The organic arsenicals are not much toxic to mammals than the inorganic forms, are crystalline soils, and are relatively soluble in water. The arsonate, or pentavalent arsenic, acts in a manner different from the trivalent form of the inorganic arsenicals just described. The arsonates upset plant metabolism and inter act with normal growth by entering into reactions in place of phosphate. They not only substitute for essential phosphate but also are absorbed and translocated in a fashion similar to that of phosphates.

The trivalent arsenates are exclusively contact herbicides, whereas the pentavalent arsonates are translocated to underground tubers and rhizomes, making them highly useful against Johnson-grass and nut sedges. They are normally applied as spot treatments.

2,4-D

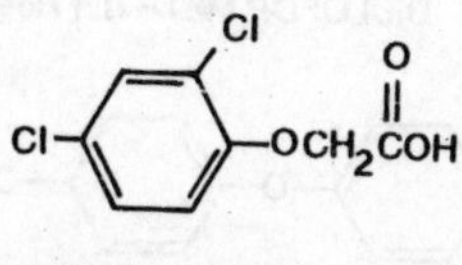

(2,4-dichlorophenoxy) acetic acid

2.4,5-T

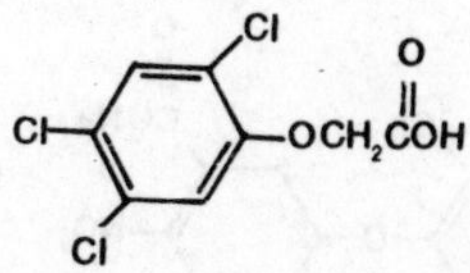

(2'4'5- trichlorophenoxy) acitic acid

SILVEX

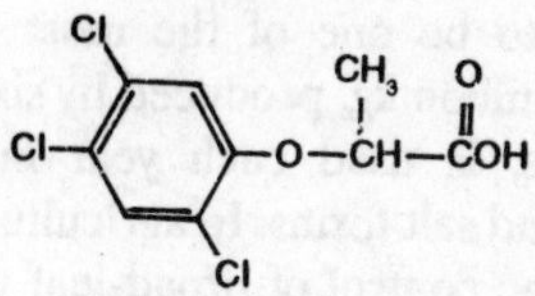

(2,4,5,- trichlorophenoxy) propionic acid

MCPA

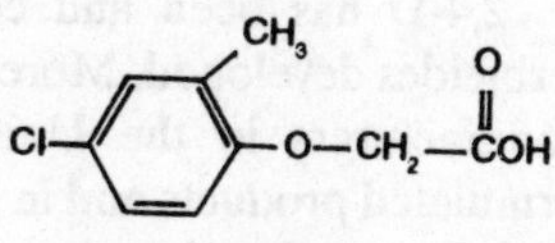

[(4-chloro-o-tolyl) OXY] acetic acid

Phenoxyaliphatic Acids

An organic herbicide introduced in 1944, later to be known as 2,4-D, was the first of the "phenoxy herbicides," "phenoxyacetic acid derivatives,", or "hormone" weed killers. These largely selective for broad-leaf weeds and were translocated throughout the plant, 2,4-D provided most of the impetus in commercial search for other organic herbicides in the 1940s. Several compounds belong to this group, of which 2,4-D and 2,4,5-T are the most commonly used. Other important compounds in this group are 2,4-DB, MCPA, and silvex.

The phenoxy herbicides have complex mechanisms of action similar to those of auxins (growth hormones) in some. They affect cellular division, activate phosphate metabolism, and modify nucleic acid metabolism.

2,4-D, MCPA, and 2,4,5,-T have been used for years in very large volume worldwide with no harmful effects on human or animal health. However, 2,4,5-T, used mainly to control woody perennials, became the subject of extended investigation, particularly because of its use in Vietnam in combination with 2,4-D as Agent Orange. Certain samples were found to contain excessive amounts of a highly toxic impurity, 2,3,7,8- tetrachlorodibenzo-p-dioxin, commonly known to as tetrachlorodioxin, or dioxin. Subsequent slight variations in manufacturing procedures have brought the dioxin content down to tolerable levels.

ACIFLUORFEN (Blazer)

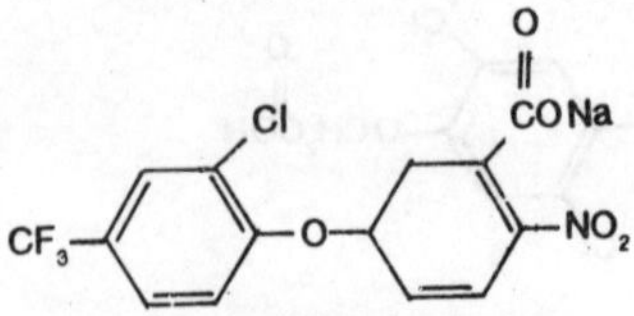

sodium 5-[2- chlora- 4-(trifluoro-
methyl)-phenoxy]- 2- nitrobenzoate

DICLOFOP METHYL (Hoelon)

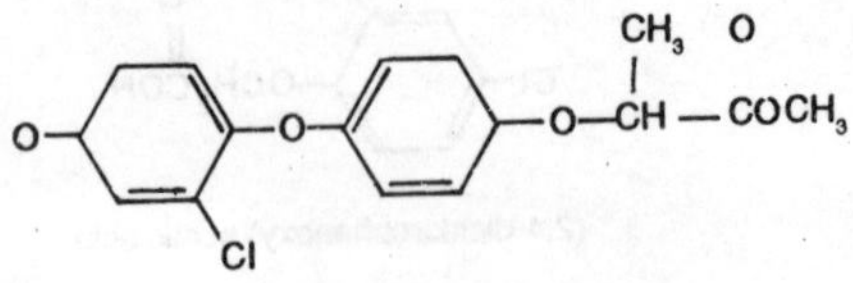

2-[4-(2,4-dichlorophenenoxy)-
phenoxy] -methyle- propanoate

2,4-D has been and continues to be one of the most useful herbicides developed. More than 32 million kg, produced by six basic manufacturers in the United States, is used each year in 1500 formulated products and in 35 ester and salt forms. In agriculture it is used on cereal and grain crops for the control of broad-leaf weeds, and on rights-of-way, turf and lawns, and in forest conservation programmes. The manufacturing process for 2,4-D used in the United States does not result lead to any level of tetrachlorodioxin contamination.

Aciflurofen (sodium salt) is a latest addition to the phenoxys, as is diclofop methyl. Aciflurofen controls annual broad-leaf and grass weeds in soybeans, peanuts, and other legumes. Diclofop methyl is sold as Illoxan in other countries and is used for post-emergence in wheat, barely, soybeans, sugar beets and vegetables.

Substituted Amides

The amide herbicides have different biological properties. These are simple molecules that are easily degraded by plants and soil. One of the earliest was allidochlor, which is used for the grasses. Allidochlor steps the germination or early seedling growth of most annual grasses probably by alkylation of the SH groups of proteins.

DIPHENAMID

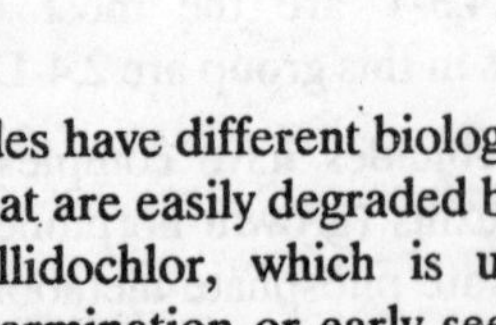

N,N -dimethyl-2,2 - diphenytlacetamide

Diphenamid is used as a preemergence soil treatment and has little contact effect. Most established plants are resistant to

PROPANIL

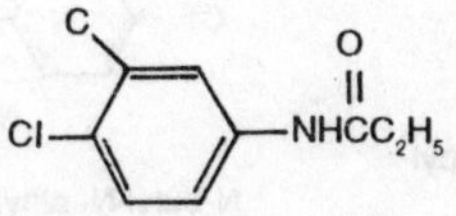

3,4 dichloropropionanilide

NAPROPAMIDE (Devrinol)

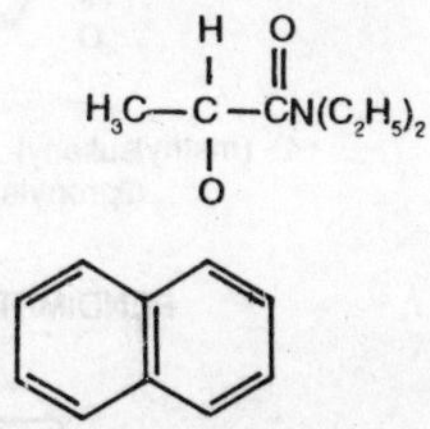

2- (a-naphthoxy)- NN-
diethylpropionamide

diphenamid, because it affects only seedlings. It persists from 3 to 12 months in soil.

Propanil has been used widely on rice fields as a selective post-emergence control for a broad spectrum of weeds. Napropamide is used in the control of grass and broad-leaf weeds in vineyards and orchards, and in direct-seeded tomatoes, strawberries, ornamentals, and tobacco.

There is no consistent pattern in the mechanism of action amide herbicides. Some are applied only to the soil and are active through the root system or seeds, while others are applied only to foliage.

Nitroanilines

The nitroanilines are probably the group of herbicides most heavily used in agriculture. They are used almost exclusively as soil-incorporated, preemergence selective herbicides in many field crops. Examples of the substituted dinitrotoluidines are triflurlin, benefin, nitralin, and isopropalin. Triflurlin has very low water solubility, which reduces leaching and movement from the target. The

TRIFLURALIN (Treflan)

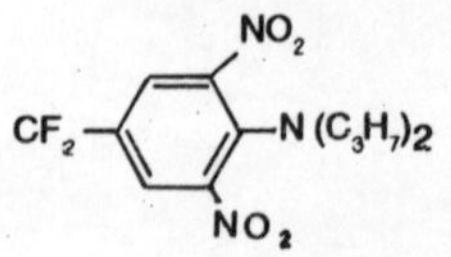

a,a,a, - trifluoro-2,6 dinitro-N-N dipropyl-para-toluidine

BENEFIN

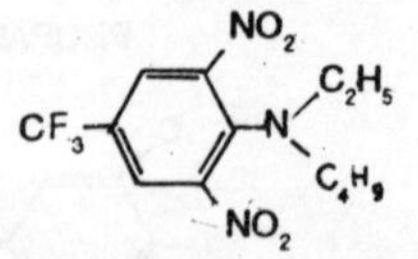

N-butyl-N- ethyl a,a,a,- trifluoro-2,6-dinitro-p-toluidine

ISOPROPALIN

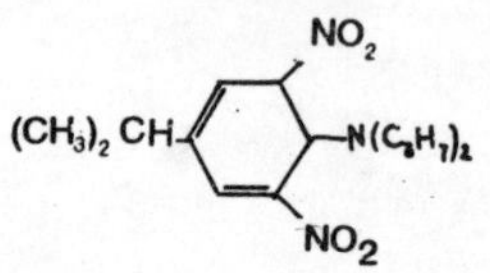

2,6-dinitro-N,N-dipropylcumidine

NITRALIN

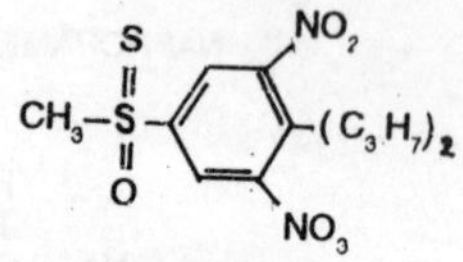

4 - (methylsulfonyl- 2,6- dinitro-N,N-dipropylaniline

ORYZALIN (Surflan)

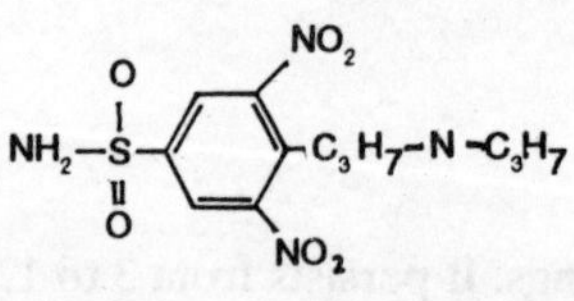

3,5 dinitro-N^4 N^4-dipropylsulfanilamide

PENDIMETHALIN (Prowl)

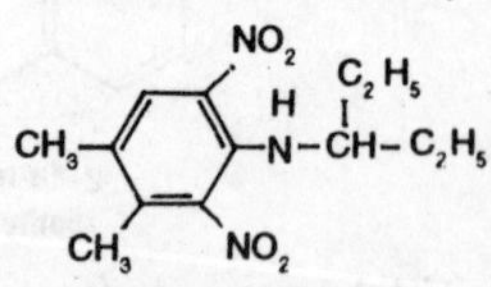

N-(1-ethylpropyl_-3,4-dimethyl-2,6 -dinitrobenzenamine

FLUCHLORALIN (Basalin)

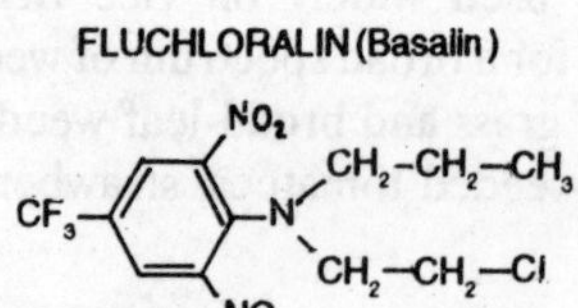

N- (2- chloroethyl- a,a,a, - trifluoro-2.6- dinotro -N- propyl-p-toluidine

nitroanilines inhibit both root and shoot growth when absorbed by roots, but they have an involved mechanism of action, which includes inhibiting the development of several enzymes and the uncoupling of oxidative phosphorylation.

The Broad leaf weeds in cotton, soybeans non-beaging fruit trees, nut trees, vineyard's & ornamentals etc. are controlled by Oryzalin. Pendimethalin checks annual grass and certain broad-leaf weeds in corn and is preplant incorporated in cotton, soybeans and tobacco. Fluchlorin is preplant and preemergence incorporated for control of annual grasses and certain broad-leaf weeds in soybeans and cotton.

Substituted Ureas

Start from large number of substituted areas have been known a function as herbicides and are therefore being used today.

The substituted ureas, H_2NCNH_2, have their hydrogen atoms replaced (substituted) by various carbon chains and rings, conceding a utilitarian group of compounds. Illustrated are monuron, diuron, fenuron-TCA, linuron, siduron, and tebuthiuron. Most of the ureas are relatively nonselective and are ordinarily applied to the soil as preemergence herbicides; some have post-emergence uses, while others are active when foliar applied. The ureas are strongly adsorbed by the soil, then absorbed by roots. Their mode of action is to inhibit photosynthesis known is the production of plant sugars-and,indirectly, to hinder the Hill reaction.

MONURON

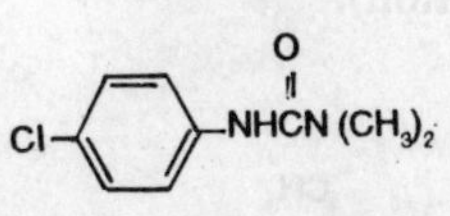

3-(P-chloropnenyl)- 1,1- dimethylurea

DIURON

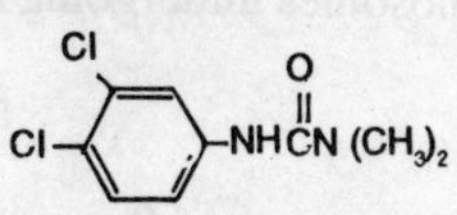

3 (3 4-dichlorophenyl)-1,1-dimethylurea

FENURON - TCA

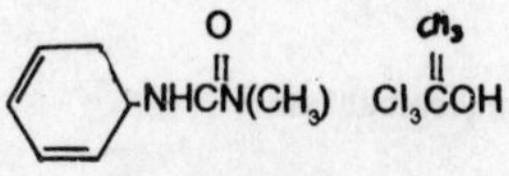

1,1-dimethyl-3- phenylurea
mono (trichloroacetate)

LINURON

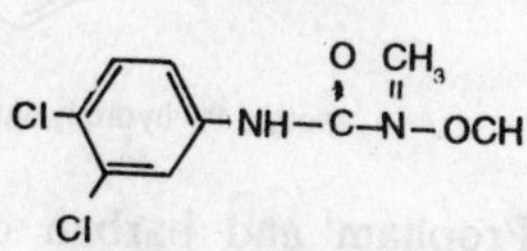

3-(3-4 dichlorophenyl)-1-methoxy-1-methylurea

SIDURON

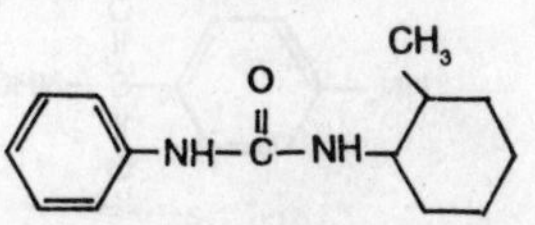

1-(2-methylcyclohexyl)-3-phenylurea

TEBUTHIURON (spike)

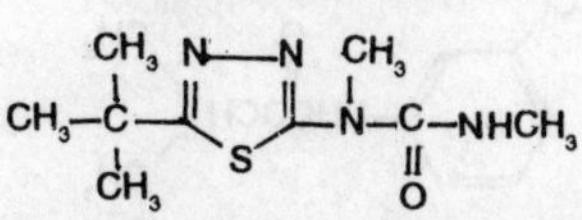

N-[5-(1,1-dimethyl)-1,3,4 -thiadiazol
2-yl]N,N dimethylurea

Carbamates

Carbamates, as the name indicates are the estees of Calls genic acid. These are physichologically very active. Some of these carbamates, as we shall observes later are insecticidal while others are

PROPHAM

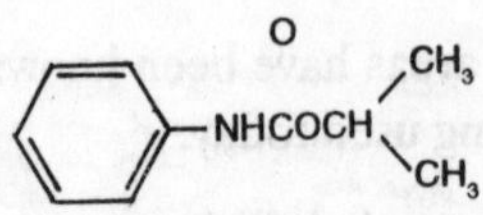

isopropyl carbanilate

BARBAN

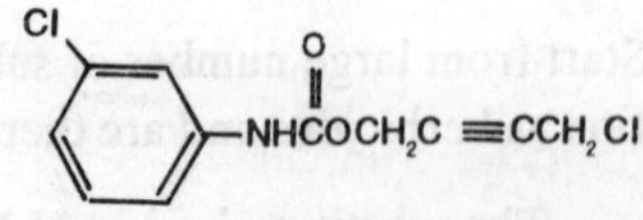

4- chloro-2-butynyl m-chlorocarbanilate

furgicidal. Still other carbamates are herbicidal, and we shall deal with that group in this section. Discovered in 1945, the carbamates are used primarily as selective preemergence herbicides, but some are also effective in post- emergence use.

Propham was the first herbicidal carbamates followed first by chlorpropham, then by barban and terbucarb. These carbamates kill plants by stopping cell division and plant tissue growth. Two effects are noted: cessation of protein production and shortening of chromosomes undergoing mitosis (duplication).

PHENMEDIPHAM

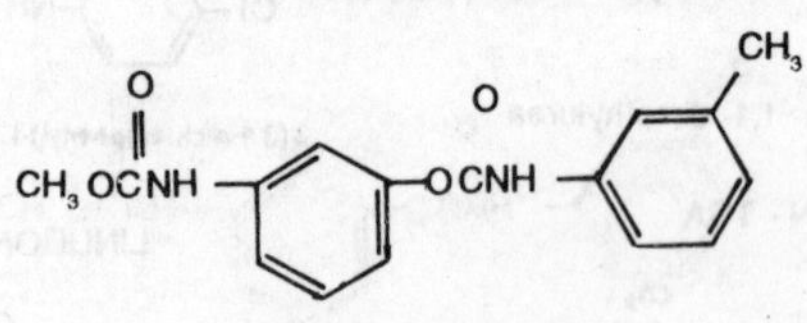

methyl m- hydroxy carbanilate m-methylcarbanilate

Propham and barban contain an aryl group which is a six membered carbon ring (normally the pnenyl ring) hence they are

CHLORPROPHAM

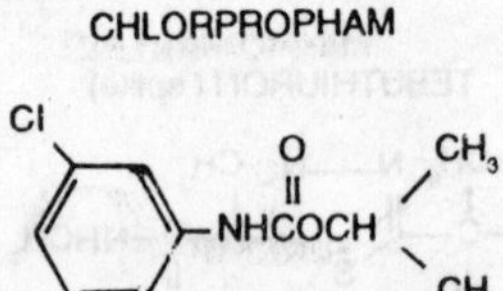

isopropyl m - chloracarpanilate

ASULAM (Asulox)

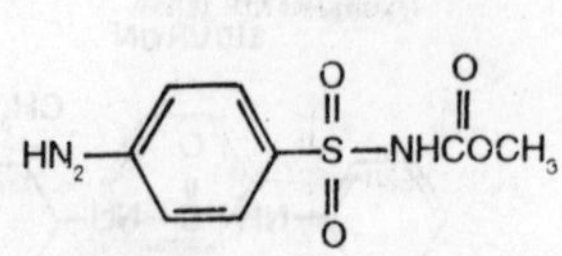

methyl sulfamilylcarbamate

called aryl carbamates. The alkyl group is the carbon chain or hydrocarbon radical with one hydrogen displaced by the attachment of the alkyl group to the rest of the molecule.

Phenmedipham is used as a pest emergence herbicide in sugar beet production. It contains two carbamate radicates in its molecule. Chlorpropham has largely replaced propham for field crops and is basically useful for alfalfa, clovers, tomatoes, safflower, and soybeans. Potatoes placed in storage can be treated with chlorpropham to inhibit sprouting. Asulam is used in most instances for grass control, such as crabgrass, Johnsongrass, goosegrass, foxtail, and

EPTC

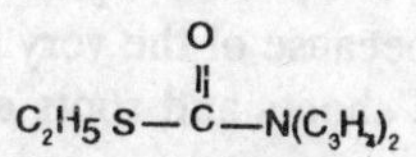

S -ethyl diprophylthiocarbamate

PEBULATE

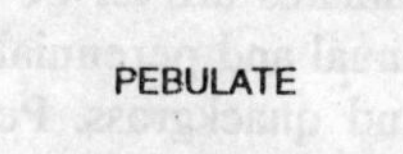

S-propyl butylethylthicarbanmate

CDEC

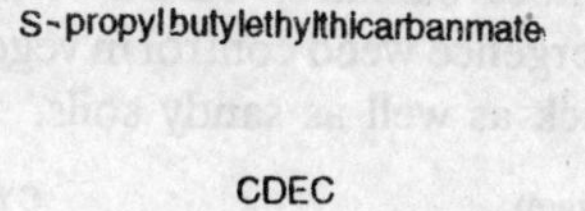

2 - chloroallyl-
diethy ldithiocarebamate

METHAM

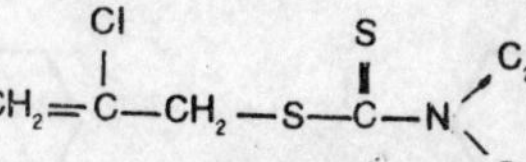

Sodium N- methyl -
dithiocarpamate

barnyardgrass in sugarcane. It is also useful in reforestation and Christmas-tree plantings.

Thiocarbamates

The thiocarbamates as the name indicates contains *sulfur* (from the Greek word theion) and are obtained from the hypothetical thiocarbamic acid.

The thiocarbamates are selective herbicides sold for weed control in croplands. The thio-and dithiocarbamates have exceptionally vapor pressures. Therefore these are quite volatile, and mist be incorporated in the soil at time of application. The phio carbamates must essentially be used either as preplant or pre-emergence soil incorporated herbicides because of the very fact that they inhibit the development of seedlings shoots and roots as soon as there emerge from the weed seeds.

Example of thiocarbamates are EPTC and pebulate. EPTC is used mostly to control annual and perennial grassy weeds, including nutgrass, Johnsongrass and quackgrass. Pebulate is useful against both grassy and broad-leaf weeds in sugar beets, tobacco, and tomatoes. Butylate, cycloate, diallate, molinate, triallate and vernolate are also included among the Thiocarbamates.

Among the dithiocarbamates, CDEC is an exceptionally good herbicide for preemergence weed control in vegetable and field crops. It is effective in muck as well as sandy soils. Metham is a general

MOLINATE ordram)

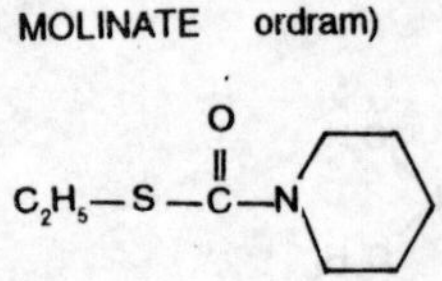

S ethyl hexahydro-1H azepine-1-carbothioate

CYCLOATE (Ro Neet)

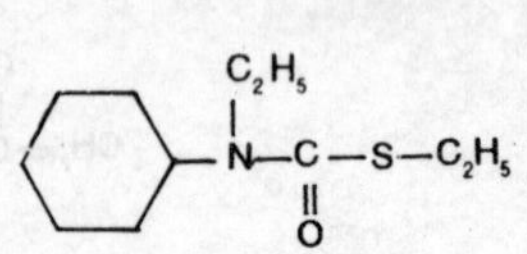

S - ethylcyclohexylethy ethythiocarbamate

VERNOLATE (Vernam)

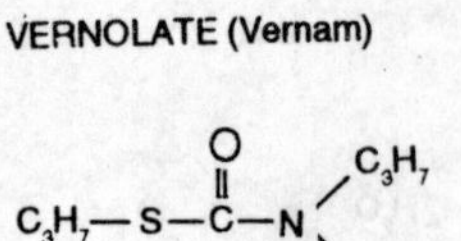

S- propylidpropylthiocarbamate

BUTYLATE (Sutan)

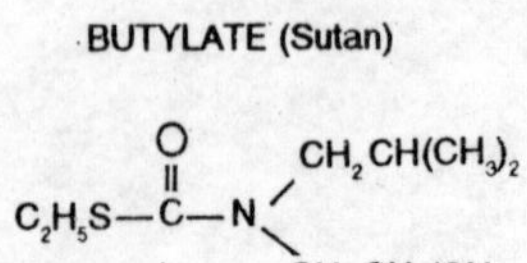

S ethyl diisobutylthiocarbamate

TRIAZINE NUCLEUS

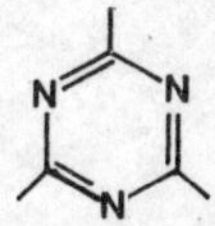

purpose soil fumigant for weeds, weed seeds, nematodes, insects, and soil-borne disease microorganisms. Molinate is effective for water-grass control ink rice. Cycloate is used mostly in sugar beets, table beets, and spinach to control annual and perennial grasses and broad-leaf weeds. Vernolate is effective for control of most grass weeds and some broad-leafs in soybeans and peanuts. Butylate is incorporated preplant to control grass weeds in corn.

Heterocyclic Nitrogens

The *triazines,* which are six-member rings containing three nitrogens (the prefix tri-means "three") and azine (a nitrogen-

ATRAZINE

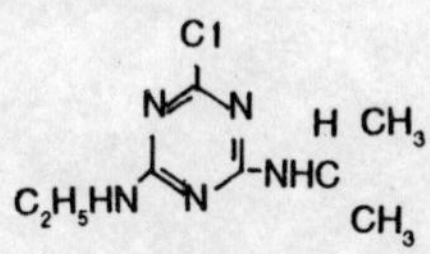

2-Chloro-4-(ethylamino)-6 (Isoprophyla amina)-s-triazine

SIMAZINE

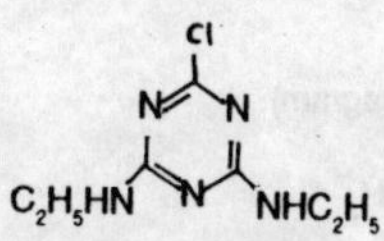

2-chlaro-4.6-bis-(ethyla mino)-s-triazine

CYANZINE

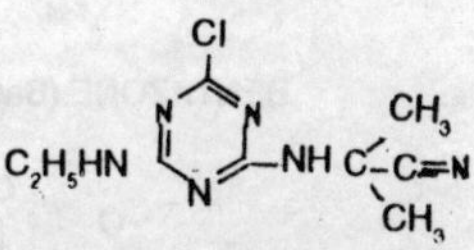

2 [[4-chloro-6-(ethylamino)- 2yl [amino]2 methylpropionitrile

PROMETON

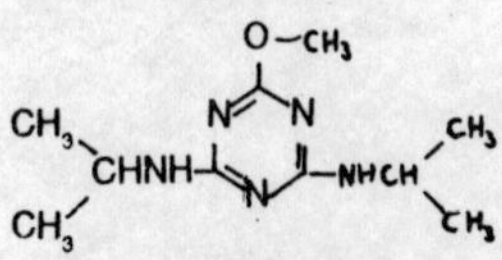

2,4-bis (isoprophylamino) -6-methoxy-5-triazine

PROPAZINE

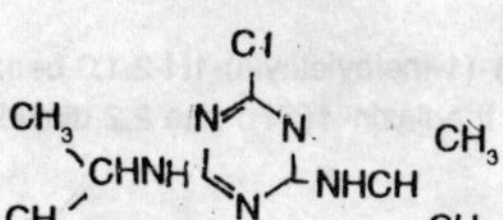

METRIBUZIN (Lexone Sencor

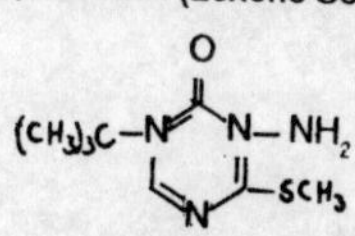

2-chloro-4.6-(1)-dimethylethyl)-3- (methylthio)-1,2,4-triazin-5(4H)-One

HEXAZINONE (Velpar)

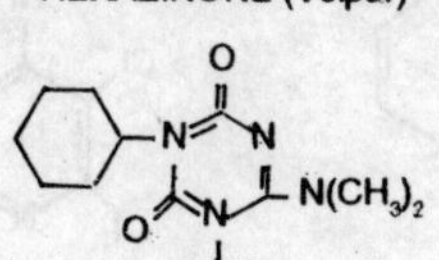

3-cyclohexy-6-(dimethylamino -1 -methyl-1,3,5-triazine-2,4- (1H,3H)- dione

AMITROLE

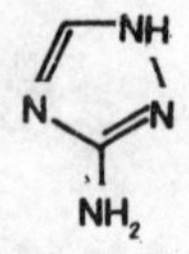

3-amino S-triazole

OXADIAZON (Ronstar)

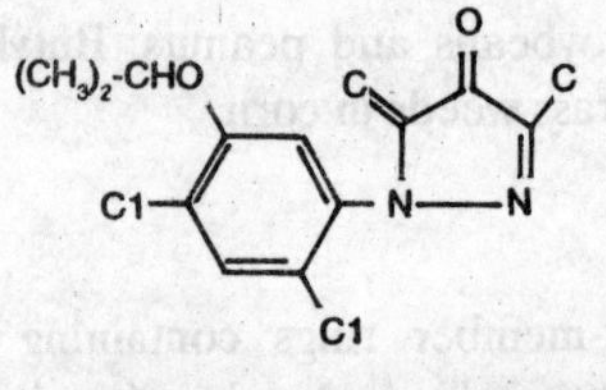

2- text- butyl-4-(24- dichloro -5- isopropoxy - phenyl- . Δ²-1,3,4 - oxadiazolic -5-one

BENTAZONE (Basagram)

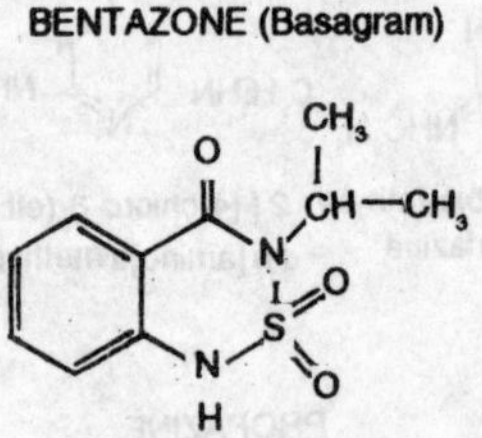

3.-(1-methylethyle)-1H-2,1,3-benzo thadiazin-4(3H)-one 2,2-dioxide

PYRIDINE

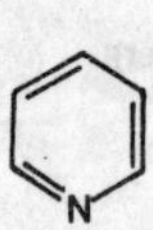

PICLORAM

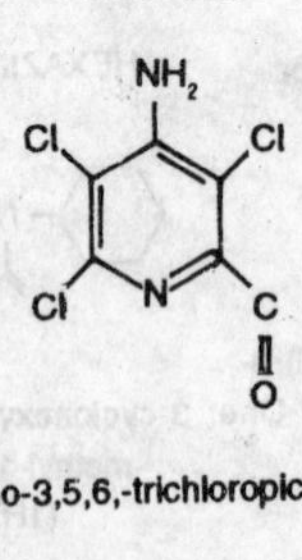

4-amino-3,5,6,-trichloropicalinic acid

URACIL NUCLEUS

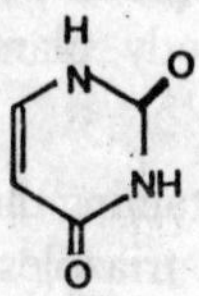

LENACIL

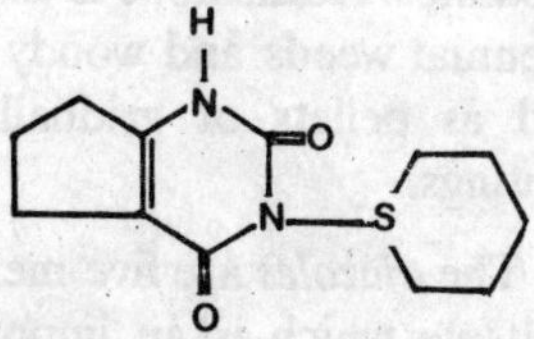

3-cyclohexyl-6,7-dihydro-1H-cyclopentapyrimidine-2,4 -(3'H',5'H)-dione

BROMACIL

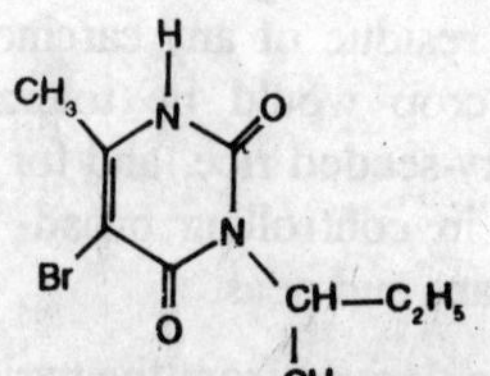

5-bromo-3-Sec-butyl-6methyluracil
[5 - bromo-6- methyl -3- (1- methylpropyl) uracil]

TERBACIL

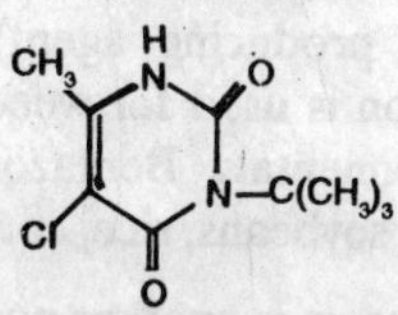

3-tert-burty-5 chloro- 6-methyluracil

containing ring) make up the heterocyclic nitrogens. The fundamental triazine nucleus is illustrated, showing the positions of the three nitrogens in the ring.

Most likely Tirazines which are strong unhibitors of photosphthesis, seems to be the most familiar group of netesocyclic nitregens because of their heavy use and notoriety. Their selectivity depends on the ability of tolerant plants to degrade or metabolize the parent compound (the susceptible plants do not). Triazines are applied to the soil primarily for their post-emergence activity. There are a number of triazines in the market today, some of which are illustrated.

The triazines are used selectively in certain crops and nonselectively in still others. They are used extensively in corn production and nonselectively on industrial sites.

Metribuzin is used on soybeans, wheat, sugarcane, and a few vegetables. Hexazinone is used in the control of annual, biennial, and perennial weeds and woody plants on noncroplant. Sometimes it is used as pellets or gridballs to control woody plants in conifer plantings.

The *triazoles* are five-membered rings containing three nitrogens. Amitsole which is an important member of triazoles functions in manner similar to triazines that is by inhibiting metosmthesis.In the form of aminotriazol, it became embroiled in the historic "cranberry incident" of 1959. This event culminated in the addition of the Delanely Amendment to the Pure Food and Drug Act in 1960.In essence, this amendment stated that no residue of any carcinogen (cancer- producing agent) in a food crop would be tolerated. Oxadiazon is used for weed control in dry-seeded rice, and for turf and ornamentals. Bentazon is effective in controlling broad- leaf weeds in soybeans, rice, corn, peanuts, beans and peas.

Picloram is another popular herbicide derived from the pyridine molecule.Picloram is a readily translocated herbicide used against broad-leafed and woody plants; it may be taken up from either the roots or the foliage. It has also been used experimentally as a growth regulator at extremely low dosages on apricots, cherries. The regulation of protein and enzyme synthesis through its effects on nucleic acid synthesic and metabolism is probably its mode of action.

The substituted uracils give a wide range of grass and broad-leaf control for an extended period by inhibition of photosynthesis Illustrated are terbacil, bromacil and lenacil.

Aliphatic Acids

TLA and Dalapen are the two aliphatic acids heavily used as herbicides. These are used against grasses mainly quackgrass and Bermudagrass.

TCA DALAPON

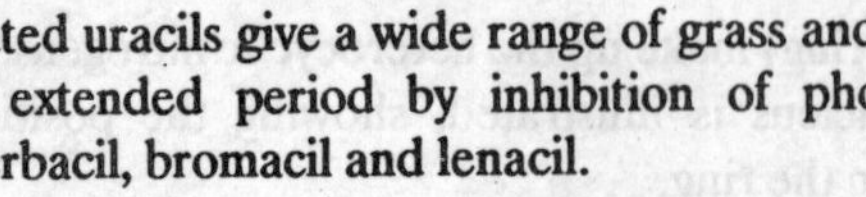

trichloroacetic acid 2,2,2-dichloroproprone acid

Arylalikphatic Acids
(Substituted Benzoic Acids or "Benzoics")

A number of these materials are employed as herbicides and are applied to the soil against germinating seeds and seedlings. The mode of action for dicamba and fenac is quite vague ; however, it is probably similar to that of the phenoxy herbicides (2,4-D, etc.), which interfere with the metabolism of nucleic acids. DCPA and chloramben produce auxinlike growth effects in plants but the available research data are insuficient to propose a mechanism of action.

Phenol Derivatives

The phenols which have been in use since a long time are still used in great quantity as general contact herbicides, usually against broad-leaf weeds, and selectively in cereal crops.

DICAMBA

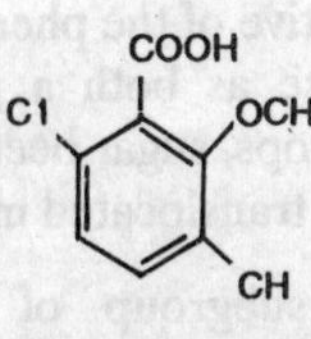

(2,3,6-trichlcropenyl) acctic acid

FENAC

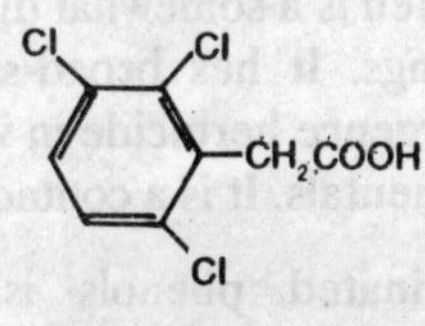

(2,3,6 - trichlorophenyl) acetic acid

DCPA

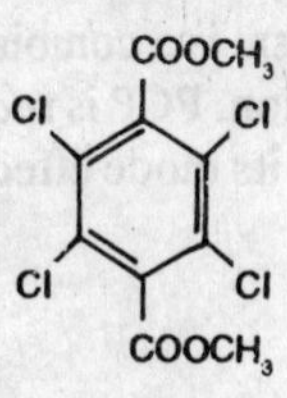

dimethyl tetrachloraterephthalate

CHLORAMBEN

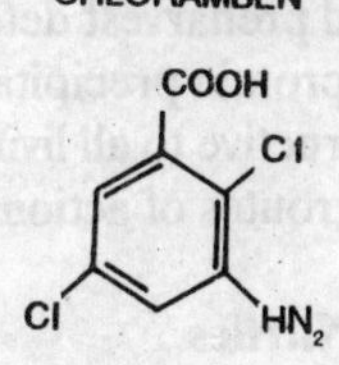

3-amino-2,5- dichloorobenzoic acid

DNOC

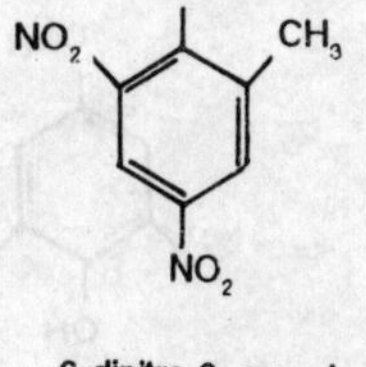

6-dinitro-0- cresol

DINOSEB

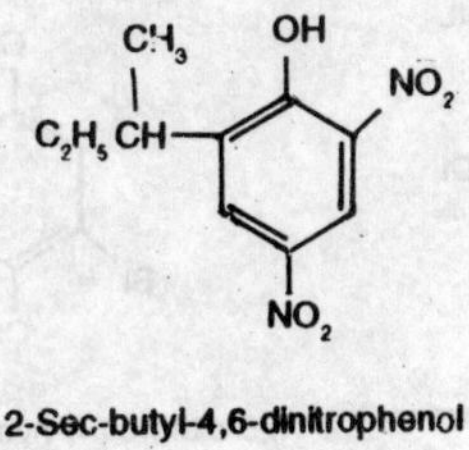

2-Sec-butyl-4,6-dinitrophenol

NITROFEN

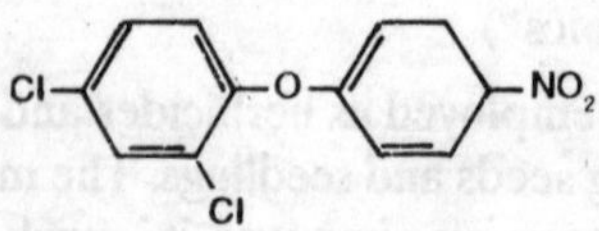

2,4-dichlorophenyl P-nitrophenyl ether

PCP

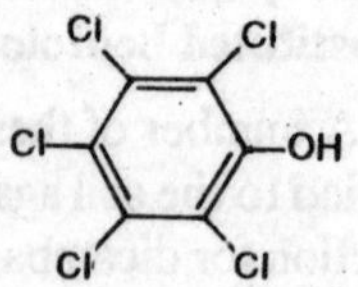

pentachlorophenol

Phenol derivatives are highly toxic to humans whatever be the route of entry into the body, and are nonselective foliar herbicides that are most effective in hot weather. The knitrophenols, represented by DNOC, were first introduced as herbicides in 1932. Dinoseb is the besst-known representative of the nitrophenols, which act by uncoupling oxidative phosphorylation. The dinitrophenols have also been used as ovicides, insecticides, fungicides and blossom-thinning agents.

Nitrofen is a somewhat different relative of the phenols, with two phenyl rings. It has broad-spectrum use as both a pre-post and post-emergence herbicide in vegetable crops, sugar beets, rice, and a few ornamentals. It is a contact and not a translocated material.

Chlorinated phenols is another subgroup of the phenol derinatines, of which only one member is recognised as a herbicide, namely PCP or pendachlorophenol. It is also used for termite protection and wood treatment for fungal rots, and as a noselective herbicide and preharvest defoliant. Its mechanism is a combination of plasmolysis, protein precipitation, and desiccation. PCP is recognized as being destructive to all living cells because of its mode effectiveness and multiple routes of action.

Substituted Nitriles

Nitriles are *organic* compounds containing the C≡N or cyanide grouping. There are three substituted nitrile herbicides, which have a

DICHLOBENIL

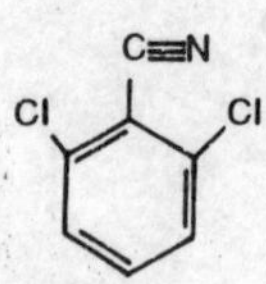

2,6- dichlorobenzonitriti

BROMOXYNIL

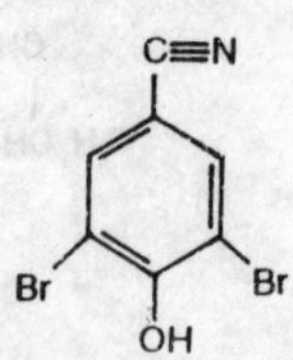

3,5-dibromo-4-hydroxybenzonitrile

10XYNIL

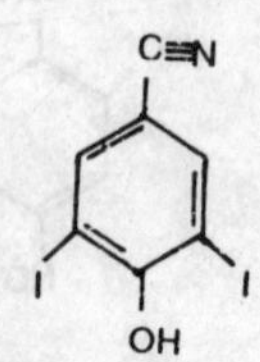

4-hydroxy-3.5-diiodobenzonitrile

wide range of uses against grasses and broad-leaf weeds. Their mechanisms of action are extensive, involivng seedling growth inhibition, lkpotato sprout inhibition, and gross disruption of tissues by inhibiting oxidative phosphorylation and preventing the fixation of CO_2. These effects, however, do not explain their rapid action. Ioxynil is not registered for use in the United States.

Bipyridyliums

The name bipyridylium suggests the attachment of two pyridyl rings. The group involves two important herbicides namely diqual and paraquat. Both are contact herbicides that damage plant tissues quickly, causing the plants to appear frostbitten due to the destruction of the cell membrane. This rapid wilting and desiccation occur within hours, making these novel herbicides also useful as preharvest desiccants for seed crops, cotton, soybeans, sugarcane, and sun-flowers. Diquat is also used in aquatic weed control. Neither material is active in soils. They are available only to professional weed control specialists, who can achieve spectacular results for home-owners.

DIQUAT

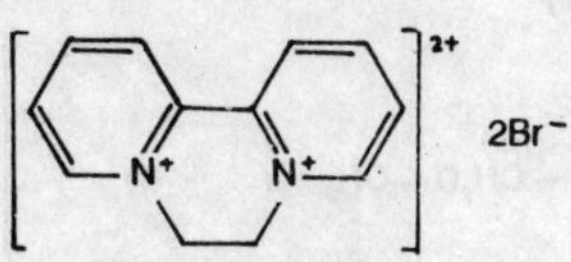

6,7-dihydrodiphyrido (1,2-α:2 ,1 -c) pyrazidinium
(dibromide

PARAOUAT

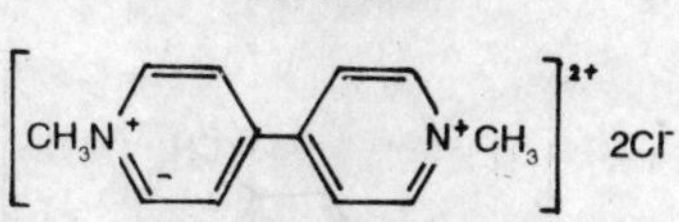

1,1-dimethyl-44 bipyridyiumion
(dichloxide)

Myco-Herbicides

A new approach to need control is the use of disease micro-organisms useful, sometimes self-perpetuating pathogens. The first to be registered in this field is *Phytophthora Palmivora* (Devine®), a naturally occurring, highly selective disease of milkweed vine *(Morrenia odorata),* a serious pest in citrus groves. Properly applied to the soil beneath citrus trees. *P. palmivora* kills existing milkweed vines. The microorganism exists in the vine root debris and conti- nues to control germinating vines for more than one year after a single treatment. This pathogen is selective and does not infest citrus tree

INDOTHALL

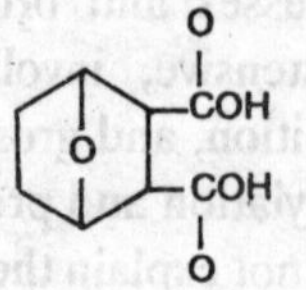

7- ox abicyclo (2,2,1) heptane-2.3-di carbozylic acid (sodium salt)

BENSULIDE

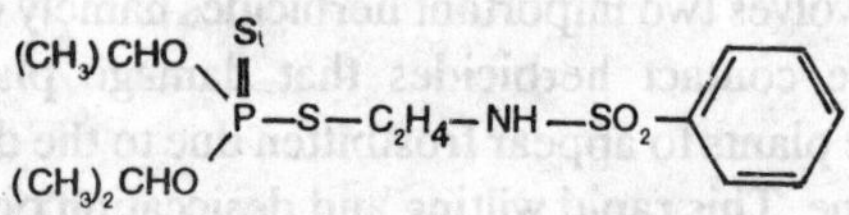

0,0-disopropyl phosphorodithioate S-ester with N-(2-mercaptoethyl) benzesulfonamide

ACROLEIN

$CH_2 = CH — CHO$

METOLACHLOR (Dual)

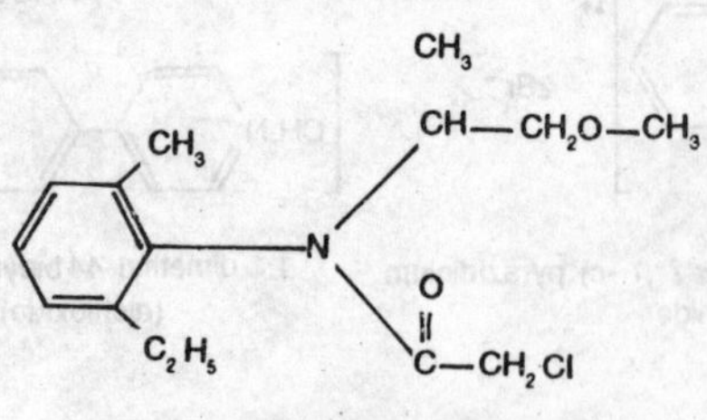

2-chloro-N--(2-ethyl 6-methyl phenyl)-N-(2-methoxy-1 methylethyl) acetamide

GLYPHOSPATE (Roundub

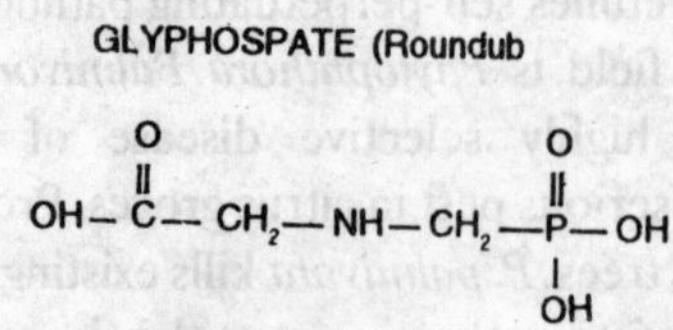

N-(Phosphonamethylo) glycine (isopropylamine salt

roots, fruit, or foliage. Precautions must be taken in its use since certain ornamental crops are susceptible to P. Palmivora.

Miscellaneous Herbicides

Methyl Bromide which comes under miscellaneous herbicides is used as a funigant for any known organism in soilnematodes, fungi, seed, insects and other plant parts. Allyl alcohol ($CH_2 = CHCH_2OH$) is a volatile, water-soluble fumigant used for the same purposes as methyl bromide.

Endothall which acts by interfering in RNA synthesis is used as acquatic weed and a selective weed crop herbicide. It's advantage over other acquatic weed killers is its low toxicity to fish. This is an outstanding example of environmental protection through persiticide selectivity.

Bensulfide is one of the better turf herbicides, especially for the control of crabgrass. It is an organophosphate, but is considered one of the less toxic herbicidal materials. Bensulide acts by inhibiting cell division in root tips. It is used as a pre-mergence herbicide in lawns to control certain grasses and broad-leaf weeds because of the fait that it is not translocated, it cannot be used as a foliar spray.

Acrolein which is an extremely useful acquatic weed herbicide has extensive tear gas effect, hence can only be used by licensed operators. Plants exposed to acrolein disintegrate with a few hours and float downstream. It is general plant toxicant, destroying plant cell membranes and reacting with various enzyme systems. Spot treatment in lakes will effectively kill weeds without destroying the fish population.

In 1974 Metolachlor was introduced it belongs to the chloroacetamide herbicides (or acetanilides). It is soil- incorporated as pre-plant and pre-mergence herbicide for annual weed grasses and some broad-leaf weeds in corn, soybeans and peanuts. It's presumed mode of action is the inhibition of nucleic acid metabolism and protein synthesis.

Glyphosate which is a nonselective, nonresidual pot emergence material was discovered in 1971 and belongs to the glyphosphate herbicide classification. Glyphosate is recognized for its effectiveness against perennial, deep-rooted, grass and broad-leaf weeds, as

OXYFLUOREEN) (Goal)

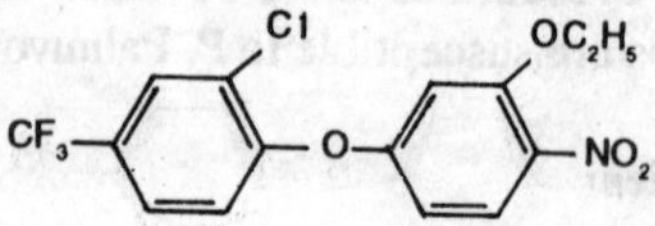

2-chloro-1-(3-ethoxy-4-nitrophenoxy)
-(trifluoromethyl) benzone

FOSAMINE AMMONIUM (Krenite)

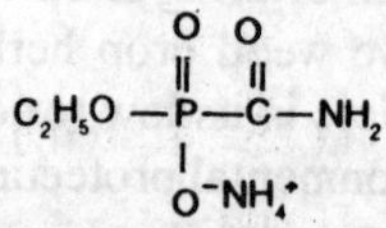

ammonium ethyl carbamylphosphonate

well as woody brush problems in crop and noncrop areas. It is a translocated, foliar-applied herbicide that can be applied at any stage of plant growth or at any time of year, with most types of application equipment, including the new wick, roller, and wiper devices. Its mechanism of action seems to be the inhibition of the synthesis of aromatic amino acids, which results in the inhibition of nucleic acid metabolism and protein synthesis. Oxyfluorfen belongs to the diphenyl ether herbicides and controls annual broad-leaf and grass weeds in soybeans, corn (witchweed control), cotton, deciduous fruits, grapes and nuts.

Fosamine ammonium belongs to the organophgosphate herbicides and is used as a foliar spray for brush control in non-cropland, principally on rights-of-way.

It has been seen that even the herbicides which are members of the same chemical class have sometimes different modes of action. Those discussed and illustrated here are but a cross section of the existing herbicides. We can expect that diferent classes will be developed in the future and their mechanisms of action will be more clearly delineated, following intensive research on this complex subject. The wealth of material precludes a brief summary. Rather, the reader is urged to frequently review the groups, to establish generalizations regarding herbicide use and action classes.

4

Herbicides and their Actions

This chapter as the name indicates talks about the mode of action of the herbicides, that is the biochemical responses of the plant appear to be associated with the herbicidal action. Because of the complex interrelations between cellular metabolism and growth development, responses to chemicals may take place at sites removed from the region of application or the original site of action. In many cases, the mode of action has not been worked out. However, it may be at possibly many herbicides are metabolically non-specific; that is, rather than having a single site of action, there could several sites and mechanisms involved in inhibition.

In presenting the modes of action for herbicides in a simple and straightforward way, I have made use of *Mode of Action of Herbicides*, by Floyed M. Ashton and Alden S. Crafts (1973), *Herbicide Handbook*, 4th ed., by the Weed Science Society of America (1979), and *The Biochemical Mode of Action Pesticides* by J. R. Corbett (1974). Readers desiring more explict information should consult one of these books.

PHYSICAL TOXICANTS

Petroleum oils

The action of oils on plants is leads to disruption of cellular membranes. Due to leakage of the cell contents into the intercellular spaces. by the use of oil, the plant undergoes darkening a loss of turgor as if frostbitten. The effects of oils are similar to the symptoms produced by the bipyridylium (diquat and paraquat) herbicides, which are also thought to disrupt membranes. The actual molecular mechanism by which oils disrupt membranes is not known, though it is suggested that they act by dissolving part of the lipid membrane.

Bipridyliums

Any ploat on which paraquat or diquat is sprayed is pilled within hours, sometimes, the effects being in visible within minutes.Paraquat and diquat kill within hours any and all plants onto which they are sprayed, with their effects sometimes visible in minutes. They are highly water soluble and are so tightly bound by soil that they become unavailable to plant roots pratically as soon as they reach the ground. Bipyridylium herbicides are almost completely dissociated in solution into positive and negative ions. The herbicidal action is due to the positive ion, which is reduced by photosynthesis to from a relatively stable free redical. The bipyiridyl free radical is easily oxidized in the presence of oxygen to reform the original ion and hydrogen peroxide, which destroys the plant tissue. They cause rapid desiccation of the foliage, with wilting as an early smyptom. At the cellular level they cause cell and chloroplast membranes to rupture, leading to frostbitten appearance. Either or both the hydroxyl radical (OH^-) and hydrogen peroxide (H_2O_2) are the primary toxicants.

AUXINLIKE EFFECTS

Phenoxyaliphatic Acids

The most extensively used phenoxyaliphatic acids are 2,4-D, 2,4,5-T, and MCPA. The effect of these herbicides on broad-leaf plant is auxinlike (auxin is a growth hormone) : elongation of the growing terminals, distortion, and in 7 to 10 days colapse, withering, and death, 2,4-D is indeed used as an auxin or growth regulator in traces. The phenoxys increase cellular division, activate phosphate metabolism and usually increase ribonucleic acid (RNA) synthesis, which increases protein synthesis. We can rule out both intgerference with oxidative phosphorylation in mitochondria and inhibition of the Hill reaction in chloroplasts as mechanisms to explain the herbicidal action of 2,4-D. Hence the mode of action of the phenoxy herbicides undues of a large number of structural and biochemical reactions involving prolonged abnormal growth and the failure of those changes characteristic of maturity and senescence.

Arylaliphatic Acids (Substituted Benzoic Acids)

Arylalphatic acids such as 2,3,6-TBA, dicamba, fenac, chloramben, and DCPA are related to benzoic acids, act in the same way as the phenoxyaliphatic acids (2,4-D and relatives); that is, they resemble auxins (growth hormones) and are persistent in their actions. For example 2,3,6-TBA cause cell elong- ation and tissue proliferation, induces adventitious roots, modifies and the aggangement or leaves and other organs, and causes fruit to develop in the absence of fertilization. Picloram, though not a member of this class but rather a heterocyclic nitorgen, also possesses the typical properties of auxins and herbicidal action similar to that of 2,4-D. The specific mode of action of the benezoic acid herbicides is unknown.

METABOLIC INHABITORS

Inhibitors of the Electron Transport Chain

The mode of action of phenols such as DNOC, dinoseb, DNAP, and pentachloro- phenol (PCP) is two fold. In high concentrations they destroy cell membranes of treated foliage resulting in fluid loss, cell collapse, and decissation. When present in low concentration the phenols block the formation of ATP by uncoupling oxidative phosphorylation. Their actions in plants are the same as in animal tissues.

Nitroanilines, also known as to as dinitroanilines, include trifiluralin, benefin, nitralin, and isopropalin. They inhibit growth of the entire plant, apparently by limiting root growth, especially the lateral or secondary roots. They are absorbed mainly by both roots and shoots and but are only slightly translocated to other plant parts. Generally, the nitroanilines seem to stimulate increased production of nucleic acids in both roots and shoots, the most relevant effect related to the phytotoxic sympoatoms. The uncoupling of exidative phosphorylation and inhibition of the production of several hormone-induced enzymes have also been demonstrated with tripluratin.

Bromoxynil, dichlobenil, and ioxynil (which is not available in the United States) belong to the nitrile herbicides. Ioxynil and bromoxynil which are contact herbicides are used to control hard-to-kill weeds in cereals. They are strong inhibitors of oxidative phosphorylation and also inhibit CO_2 fixation. These effects do not entirely explain the rapid contact effect of these compounds, but they may be indicatgors of more drastic effects on cell membrane systems. Dichlopenil which is a pre-emergence herbicide is applied by incorporation into the soil, and its site of action is in the meristems of embryos as they begin to germinate. It is strongly adsorbed by the soil, with almost no leaching, making it useful in no-till or minimum tillage vineyard and orchard programes. The precise mode of action of dichlobenil not known. It is known that the substituted nitriles release CN^-, which binds to cyhtochromec oxidase, a respiratory enzyme intermediate in the electron transport chain.

Inhibitors of Phosphorus Metabolism

The arsenicals (cacodylic acid, MMA, MSMA, DSMA, MAMA) act through enzyme systems to inhibit growth. Their actions in plant systems are no different than in animal systems. Arsenicals kill plants relatively slowly, with the first symptom being chlorosis, the appearance of bleached or yellow spots on leaves as a result of the loss of chlorophyll. Arsenic occurs in herbicides in various forms from the inorganic to arsenites arsenates, and arsonates. Arsenic is known to uncouple phosphorylation, react with – SH groups of enztmes, and act like phosphorus in all the important roles of phosphorus. All the above mentioned effects can without any doubt, be put into the underadge of mechanisms of Arsenical herbicides.

Enzyme Inhibitors

The substituted amides contains the herbicides CDAA, dephenamid, propachlor, and propanil. Apparently their chemical structures are more closely related than their modes of action, For mechanism-of-action studies on the amides have not produced a consistent pattern. The various known actions include inhibition of photosynthesis, respiration, RNA and protein synthesis, amylase, proteinase, dipeptidase, and other enzymes.

Glyphosate is a very broad spectrum herbicide that is relatively nonselective. It appears to inhibit the aromatic amino acid biosynthetic pathway and may inhibit certain enzymes whose role is unclear.

Inhibitors of Nucleic Acid Matabolism and Protein Synthesis

Carbamates. Barban, chlorpropham, phenmedipham, propham, swep and terbucarb are all included in carbamates. Their chemical structures or derivations are more closely related than their modes of action, for mechanism-of-action studies on the carbamates have not produced a homogeneous patter. They have been shown to lower down the ATP content of tissue sections, and to inhibit RNA and protein sysnthesis oxidative phosphorylation, and the Hill reaction of photo- syntheses. Basically, they interfere with cell division, which must relate to the involvment of protein synthesis and associated nucleic acid synthesis and the energy needs for these reactions are in the form of ATP.

Thiocarbamates. All thio-and dithiocarbamates have low vapour pressure, are volatile, and must be soil-incorporated for this reason. They inhibit the growth of shoots of germinating seedlings to a larger extent degree than they inhibit the roots, and they are absorbed by the shoot more readily than by the roots. Collectively they alter several aspects of plant metabolism : photosynthesis, respiration, oxidative phos- phorylation, protein synthesis, and nucleic acid metabolism. The primary site of action may be nucleic acid metabolism and protein synthesis, because CDEC and EPTC inhibit L-lecucine incorporation into protein in segments of barley coleoptile or sesbania hemp hypocotyls. Additionally, EPTC inhibits growth and RNA synthesis in soybean tissue.

Aliphatic Acids. Two examples of the chlorinated aliphatic acids are trichoroacetic acid (TCA) and dalapon. The specific mechanism of action for these herbicides are not known. It has been suggested that the primary site of action for dalapon is associated with modification of protein structure, including enzymes, not by precipitation but probably by subtle conformational changes, perhaps by combining with proteins. Dalapon is known to increase ammonia in the cells of treated plants and susceptible species do not seem to be able to form amides fast enough to prevent toxic levels of ammonia. Even less is known about TCA, but it is assumed that it acts like dalapon.

Photosynthesis Inhibitors

Ther are several modes of actions for herbicides, one of these which is of great importance is the inhibition of the Hill reaction. The

Hill reaction is a light-initiated reaction that splits water (photolysis), resulting in the production of free oxygen (O_2) by plants. Chlorophyll, which is a green pigment is plants, is an essential ingredient in the reaction, since it catalyzes the production of oxygen from water and the transfer of the hydrogen to a hydrogen acceptor. A simplified reaction for the above process is written below:

$$2H_2O + 2A \xrightarrow[\text{chlorophyll}]{\text{light}} 2AH_2 + O_2$$

where A is some unidentified hydrogen acceptor. The hydrogen and acceptor complex (AH_2) continue or in reactions with CO_2 to form plant sugars and cellulose, while free O_2 is released into the atmosphere.

Triazines

Altrazine and simazine are two of the most prominent of the numerous triazines is the market today. The triazines inhibit the growth of all organs of intact plants, which results due to a defi- ciency of photosynthate, necessary for growth, caused by inhibition of photosynthesis. The tolerance of certain species of higher plants to the triazines probably results from their ability to rapidly degrade the herbicide to nontoxic matabolites. The mechnism of action of the triazines is blockage of photosynthesis, specifically the photolysis of water. Since this along cannot account for the great effects of the triazines, it appears that a secondary phytotoxic agent is formed by a reaction that is probably coupled to the water photolysis and carries the burden of triazine activity.

Triazoles

Amitrole is the only member of the unique, 5-member-ring class of triazoles. It stimulates growth at low concentrations and inhibits growth at high concentrations. Amitrole arts by interfering with the development of chloroplasts which results in the loss of green pigment in the treated plants, which therefore, appear to become alberes. Amitrole is absorved by leaves and roots and translocates throughout the plant, causing biochemical alterations that affect carbohydrate, lipid, and nitorgen metabolism. Its primary site of action seems to be the biochemistry of chloroplast development. This, however, does not explain the phytotoxic symptoms beyond albinism.

Substituted Ureas

The dermatives of urea involves monuron, diuron, fenuron, linuron, and siduron. They are not all equally phytotoxic. It is common to find a four-to six fold difference in toxicity to some species. Most are readily absorbed by the roots and repidly translocated to the upper plant parts, producing phytotoxic symptoms that are most visible in the leaves. The primary site of action for the urea herbicides is the inhibition of the Hill reaction of photosynthesis. But since the plants do not merely starve from lack of photosynthate, it is theorized that a secondary toxic substance is formed in the oxygen-releasing of photosynthesis.

Substituted Uracils

Bromacil, isocil, and terbacil, are three widely used uracils. All are soil-borne herbicides, readily absorbed by roots and translocated to the leaves. They inhibit the Hill reaction and interfere with a step in the photosynthetic pathway near that of oxygen release. In this way the mechanism of action of substituted uracils resemble those of substituted ureas.

5

Fungicides and Bactericides

Strictly speaking chemicals used to kill or halt the development of fungi. However, for our purposes we shall consider them as chemicals used to control bacterial as well as fungal plant pathogens, the causal agents of most plant diseases. Other organism like viruses, rickettisias, alge, nematodes, mycoplasma like organism and parasitic seed plaints causes plant diseases. In this chapter we discuss only those chemicals used for control of fungi and bacteria.

Of plant diseases here are hundreds of examples which include storage rots, seedling diseases, root rots, gall diseases, vascular wilts, leaf blights, rusts, smuts, mildews, and viral diseases. These can, in many instances, be controlled by the early and continued application of selected fungicides that either kill the pathogens or inhibit their development.

With today's fungicides most plant diseases can be controlled to some extent. Phytophthora and phizoctonia rots Fusarium, verticillium and bacterial wilts and the viruses. are among those that are only beginning to be controlled with chemicals. The difficulties with these diseases are that they either occur below ground, and thus beyond the reach of fungicides, or they are systemic within the plant.

Fungus is a plant living in close quarers with its host so fungal diseases are basically more difficult to control with chemicals than are insects. This explains the difficulty of finding selective chemicals that kill the fungus without harming the plant. Also, fungi that can be controlled by fungicides may undergo secondary cycles rapidly and produce from 12 to 25 "generations" during a 3-month growing season. Consequently, repeated applications of protective fungicides may be necessary, due to plant growth dilution and removal by rain and other weathering.

Before there is any evidence of disease Fungicides must be applied to plants during stages when they are vulnerable to inoculation by pathogens, before there is any evidence of disease. Those fungicides referred to as *chemotherapeutants* can help to control certain disease after the symptoms appear. Also, protective fungicides are commonly used, even after symptoms of diseaese have appeared. Eradicant fungicides are usually applied directly to the pathogen during its overwintering stage, long before disease has begun and symptoms have appeared. However, the fungicide must be applied as protectant spray, in advance of the pathogens, to prevent the disease are in the case of crops, whose sale depends on appearance, such as lettuce and celery.

In our present arsenal there are about 225 fungicidal materials, most of which are recently discovered. Most of these act as protectants, preventing spore germination and sub- sequent fungal penetration of plant tissues. Protectants are applied repeatedly to cover new plant growth and to replenish the fungicide that has deteriorated or has been washed off by rain.

The application principle for fungicides differs from that of herbicides and insecticides. Only that protion of the plant that has a coating of dust or spray film of fungicide is protected from disease. Thus a good uniform coverage is essential. Fungicides, with several new exceptions, are not systemic in their action. They are applied as sprays or dusts, but sprays are preferable, since the films stick more readily, remain longer, can be applied during any time of the day, and result in less off-target drift.

Thanks to modern chemistry, many of the serious diseases of grain crops are controlled by treating the seeds with selective materials. Other are controlled with resistant varieties. Diseases of fruit and vegetable are often controlled by sprays or dusts of fungicides.

Historically, fungicides have relied on sulfur, copper, and mercury compounds, and even today most of our plant diseases could be controlled by these groups. The development of organic fungicides vests in the reason, that sulfur and copper compounds can retard grwoth in sensitive plants. These sometimes have greater fungicidal activiy and usually have less phytotoxicity.

In organic forms of copper, sulfur, land mercury, until recently and metallic complexes of cadmium, chromium and zinc along with a variety of organic compounds are inclusions of the general purpose fungicides for agriculture whereas the general purpose lawn and garden fungicides are few in number and are usually organic compounds. The general-purpose lawn and garden fungicides are few in number and are usually organic compounds.

INORGANIC FUNGICIDES

Sulfur

Sulfur in many forms is probably the oldest effective fungicide known and is still a very useful garden fungicide. There are three physical forms or formulations of sulfur used as fungicides. The first is

TABLE 4.1.

A sampling of the vast number of inorganic copper compounds used as fungicides.

Name	*Chemical formula*	*Uses*
Cupric sulfate	$CuSO_4\ 5H_2O$	Seed treatment and preparation of Bordeaux mixture
Cupric hydroxide (Kocide®)	$Cu(OH)_2$	Seed treatment, foliage spray; many fungal diseases
Copper oxychloride	$3Cu(OH)_2\ CuCl_2$	Powdery mildews
Copper oxychloride sulfate	$3Cu(OH)_2\ CuCl_2$ $3Cu(OH)_2\ CuSO_4$	Many fungal diseases
Copper ammonium carbonate (Copper-Count-N®)	Chemical complex (formula not known)	Many citrus,deciduous fruit, and vegetable diseases
Cuprous oxide	Cu_2O	Powdery mildews
Basic copper sulfate	$CuSO_4\ Cu(OH)_2\ H_2O$	Seed treatment and preparation of Bordeaux mixture
Cupric carbonate (malachite®)	$Cu(OH)_2\ CuCO_2$	Many fungal diseases
Copper resinate (Citcop®)	Salts of fatty and rosin acids	Bacterial and fungal diseases of grapes, citrus, vegetables

finely grounds sulfur dust that contains 1 to 5 per cent clay or talc to assist in dusting qualities. The sulfur in this form may be used as a carrier for another fungicides or an insecticide. The second is flotation or colloidal sulfur, which is so very find that it must be formulated as a wet paste in order to be mixed with water. It would be impossible to mix this with water when it is in its original dry, microparticle size, as it would merely flow but would merely float. Wettable sulfur is the third form; it is finely ground with a wetting agent so that it will mix readily with water for spraying. The easiest to use, of course, is dusting sulfur, applied when plants are slightly moist with the morning dew.

The particle size of wettable sulfur should be no larger than 7 7μm for controlling the most effective disease. A good grade of dusting sulfur should pass through a 32.5-mesh or finer screen.

Copper

The majority of inorganic copper compounds are practically insoluble in water and are pretty blue, green, red, or yellow powders sold as fungicides. The various forms include Bordeaux mixture, named after the Bordeaux region in France, where it originated. Bordeaux is a chemically undefined mixture of copper sulfate and hydrated lime, which was accidentally discovered when sprayed on grapes in Bordeaux to scare off "freeloaders." It was soon observed that downy mildew, a disease of grapes, disapeared from the treated plants. From this unique origin began the commercialization of fungicides. The fungicidal as well as phytotoxic and properties are provided by the copper, ion, which becomes available from both the highly soluble and relatively insoluble copper salts. A few of the many inorganic copper compounds used over the years are presented.

The EPA has determined that no tolerance level need be set for a large number of copper compounds : Bordeaux mixture, copper acetate, copper carbonate – basic, copper-lime mixtures, copper oxychloride, copper silicate, copper sulfate, copper sulface – basic copper – zinc-chromate, cuprous oxide, cupric oxide, and copper hydroxide.

A comment on solubility is appropriate at this point. In general, protective fungicides have low ionization constant, but in water some toxicant does go into solution. That small quantity absorbed by the fungal spore is then replaced in solution from the residue. The spore continues to accumulate the toxic ion in sublethal doses, whose cumulative effect is lethal. Except for powdery mildews, water in the penetration court – the place on plant tissue softened by fungal mycelium, which permits it to penetrate the plant cutlicle – thus permits spore germination and makes soluble the toxic portion of the fungicidal residue.

To discourage killing all or portions as the host plant, the copper,ion which being toxic to all plant cells must be used in dicrete dosages or in relatively insoluble forms. This is the basis for the use of relatively insollluble or "fixed" copper fingicides, which release only very low levels of copper, adequate for fungicidal activity, but not enough to affect the host plant.

Copper compounds are not easily washed from leaves by rain, since they are relatively insoluble in water, and thus give longer protection against disease than do most ofd the organic meterials. During spraying no special precaution is required as they are relatively safe to use. Although copper is an essential element for plants, there is some danger in an accumulation of copper in agricultural soils resulting from frequent and prolonged use. After using fixed copper for disease control a serious problem, in fact, of copper toxicity is being experienced by certain, citrus growers in Florida.

The currently accepted theory for the mode of action of copper's fungistatic action is its nonspecific denaturation of protein. The Cu^{++} ion reacts with enzymes having reactive sulfhydryl group which would explain its toxicity to all forms of plate life.

Mercury

The inorganic mercurial fungicides are probably the most toxic of the fungicides. Mercury's fungicidal properties and toxicity to animals are due in part to the degree of association of divalent mercury ions, which are toxic to all forms of life. As a result, no mercury residues, are permitted in foods or feed.

Over the past 30 dears, many organic mercury compounds were developed, but they have been replaced by other organic fungicides. Phenyl- mercury acetate (PMA) was useful for turf diseases, as a seed treatment, and as a dormant spray for fruit trees whereas ceresan is typical of those used as seed treatment. The mode of action for the mercurials is the nonselective or nonspecific inhibition of enzymes, especially those containing iron and sunfhydryal sites.

CERESAN

$CH_3OCH_2CH_2HgCl$

2-methoxyalhylmercuric chloride

PMA

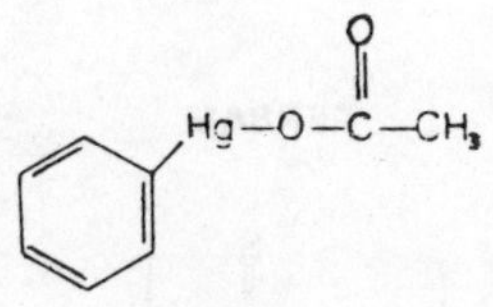

phenylmercury acetate

With one or two exceptions, both organic and in organic mercurial Fungicides have been banned from home and agricultural use by the EPA. This decision hinged on their toxicity to warm-blooded animals and accumulation of mercury in the enviornment.

Organic Fungicides

To replace the more harsh, less selective in organic materials many synthetic sulfur and other organic fungicides have been developed over the past 35 years. Most of them had no measurable build up effect on the enviornment after many years of use. Thiram, the first of the organic sulfur fungicides, was discovered in 1931; it was followed by many others. On 1943 and 1949 respectively, other new classes, the dithiocarbamates and dicarboxmides (Zineb and captan) was introduced. Now more than 200 fungicides of all classes are in use or in various stages of development.

Several outstanding qualities are passessed by newer organic fungicides, that is being extremely efficient, smaller quantities are required than of those used in the past, usually lasts longer and are safe for crops, animals and the environment. Most of the newer fungicides also have very low phytotoxicity, many being at least ten times safer than the copper meterials. And most of them are readily degraded by soil microrganisms, thus preventing their accumulation in soils.

THIRAM

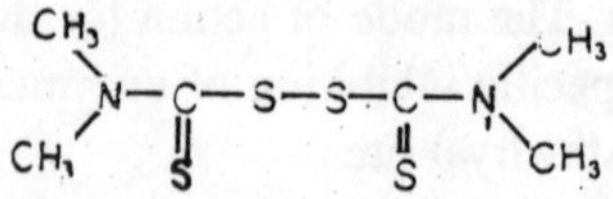

bis(dimethylthio-carbamoyl) disulfide

MANEB

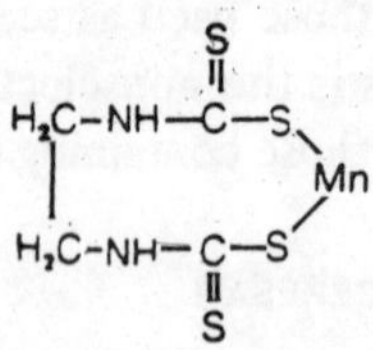

manganese ethylenebisdithiocarbamate

FERBAM

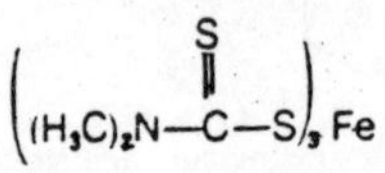

ferric dimethyldithiocarbamate

ZINEB

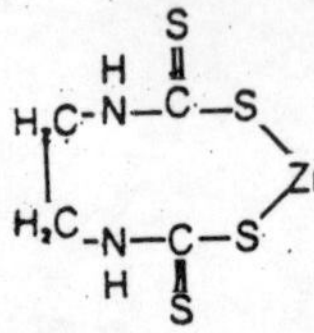

zinc ethylenebisdithiocarbamate

TRICYCLAZOLE (Beam®)

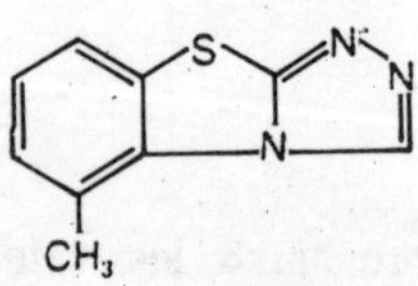

5-methyl-1,2,4-triazolo(3,4-b)-benzothiazole

Dithiocarbamates

Among the dithiocarbamates we find the "old reliables" – thiram, meneb, ferbam, ziram, Vapam® (SMDC), and zineb all developed in the early 1930s and 1940s. Such fungicides probably have greater popularity and use than all other fungicides combined. Except for systemic action, they are employed collectively in every use known for fungicides. The dithiocarbamates probably act by being matabolized to the isothiocyanate radical (–N = C = S), which inactivates the –SH groups in amino acids contained within the individual pathogen cells.

Thizoles

The thirzoles, a class of compounds that offers a surprising chemical disposition, contains among others, ethazol. The five-

ETHAZOL (Terrazole®)

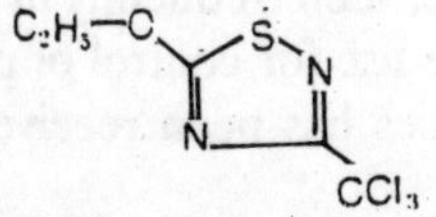

5-ethoxy-3-trichloromethyl-1,2,4-thiadiazole

membered ring of the thiazoles is claved rather quickly under soil conditions to form either the fungicidal $-N = C = S$ or a dithiocarbamate, depending on the structure of the parent molecule. Ethazol is used only as a soil fungicide and, as such, is exposed to the ring cleavage just mentioned. The probable mode of action is similar to that of the dithiocarbamates.

PCP

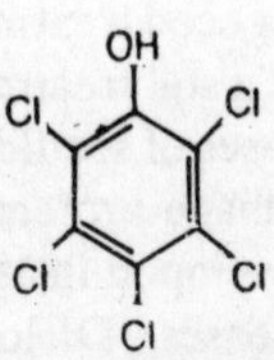

pentachlorophenol

CHLOROTHALONIL

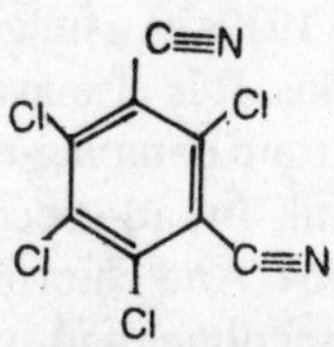

tetrachloroisophthalonitrile

CHLORONEB

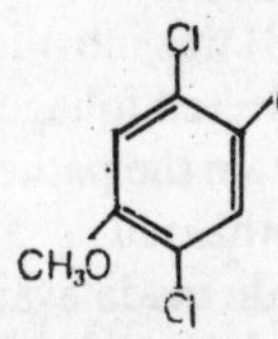

1,4-dichloro-2,5-dimethoxybenzene

ANILAZINE

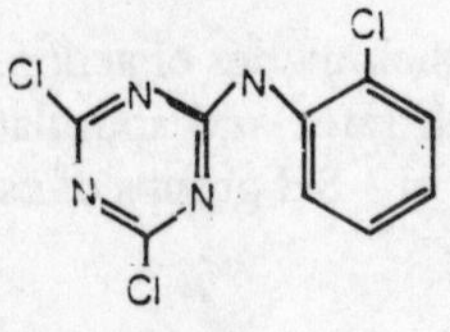

2,4-dichloro-6-(o-chloroanilino)-s-triazine

HEXACHLOROBENZENE

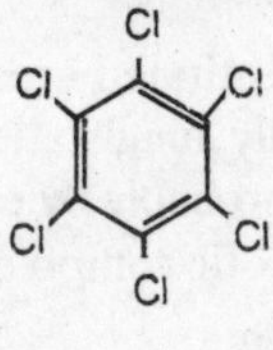

1,2,3,4,5,6-hexachlorobenzene

PCNB

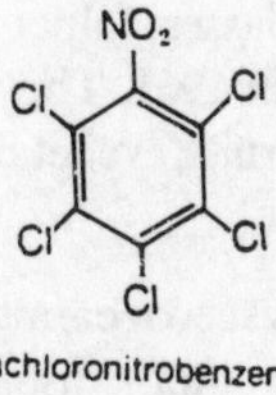

pentachloronitrobenzene

DICLORAN (DCNA, Botran®)

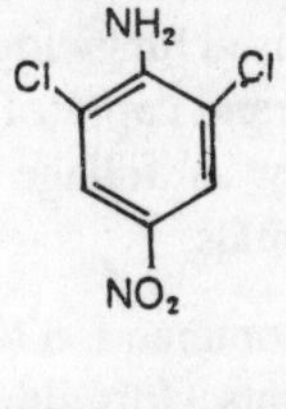

2,6-dichloro-4-nitroaniline

Triazines

The triazines structure, seen frequently in herbicides, is found in only one fungicide. A wide use for control of potato and tomato leaf spots and turf-grass diseases has been received by Anilazine which was introduced in 1955.

Substitued Aromatics

The substituted aromatics belong in a somewhat arbitrary classification assigned to the simple benzene derivatives that possess long-recognized fungicidal porperties.

For seed treatment and as soil treatment to contract stinking smut of wheat, Hexachlorobenzene, was introduced in 1945. Pentachlorophenol (PCP) has been used since 1936 as a wood preservative as a seed treatment, and, as a herbicide. Pentachloronitrobenzene (PCNB) was introduced in the 1930s as a fungicide for seed treatment and selected foliage applications. It is also availed as a soil treatment to restrain the pathogens of certain damping-off diseases of seedlings. Chlorothalonil is a very useful, broad-spectrum foliage-protectant fungicide made available in 1964. And chloroneb, developed in 1965, is heavily used for cotton seedling and turf diseases. Dicloran (DCNA) is a highly useful fungicide against *Botrytis, Monilinia, Rhizopus, Sclerotinia*, and *Sclerotium* species, on a wide range of fruits and vegetables.

Substituted aromatics are diverse in their modes of action. Being generally fungistatic, they reduce growth rates and sporulation of fungi, probably by combining with $-NH_2$ or $-SH$ groups of essential meta- bolic compounds.

Dicarboximides (Sulfenimides)

Dicarboximides are three extremely useful foliage protectant fungicides. Captan appeared in 1949 and is undoubtedly the most heavily used fungicide around the home of all classes; folpet appeared in 1962; and captafol (Difolatan®) appeared in 1961. They are used primarily as foliage dusts and sprays on fruits, vegetables, and ornamentals.

Recommended for lawn and garden use, as seed treatment and as protectants formildews, late blight, and other diseases, the

CAPTAN

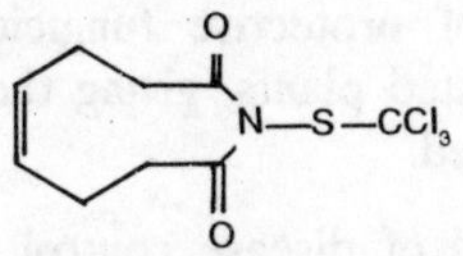

N- (trichloro methylio)-4- cycloherene-
1,2- dicarboximide

FOLPET

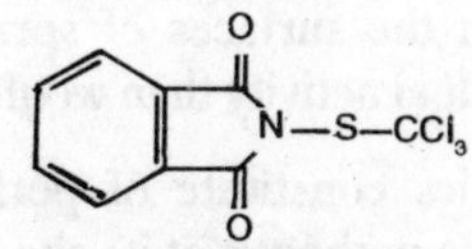

N- trichloromethylthioPhthalimide

Dicarboximides are some of the safest of all pesticides available. Remember the garden adage, "When in doubt, use captan."

CAPTAFOL (Difolatan)

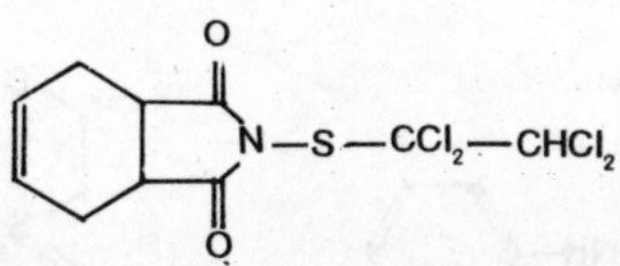

tetrachloroethylmercaptocyclohexenedicarboximide

Many compounds containing the $-SCCl_3$ moiety are fungitoxic, a fact that indicates that group as a toxophore (molecular unit that account for the molecule's toxicity). Fungitoxicity of the dicarboximides is aparently nonspecific and is not a result of a single mode of action. The dicarboximides' lethal effect on disease organisms is probably due to the inhibition of the synthesis of amino compounds and enzymes containing the radical.

Systemic Fungicides

Only in recent years have successful systemic fungicides been marketed, and very few are available. Systemics are absorbed by the plant and carried by translocation through the cuticle and across leaves to the growing points. Most systemic fungicides have eradicant properties that stop the progress of existing infections. They are therapeutic in that they can be used to cure plant diseases. To give prolonged disease control a few of the systemics can be applied as soil treatment and are slowly absorbed assimilated through the roots.

These systemics offer much better control of diseases than is possible with a protectant fungicide that require uniform application and remain essentially where it is sprayed onto the plant surfaces.

There is, however, some redistribution of protective fungicidal residues on the surfaces of sprayed or dusted plants, giving them longer residual activity than would be expected.

Systemics constitute of perfect method of disease control by attacking the pathogen at its site of entry or activity, and they reduce the risk of contaminating the environment by frequent broadcast fungicidal treatments. Undoubtedly, as newer and more selective systemic molecules are synthesized, they will gradually replace the protectants that compose the bulk of our fungicidal arsenal.

Those systemics currently in commerical use are mentioned in order of their appearence. None are available for home use.

CARBOXIN

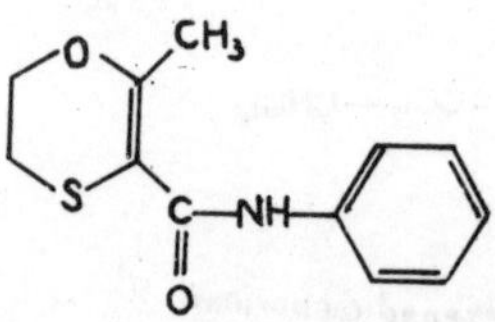

2,3-dihydro-5-carboxanilido-6-methyl-1,4-oxathiin

OXYCARBOXIN

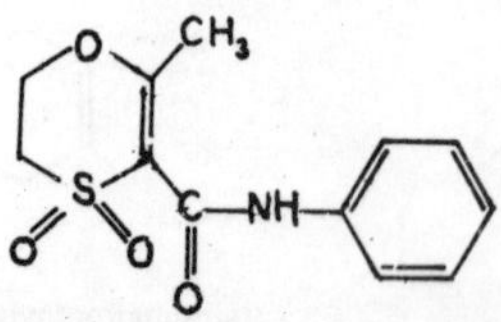

2,3- dihydro-5 carboxanilido -6- methyl-1,4 oxathiin -4,4- dioxide

Oxathiins. The first of the systemics to suceed in practice, introduced in 1966, were the Oxathiins, represented by carboxin and oxycarboxin. They are used as seed treatments for the cereal crop, particularly those affected by embryo-infecting smut fungi, and they have potential for other used. They are selectively toxic to the smuts, rusts, and to *Rhizoctonis (Thanatephorus),* organisms belonging to the Basidiomycetes. The apparent mode of action of the oxathiins beings with their selective concentration in the fungal cells, followed by the inhibition of succinic dehydrogenase, an important enzyme to respiration in the mitochondrial systems.

Benzimidazoles. The benzimidazoles, represented by benomyl and thiabendazole (TBZ), were introduced in 1968 and have received wide acceptance as systemic fungicides against a broad spectrum of diseases. Benomy has the widest spectrum of fungitoxic activity of all the newer systemics, including the *Sclerotinia, Botrytis*, and *Rhizoctonia species*, and the powdery mildews and apple scab. Thiabendazole has a similar spectrum of activity to that of benomyl. Introduced in 1969, thiophanate, although not a benzimidazole in its original structure, is converted to that group by the host plant land the

BENOXYL

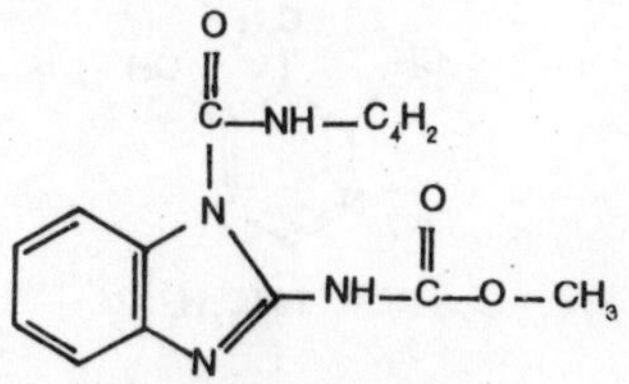

methyl 1- (butylcarbamoyl) -2-
benximidazolecarbamate

THIABENDAZOLE

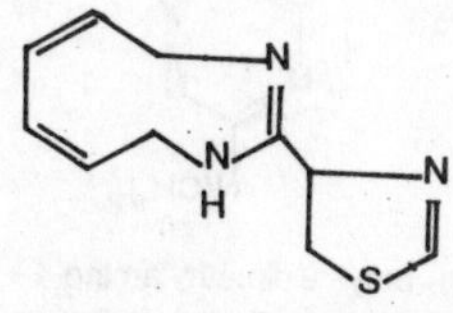

2-(4-thiazoyl) benzimidazole

THIOPHANATE

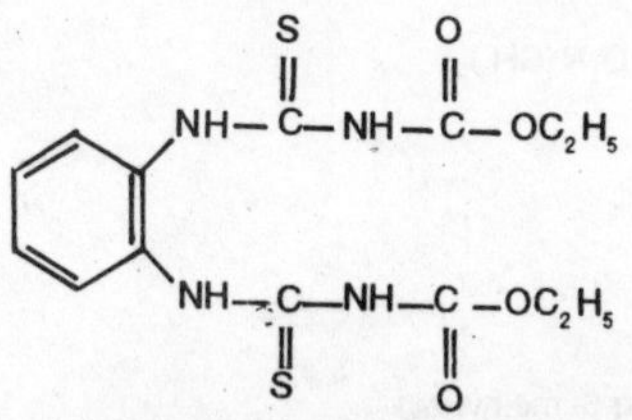

1,2-bis(3-ethoxycarbonyl-2-thioureido)benzene

CARBENDAZIM (Lignasan)

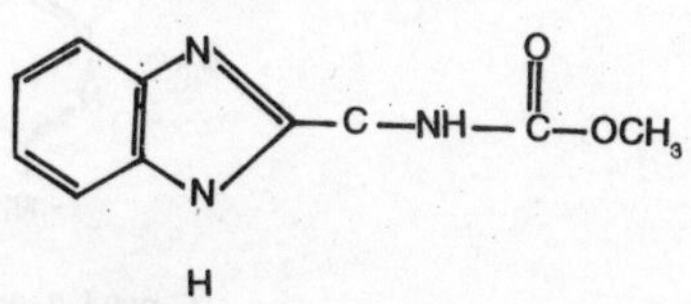

2- (methoxycarbonylaminq) - benzimidazole

fungus through their metabolism. Thiophanate has a fungitoxicity similar to that of benomyl. All three compounds have been used in foliar application, seed treatment, dipping of fruit or roots, and soil application. Their mode of action appears to be the induction of abnormalities in spore germination, cellular multiplication, and growth, as a result of their interference in the systhesis of that vital nucleic meterial, DNA.

A later addition to the benzimidazoles is carbendazim, introduced in 1973. The interesting quality of this systemic is its proved usefulness in the control of the formidable Dutch elm disease, in the formulation of Lignasan®, hydrochloride salt, it is injected into the trunks of diseased trees, and results in a slow curative action.

Pyrimidines. The pyrimidine systemic fungicides appeared in the late 1960s, and include dimethirimol, ethirimol and bupirimate. They are very active against specific types of powdery mildews : for instance, dimethirimol works well on cucurbits. Ethirimol is for cereals and other field crops, and bupirimate controls powdery mildews on apples and greenhouse roses.

DIMETHIRIMOL (Milcurb)

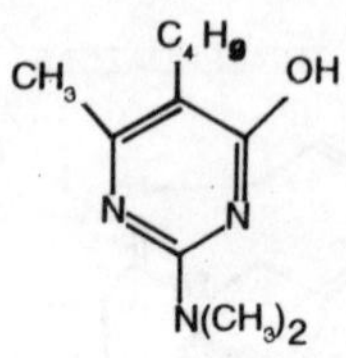

5-*n*- butyl- 2 dimethylaminq-4-
hydroxy-6-methylprimedine

ETHIRIMOL (Milcurb Super)

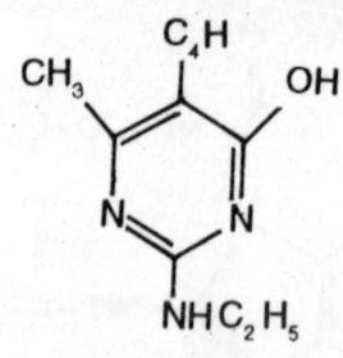

-butyl-2- ethylaminq-4- hydroxy-
6-methylprimidine

BUPIRIMATE (Nimrod)

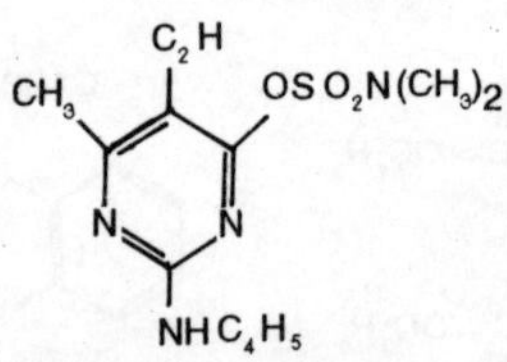

5- butyl- 2 ethyl aminq-6- methyl-
primidin-4 yl- dimethysulfamate

Organophosphates. Among the newer systemic fungicides are two organophosphorous meterials, IBP and Conen ®. Developed in Japan and intorduced in 1965, both are effective against rice blast, stem rot, and rice sheath blight.

Conen

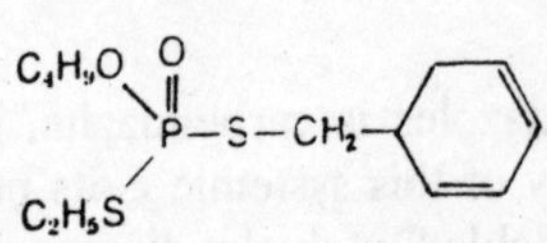

O-butyl-S-benzyl-S-ethyl pr osphorodithioate

IBP (Kitazin)

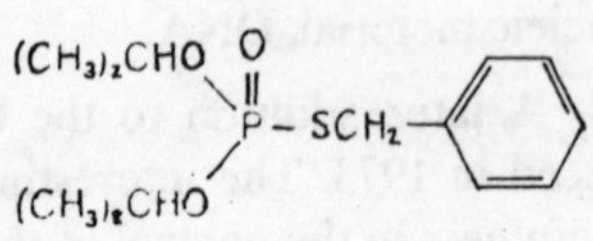

O O-diisopropyl-S-benzyl thiophosphate

Acylalanines. Acylalanines are another new group of systemic fungicides which includes metalaxyl and funalacyl. They are effective aganist soil-borne diseases caused by *Pythium* and *Phytophthora* and foliar diseases caused by the phycomycetes (downy mildews). They show promise as foliar, soil, and seed treatments for agricultural crops. They offer systemic and corrective or curative activity as well as residual-protectant activity. However, against the Ascomycetes, Basidiomycetes and Fungi Imperfecti they have little or no activity.

METALAXYL (Ridomil®. Dual®)

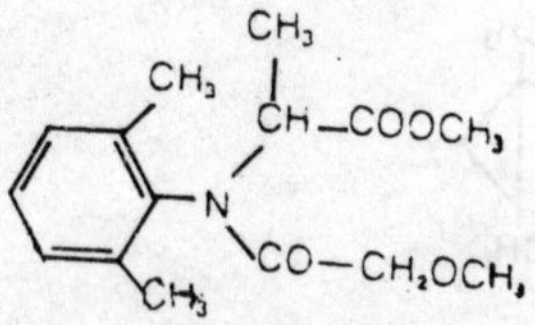

N-(2,6-dimethylphenyl)-N-(methoxy-acetyl) alanine methyl ester

FURALAXYL (Fongarid®)

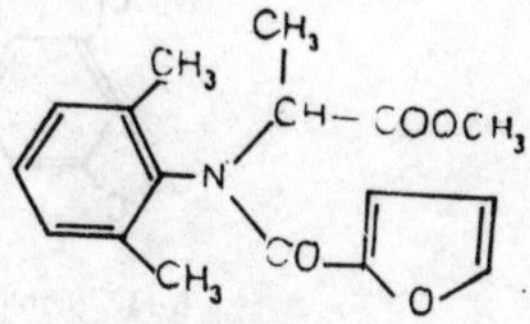

methyl N-2,6-dimethylphenyl-N furoyl-(2)-alaninate

TRIADIMEFON (Bayleton)

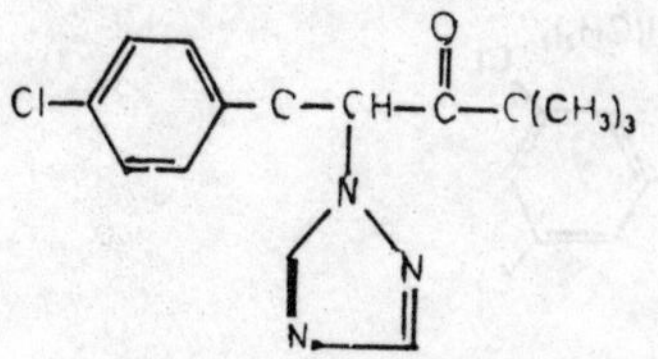

1-(4-chlorophenoxy)-3 3-dimethyl-1-(1H01 2 4-triazol-1-yl)-2-butanone

TRIFORINE

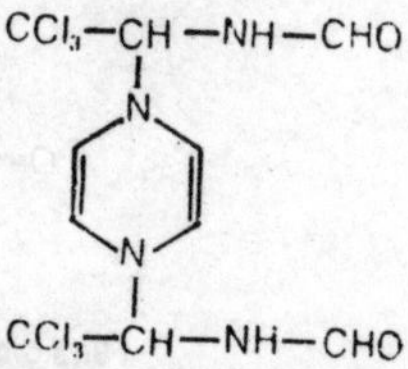

N-N -(1,4-piperazinediyl-bis 2 2,2-trichloroethylidene))-bis-(formamide)

Triazoles. The sole systemic fungicide of the triazole group is the Triadimefon which carries both protective and curative actions and is effective against mildews and rusts on vegetables, cereals, coffee, deciduous fruit, grapes and ornamentals.

Piperazines. The poperazine systemic fungicides include only one member at this time, triforine, which is amazingly effective against powdery mildew on any host. Additionally it is active against scab and other diseases of fruit and berries, rust and black spot on ornamentals, powdery mildew and other leaf diseases on cereals, rust on cereals, several diseases of vegetables, land storage diseases of fruit.

Imides. A second group of dicarboximides, the imides, are structurally significantly different from the originals. These systemic chemicals apeared during the 1970s. The group now includes procymidone, iprodione and vinclozolin. They are particularly effective against *Botrytis, Monilinia* and *Sclerotinia*. Iprodione is also active against *Alternaris, Heminthosporium, Rhizoctonia, Corticium Typhula* and *Fusarium* species.

Dinitrophenols

We have mentioned the dinitorphenols are insecticides and as herbicides. Their mode of action as fungicides is the same : the

PROCYMIDONE (Sumilex®)

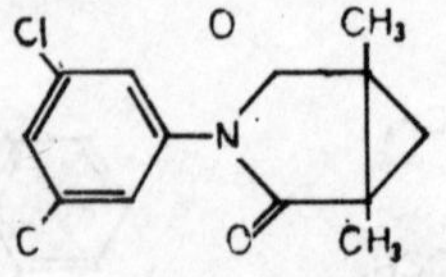

N-(3 5-dichlorophenyl)-1,2-dimethyl-
cyclopropane-1 2-dicarboximide

IPRODIONE (Rovral®)

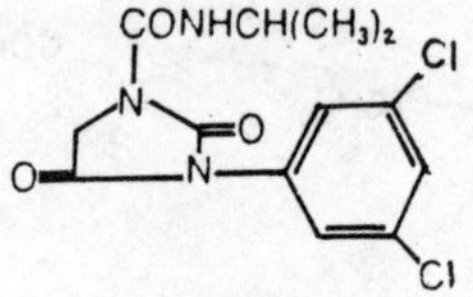

3-(3.5-dichlorophenyl)-*N*-(1-methyl-
ethyl)-2.4-dioxo-1-imidazolidine-
carboxamide

DINOCAP

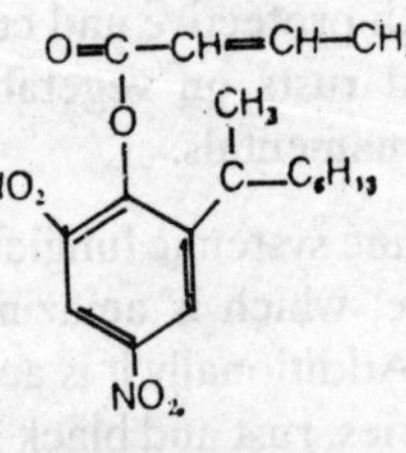

2.4-dinitro-6-(2-octyl)phenyl crotonate

uncoupling of oxidative phosphorylation in cells with an attendant upset of the energy systems within the cells.

VINCLOZOLIN (Ronilan)

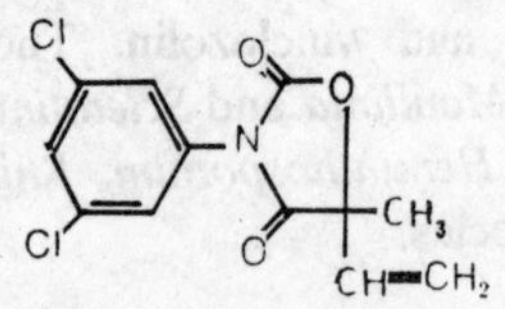

3-(3.5-dichlorophen, -5-ethenyl-
5-methyl-2 4-oxazolidinedione

Dinocap (Karathane®) has been since the late 1930s, both as an acaricide and for powdery mildew on a number of fruit and vegetable crops. Dinocap undoubtedly acts in the vapour phase, since it is quite effective against powdery mildews whose spores germinate in the absence of water. This is popular home fungicide.

Quinones

The quinones are a facinating chemical group that offers countless numbers of molecules that are potential fungicides. Chloranil is the first of these to appear (1937). Until the decorboximides it was used heavily as a seed treatment and Foliar application. The most popular of this group, however, is dichlone. It is used on a number of fruit and vegetable crops and for treatment of

CHLORANIL

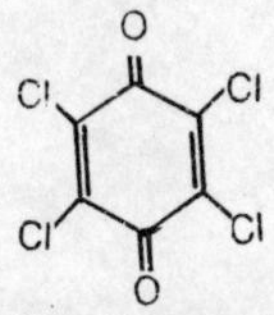

2.3.5.6-tetrachloro-1.4-benzoquinone

DICHLONE

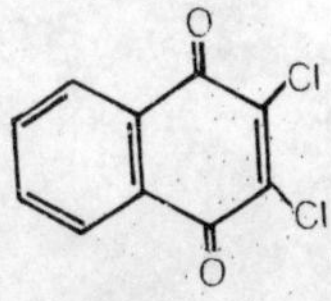

2.3-dichloro-1.4-naphthoquinone

ponds, to control blue-green algae. Dichlone affects cellular respiration in many fungi and acts by attaching to the –SH groups in enzymes, thus inhibiting their action and indirectly uncoupling oxidative phosphorylation.

Aliphatic Nitrogen Compounds

A Fungicide that has proved effective in controlling certain diseases such as apple and pear and cherry leaf spot was introduced in the mid 1950s and is known as Dodine having disease specificity and right systemic qualities. It has disease spcificity and slight systemic qualities. Its mode of action is not totally clear, but it is taken ukp repidly by fungal cells, causing leakage in these cells, possible by alteration in membrane permeability. The guanidine nucleus of dodine is also known to inhibit the synthesis of RNA.

FENTIN HYDROXIDE (Du-Ter®)

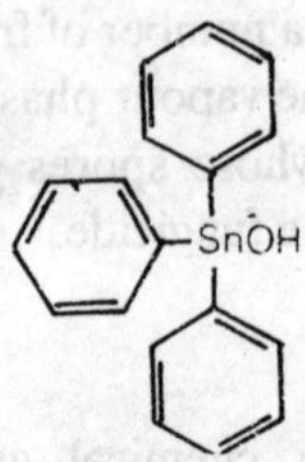

triphenyltinhydroxide

DODINE

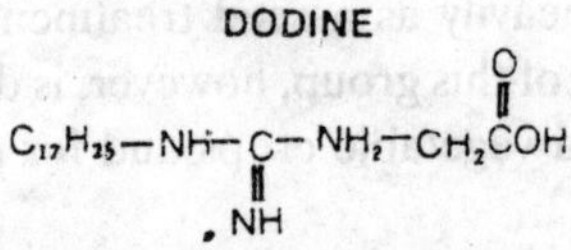

n-dodecylguanidine acetate

CHLOROPICRIN

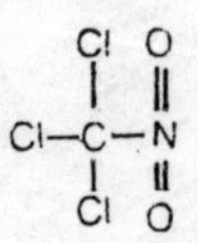

trichloronitromethane

METHYL BROMIDE

CH_3Br

bromomethane

MIT

$CH_3—N=C=S$

methylisothiocyanate

Fumigants

As with the insecticides, there are distinct highly volatile, small-molecule fungicides that have fumigant action. They are unrelated chemically but are handled alike and are dealt with as a single class in this book. Chloropicrin was mentioned in as a waring agent, in grain fumigants, but it is also an ideal fumigant itself : It controls fungi, insects nematodes, and weed seeds in the soil. Equally effective against fungi, nematodes and weeds. Methylbromide, is also listed as a fumigant insecticide. Methylisothiocyanate (MIT) is closely related to the dithiocabamates and has a similar mode of action against fungi, nematodes and weeds. SMDC was listed, but not illustrated, in the dithiocarbamate fungicide, where it belongs chemically. SMDC decomposes in the soil to yield methylisothiocyanate. None of these is available for home use.

Antibiotics

Such antibiotics are penicillin, tetracycline, and chloramphenicol are used medically against human becterial diseases. These are not involved in our present study. But we note in passing that the oxytetracylines are therapeutic against some of those mysterious mycoplasmalike diseases. As in the case of the medically important antibiotics the antibiotic fungicides are substances produced by microorganisms, which in very dilute concentrates inhibit growth and even destroy other microorganisms. To date, several hundred antibiotics have been reported to have fungicidal activity, and the chemical structures of about half of these are already known.

The largest source of antifungal antibiotics is the actinomycetes, a group of lower plants. Two antibiotics streptomycin and cycloneximide is obtained brom one amuzion species, *Streptomyces griseus,* which is found within the action mycetes.

STREPTOMYCIN

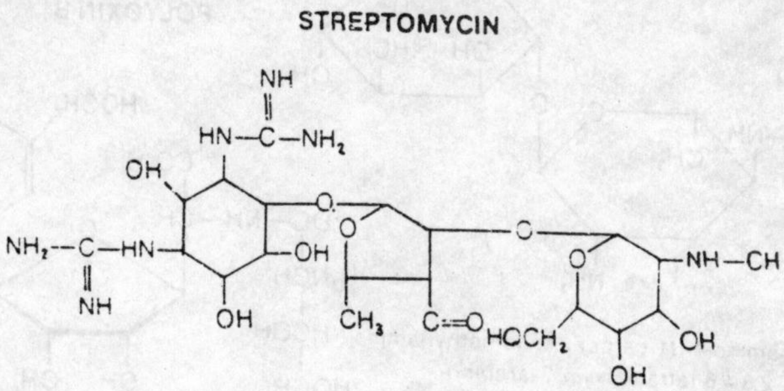

2.4-diguanidino-3.5.6-trihydroxycyclohexyl-5-deoxy-2-O-(2-deoxy-2-methylamino-α-glucopyranosyl)-3-formyl pentofuranoside

Streptomycin is used as dust, spray, and seed treatment for the prevention of mostly of bacterial diseases such as blight on apples and pears, soft rot on leafy vegetable, and some seedling diseases. It is also effective against a few fungal diseases.

Streptomycin, the mode of action of which is not clearly understood, probably interferes into synthesis of proteins. Despite the evidence of antibiotic-resistant strains, strepomycin has a place in the control of some bacterial diseases, and the tetracyclines may well play an important part in controliing some mycoplasmalike diseases of plant. Because streptomycin appears to have a single and specific site of action, resistance to it by both beceteria and mycoplasmas in inveitable. Cycloheximide is a smaller, less complicated antibiotic, about which more is understood. First, cycloheximide is toxic to a

CYCLOHEXIMIDE

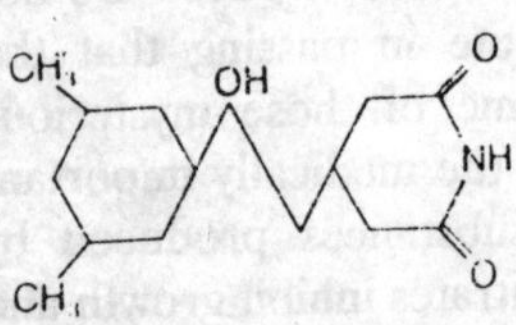

β[2-(3,5-dimethyl-2-oxocyclohexyl)-2-hydroxyethyl]glutarimide

BLASTICIDIN-S

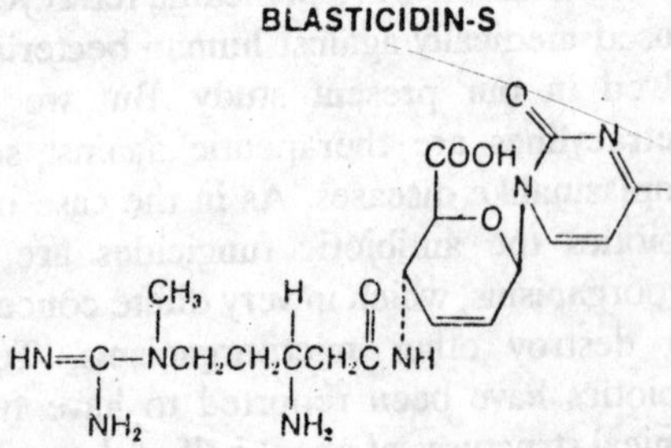

(*S*)-4[[3-amino-5[(aminoiminomethyl)methylamino]-1-oxopentyl]amino]-1-[4-amino-2-oxo-1(2*H*)-pyrimidinyl]-1,2,3,4-tetradeoxy-β-D-erythrohex-2-enopyranuronic acid

KASUGAMYCIN

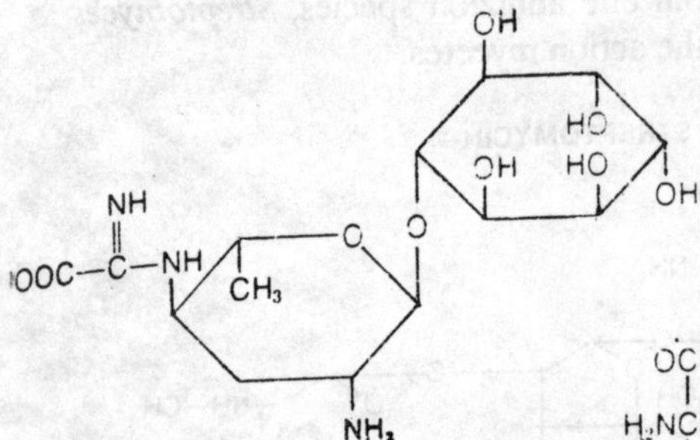

D-3-*O*-[2-amino-4-[(1-carboxyiminomethyl)amino]-2,3,4,6-tetradeoxy-α-D-arabino-hexopyranosyl]-D-chiro-inositol

POLYOXIN B

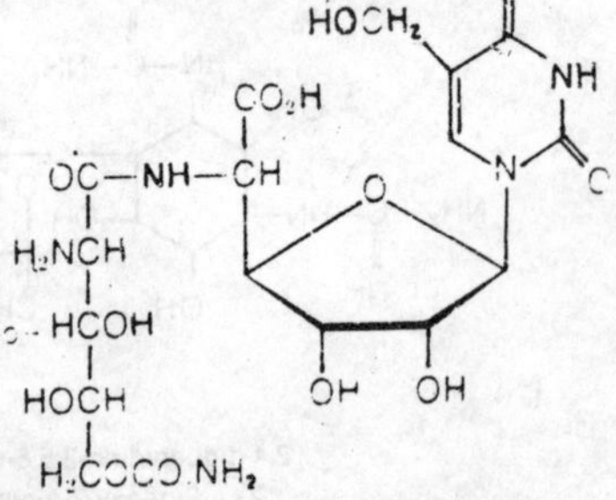

5-[[2-amino-5-*O*-(aminocarbonyl)-2-deoxy-L-xylonoyl]amino]-pyrimidinyl]-β-D-allofura-1,5-dideoxy-1[3,4-dihydro-5-(hydroxymethyl)-2,4-dioxo-1(2*H*)-nuronic acid

wide range of organisms, including yeasts, filament-forming fungi, algae, protozoa higher plant and especially mammals. Surprisingly, it is inactive against becteria, perhaps because the bacteria fail to absorb it. Cycloheximide cause growth inhibition in yeasts and filament forming fungi by inhibiting portein and RNA synthesis. Thus, with a fair amount of confidence, we can say that both streptomycin and *cycloheximide* act by inhibiting the synthesis of nucleic acids.

In the control of powdery mildew, yusts, turf diseases and certain blights a Fungicide known as Cycloheximide was intorduced in 1948 which at present has become very popular. It is best known commercially under the name Acti-Dione®. Because of its high acute toxicity, it cannot be purchased for home garden use.

Blasticidin-S, discovered in 1955, is produced by the fermentation of *Streptomyces griseochromogenses*. Kasugamycin, introduced in 1963, is formed by *Streptomyces kasugaenis*, and Polyoxins are extracted from Streptomyces cacaoi. All three are Japanese contributions to the systemic antibiotics. The first two are effective against rice, blast, and the latter against rice sheath blight.

6

Fungicides and their actions

All agro chemical show their effects on different organic of the insect pests in different ways. Fungicidal action is usually expressed in one of two physically visible ways : the inhibition of spore germination or the inhibition of fungus growth. Most fungicides prevent spore germination or kill the spore immediately following germination. Some of these chemical inhibitors or toxicants also retard or halt fungus growth when applied after the infectious stage has developed. The newer systemics fungicides have eradicant properties and stop the progress of existing infections. What happens at the cellular level to cause these readily visible results ? As currently viewed, all fungicides are metabolic inhibitors; that is, they block some vital metabolic process. For the sake of organization and simplicity, we can classify the modes of action into three broad groups : inhibitors of the electorn transport chain, inhibitors of enzymes, and inhibitors of necleic acid metabolism and protenin synthesis.

INHIBITORS OF THE ELECTRON TRANSPORT CHAIN

Sulfur

Whenever Certain pathogenic organisms and most mites come contact with sulfur and in also by its fumigant action at temmeratures above 22°C they are killed. The fumigant effect is, however, somewhat secondary at marginal temperatures and under windy conditions. It is quite effective in controlling powdery mildews of plants that are not unduly sensitive to sulfur. Unlike those of any other fungus, spores of powdery mildews will germinate in the absence of a film of water in the penetration court (a spot on the tissue that is "softend" prior to penetration by spore). Sulfur is absorbed by fungi in the vapour state, land its fumigant effect – acting at a distance – is undoubtedly important in killing spores of powdery mildews. Sulfur interferes in electron transport along the cytochromes and is then reduced to hydrogen sulfide (H_2S), a toxic entity to most cellular porteins. The way fungi respond to sulfur suggests that sulfur can stimulate enzyme activity. The reduction of sulfur to hydrogen sulfide (H_2S) by fungi is a metabolic process that diverts protons (H^+) from steps of normal hydrogenation reactions. This reduction is aerobic (requires oxygen) and involves a cytochrome syhstem, possibly cytochomeb, a respiratory lenzyme involved in the oxidation of nutritional components and foreign compounds. Sulfur presumably serves as an acceptor of hydrogen atoms from dehydrogenases, and then activates to full capacity the enzymic steps preceding sulfur reduction.

Oxathiins

Carboxin and oxycarboxin are the only two oxthiins currently in use. Carboxin inhibits glucose and acetate oxidation by intact fungi,and noncompetitively inhibits succinate, but not $NADH_2$, oxidation by mitochondria. This effect occurs in the electron transport chain of respiration. Succinate accumulates in carboxin-treated cells of fungi, suggesting that inhibition of succinate oxidation is the primary site of action. Oxycarboxin, though less toxic, has the same mode of action.

ENZYME INHIBITORS

Copper

There are several inorganic copper compounds like protectants meterials applied before the pathogen appears, to protect plants from inoculation. They prevent the spread of an infection but cannot eradicate an existing one. Thus, the site of copper fungicidal action must be the fungal spore. Copper ions are concentrated by fungal spores from the surrounding medium, some up to 100-fold over that in the immediate invironment. This high concentration of copper inside the spore supports the generally held view that the fungitoxic activity of copper ions is due to a nonspecific attraction for various groups in the cell, such as imidazole, carboxyl, phosphate, or thiol, resulting in nonspecific dentujration of protein and enzymes. The currently accepted theory for the mode of action of copper's fungistatic action is its nonspecific denaturation of protein. The cupric ion (Cu^{++}) reacts with enzymes having reactive sulfhydryl (–SH) groups, which explains copper's toxicity to all forms of plant life, and especially its toxicity to the vulnerable copper-concentrating spores and cells.

Mercury

The fungicides of mercurry compounds fungicides are purely historical and classical toxicologic interest, for they are no longer registered except for occasional use. Considerable concern over the widespread use of mercury pesticides focused on the toxic nature of mercury itself and the discovery in 1970 that methyl mercury (CH_3Hg^+)– which is more toxic to life than aryl (aromatic-ring derivatives) mercurials or inorganic salts of mercury– is synthesized in the biosphere. As a result, most registrations for mercury fungicides were canceled. The biochemical basis for mercury's toxicity is generally thougt to consist to interactions with the thiol groups of preoteins, where mercury ions (HG^{++}) form compounds attaching to one or two sulfurs. Since there are so many functional –SH groups in a living system, it is virtually impossible to select one reaction of mercury as being the most important to its toxic effect. Organic mercury is different only in that its chemical form is R-Hg^+ rather than Hg^{++}. The organic mercurials are more soluble in lipid than the inorganic Hg^{++}, enabling them to penetrate more readily into cells and tissues. In general, inorganic mercury is a more effective enzyme inhibitior than organic mercury, though the organic forms remains as potent toxicants.

Dithiocarbamates

This category includes in the dithiocarbamates, the oldest group of the organic fungicides, are maneb, ferbam, zineb, manzate, dithana, polyram and ziram. The dithiocarbamates may act by being metabolized to the isothio- cyanate radical (–N = C = S), which inactivates the –SH groups in amino acids, proteins, and enzymes contained within the individual pathogen. It is dufficult to establish exactly how they act since they are unstable, and the chemical nature of the fungitoxic agent(s) is not known for certain. A more recent theory tends to discount the isothiocyanates as toxic products, suggesting instead that ethylene thiuram disulfide. An important function in both the lower and higher plants that may help explain the mode of action for not only the dithiocarbamate fungicides but also others, including those derived from the heavy metals, is *chelation*. A chelate is an organic ring structure composed of a metal atom linked to a ring by nitrogen, oxygen, or lsulfur. Particularly as they involve enzymes, chelates are powerful and essential entities in the matabolic processes of plant. Some of the metals required by the higher plants and fungi in trace amount assist enzymes in conducting their routine duties of metabolism. The metal may be active in this role as a chelate with the biological component.

One of the generally accepted theories to explain the fungicidal activity of copper, mercury, cadmium, and other heavy metals is the formation of chelates within the fungal cells. The chelates, in turn, disrupt portein synthesis and metabolism. And since the most critical protein of cells are nezymes the metals required in trace amounts appearing in abundance or excessive quantities are equivalent to the introduction of potent poisons in the cells. If the chelation is correct,

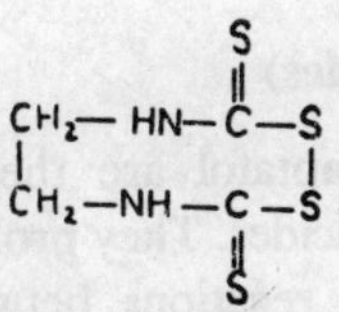

it would account for the mode of action for the organic and inorgnic heavy metal fungicides, for the formation of isothiocyanates from the dithiocarbamate molecules, and for the potency of the heavy-metal dithiocarbamates. Certain fungicides are in themselves chelating agents. They attaih to the scarce metal components, such as Fe, Mg, and Zn, within the cells, literally robbing cells of essential materials. In summary, chelation of heavy metals plays an important role in both the life and the death of cells.

Thiazoles

Ethazol, tricyclazole, and Busan-72 belong to the thiazole group. The unstable five-member ring of the thiazoles is broken rapidly under soil conditions to form either the fungicidal $-N=C=S$ or a dithiocarbamate, depending on the structure of the parent chemical.

The isothiocyanatae inactivates – SH or – SR groups in amino acids, proteins, and enzymes contained within the individual pathogen. Thus the site of action is nonspecific.

Substituted Aromatics

The substitued aromatics are diverse in their modes of action. Being generally fungistatic, they reduce growth rates and sporulation of fungi, probably by combining with amino ($-NH_2$), -SH, or SR groups in essential amino acids, proteins, or enzymes. Hexachlorobenzene possesses some fumigant qualities, especially with wheat bunt, but its mode of action is unknown. Chlorthalonil is believed to work by inactivation of thiol groups in the fungal cell. Chloroneb may act by inhibiting DNA synthesis. PCNB is effective against many species of fungi having chitin in thier cell walls, but is virtually inactive against fungi in which chitin is either absent or present in small quantities, for example, *Pythium* and Phytophthora. Thus, there is indirect evidence that PCNB may act by interfering with chitin synthesis. Pentachlorophenol (PCP) block the formation of ATP by uncoupling koxidative phosphorylation.

Dicarboximides (Sulfenimides)

Captan, folpet, and captafol are the early members of the dicarboximide class of fungicides. They probably act nonspecifically, because the nature of the reactions between the fungicides and components of the gungal cell are not definitely established. It is generally lkbelieved that – SH or – SR groups are likely reaction sites. However, folpet inhibits isolated chymotrypsin, which does not contain thiol groups, so that reactions not involving thiol groups could be involved in fungicidal action.

Quinones

Belonging to the quinones are dichlone and chloranil. Fungal spores take up a large amount of dichlone before expiring, which suggests that dichlone exerts its toxicity by a simultaneous action on a veriety of loci in major metabolic pathways. It is not clear what chemical reactions quinones undergo in the fungus. They do react readily with thiols, land they probably enter into a variety of other reactions in the cell, inculding combination with amino groups. The primary toxic effect of quinones could therefore be due to reaction with – SH or – SR and – NH_2 groups of vital enzymes.

INHIBITORS OF NUCLEIC ACID METABOLISM AND PROTEIN SYNTHESIS

Benzimidazoles

The benzimidazoles include benomyl, thiabendazole and thiophanate. They are not toxic to fungi in their original state, but must be converted to their ester metabolites, which are known to be the toxic entitles. These metabolites cause morphological distortion of germinating spores and probably act by inhibiting DNA synthesis, or by interfering with some closely related aspect of cell or nuclear division.

Antibiotics

Cycloheximide is a protein synthesis inhibitor – it inhibit the incorporation of amino acids into protein. As a consequence of its interference with protein synthesis, it may also inhibit DNA snythesis. Cycloheximide is toxic not only to fungi but also to plants, and it has an oral LD_{50} in the rate of 2.5 mg/kg, by far the most toxic of the fungicides.

Streptomycin probably inhibits protein synthesis by binding to the ribosome, one molecule of streptomycin per ribosome. It also cause misreading of the genetic code, though this is not likely the primary effect.

Aliphatic Nitrogen Compounds

Dodine has a nonspecific mode of action, but it is generally agreed that it acts by interfering with membrane sturcture. There is indirect evidence that its primary side of action is the mitochondrial membrane. The quanidine nucleus of dodine is also known to inhibit the synthesis of RNA.

Triazines

Anilazine is the only fungicide belonging to the triazine class of fungicides. The ultimate site and mechanizm of action are unknown. However, the fungicide does react readily with – NH_2 groups and less readily with –SH or –SR. Consequently, it appears likely that anilazine cause inhibition of a variety of cell processes by nonspecific combination with vital cell components.

7

Insecticides

Human beings have learned to line and compete with the insect world, even though unsects appeared long before humans oid. According to recent discoveries humanoids have existed on earth more than 3 million years, while insects are known to have existed for 250 million years. We can guess that the first materials used by our primitive ancestors that could be classed an insecticides (repellents) in the crudest definition of the word were mud and dust spread over their skin to repel biting and tickling insects, a practice that resembles the habits of water buffalo, pigs and elephants. History doesn't tell us very much about chemicals used against insects. The earliest records of insecticides pertain to the burning of "brimstone" (sulfur) as a fumigant. Pliny the Elder (AD 23-79) recorded most of the earlier insecticide uses in his *Natural History*. Gall from a green lizard to protect apples from worms and rot were also included in it a variety of materials have been used with doubtful results : extracts of pepper and tobacco, hot water, soapy water, whitewash, vinegar, turpentine, fish oil, brine, lye and many others. Even as recently as 1940, our insecticide supply was limited to several arsenicals, petroleum oils, nicotine, pyrethrum, rotenone, sulfur, hydrogen cyanide gas, and cryolite. World War II opened the Chemical Era with the introduction of a totally new concept of insect control chemical – synthetic organic insecticides, the first of which was DDT.

ORGANOCHLORINES

As the names indicates the *organochlorines* are insecticides that contain carbon chlorine and hydrogen. They are also referred to by other names : *chlorinated hydrocarbons, chlorinated organics, chlorinated insecticides, chlorinated synthetics*.

DDT and Related Insecticides

DDT affect human health agriculture and the environment. It can be considered the pesticide of greater of historical important Since EPA canceled all uses of DDT effective from January 1, The story of its rise of stardom, carrying with it the Nobel Prize, and decline to infamy is rather sensational and should be briefly narrated for the uninitiated.

DDT

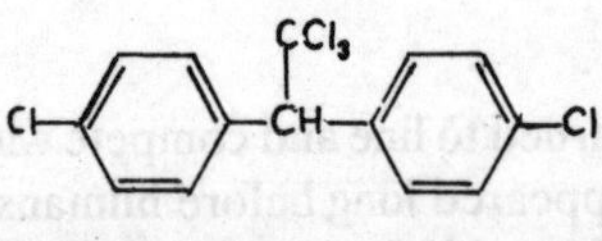

1,1,1-trichloro-2,2-bis(p-chlorophenyl)ethane

DDT which is more than 100 years old is probably the best known and most notorious chemical of this century. It is the most fascinating, and remains to be acknowledged as the most useful insecticide developed. It was first synthesized in 1873, by a German graduate student who had no idea of its tremendous insecticidal value, and after synthesis it was put on the shelf and forgotten. In 1939 a Swiss entomologist, Dr. Paul Miller, rediscovered DDT while searching for a long-lasting insecticide against the clothes moth. DDT proved to be extremely effective against flies and mosquitoes, ultimately bringing to Dr. Mller the Noble Prize in medicine in 1948 for his lifesaving discovery. We should bear in mind that its most beneficial use was in public health, for malaria control, and in many nations it still is so used.

More than 1.8 billion kg of DDT have been used throughout the world for insect control since 1940, and 80 per cent of that amount was used in agriculture. Production reached its maximum in the United States in 1961, when 73 million kg were manufactured. The greatest agricultural benefits from DDt have been in the control of the Colorado potato beetle and several other potato insect, the codling moth on apples, corn earworm, cotton bollworm, tobacco budworm, pink bollworm or cotton, and the worm complex on vegetables. It has been most useful against the gypsy moth and the spruce budworm in

TDE (DDD)

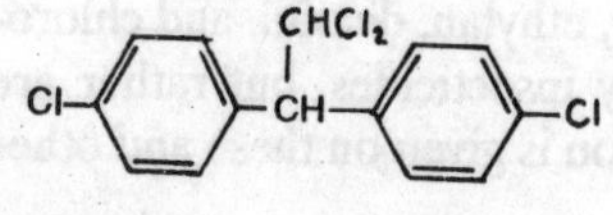

1,1-dichloro-2,2-bis(p-chlorophenyl)ethane

METHOXYCHLOR

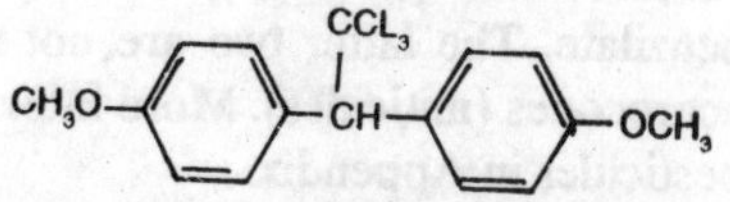

1,1,1-trichloro-2- 2-bis (p-methoxyphenyl) ethane.

forest. From the point of view of human medicine, DDT has been most successful against mosquitoes that transmit malaria and yellow fever, against body lice that can be carry typhus, and against fleas that are vectors of plague.

One of the most surprising features of DDT was its low cost. Most of that sold to the World Health Organization went for less than 22 cents pound. Without question, it was the most economical insecticide ever sold. A federal ban on the use of DDT, declared by the EPA on January 1, 1973, named DDT an environmental hazard due to its long residual life and to its accumulation, along with the metabolite DDE, in food chains, where it proved to be detrimental to certain forms of wildlife. It is no longer available to the grower or home-owner.

DDT belongs to the chemical class of diphenyl aliphatics, and as the name indicates it consist of an aliphatic, or straight carbon chain, with two (di-) phenyl rings attached, as in the illustrations. DDT was first known chemically as *dichloro diphenyl trichloroethane*, hence DDT.

The chemical structure for DDT is presented here for the reader, more as a matter of historical curiosity than any other. Five members

ETHYLAN (Perthane)

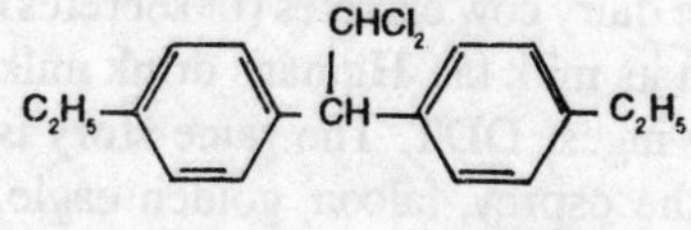

1,1-dichloro-2.2-bis (p-ethylphenyl) ethane

DICOFOL (Kelthane)

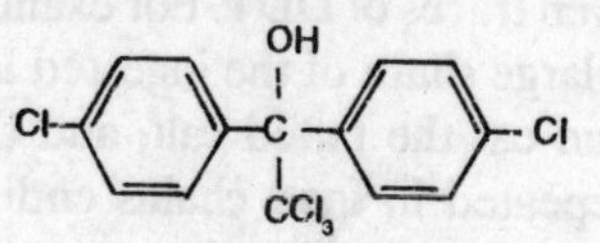

4,4-dichloro-a (trichloro methyl) benzhydrol

CHLOROBENZILATE

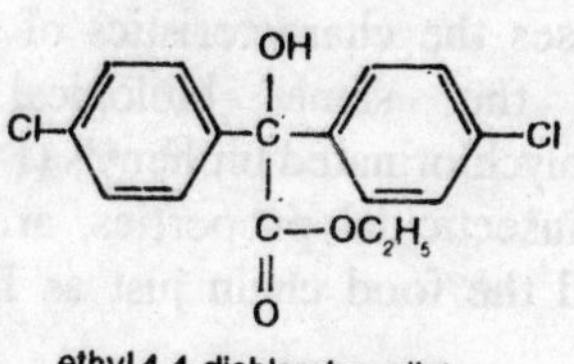

ethyl 4,4-dichlorobenzilate

of DDT are of importance because they all had an early role in pest control : TDE (or DDD), methoxychlor, ethylan, dicofol, and chlorobenzilate. The latter two are not really insecticides, but rather are acarocodes (miticides). More information is given on these and other pesticides in Appendix.

How does DDT Kill ? The mode of action is not very clear. In some complex manner it destroys the delicate balance of sodium and potassium within the neuron, thereby prevent- ing it from conducting impulses normally.

In order to understand some of the well-documented evils attributed to it let us consider a few important pents concerning DDT. The first point is DDT's chemical stability, DDT and TDE are *persistent*, that is, their chemical stability gives the products long lives in soil and aquantic environments and in animal and plant tissues. They are not readily broken down by microorganisms, enzymes, heat, or ultraviolet light. The remaining DDT relatives are considered nonpersistent.

Second, DDT has been reported in chemical literature is the most insoluble compound ever made sure that DDT's solubility in water is only about six parts per billion parts (ppb) of water. However, it is quite soluble in fatty tissue, and, as a consequence of its resistance to metabolism, it is readily stored in fatty tissue of any animal ingesting DDT along or DDT dissolved in the food it eats, even when it is part of another animal.

DDT gets accumulated in every animal that preys on the other animals because of the fait eat it is not readily metabolized and thus not exerted. It also accumulates in animals thateat plant tissue bearing even traces of DDT. For example, the dairy cow excretes (or secretes) a large share of the ingested DDT in its milk fat. Humans drink milk and eat the fatted calf, and thereby ingest DDT. The same story is repeated in food chains ending in the osprey, falcon, golden eagle, seagull, pelican, and so on.

The principle of those food chain oddities is this : Just as DDT, any chemical that possesses the characteristics of stability and fat solubility will follow the same biological magnification (*biomagnification*). The polychlorinated biphenyls (PCBs), a group of chemicals that have no insecticidal properties, are stable and fat soluble and have climbed the food chain just as DDT has. Other insecticides incriminated to some extent in biomanginification, belonging to the organochlorine group, are TDE, DDE (a major metabolite of DDT), dieldrin, aldrin, several isomers of HCH, endrin, heptachlor and mirex.

The rise and fall of DDT contains a lesson, and this is perhaps as good a place as any to moralize. Because of DDT's great success in World War II against body lice in Neples, during the typhus out-break, and in the Pacific, against mosquitoes known to carry malaria, after the war it was rapidly adopted for agricultural use with inadequate basic knowledge. And, because of its effectiveness against a host of agricultural insect pests and its ridiculously low cost, it was overused, and abused, land then—when it caused problems—was banned in rather a panic.

DDT moved to be highly effective against body lice in Neples, during world war II in typhus our break, and in the pacific against mesquitves which carried malaria. Due to the above mentioned reasons DDT was readily adopted for agricultural use without knowing much about its disadvantages. Only when it started to cause problems, it was banned, rather in a panic. A lesson that should be learnt is that no matter how effective it is against a host of agricultural insect pests, a how cheap it is, there is on absolute need for an informect caution and accoiding to available knowledge correct way of employing a specific chemical for pest control.

We need basic research performed in autonomous institutions not subjected to competition in the economic marketplace. And because this research it basic, as opposed to applied, it will of necessity be slow, expensive, long-term, and not immediately applicable. But our social policy must support such research, to enable us to cope with – if not to avoid – those chemicals that prove to pose dangers to our environment as well as our health.

HCH

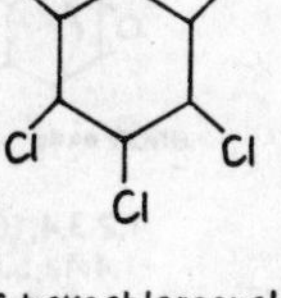

1,2,3,4,5,6-hexachlorocyclohexane

Hexachlorocyclohexane (HCH)

Hexachlorocyclohexane (HCH) – previously erroneously called benzenhaxachloride (BHC) – was first discovered in 1825. But, like DDT, it was not known to have insecticidal properties until 1940, when French and British entomologists found in the material to be active against all insects tested.

It is made by chlorinating berizence, which results in a product made up of several isomers, that is molecules containing the same kinds and numbers of atoms but differing in the internal arrangement

CHLORDANE

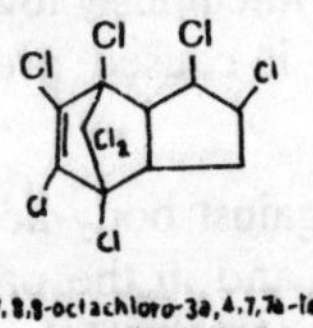

1,2,4,5,6,7,8,8-octachloro-3a,4,7,7a-tetrahydro-4,7-methanoindene

HEPTACHLOR

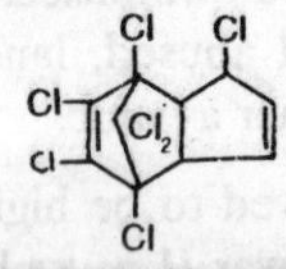

1,4,5,6,7,8,8-heptachloro-3a, 4, 7,7a tetrahydro 4,7-methanoindene

ALDRIN

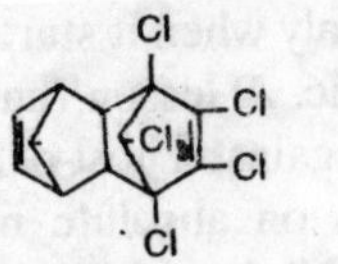

1,2,3,4, 10, 10-hexachloro-1,4,4A,5,8,8A-HEXAHYDRO 1,4-endo-exo-5, dimethanonaphthalene

ENDOSULFAN (Thiodan)

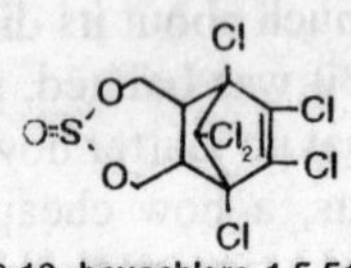

6,7,8,9,10,-hexachloro-1.5,51,6,9,a-hexahydro-6.9-methano-2,4,3-benzo dioxathiepin 3-oxide

MIREX

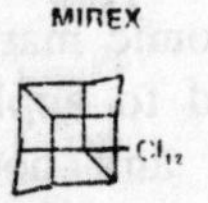

dodecachlorooctahydro-1,3,4-metheno-1H-cyclobuta[cd] pentalene

CHLORDECONE (Kepone®)

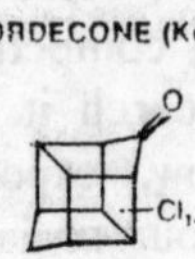

decachlorooctahydro-1,3,4-metheno-2H-cyclobuta[cd] pentalen-2-one

DIELDRIN

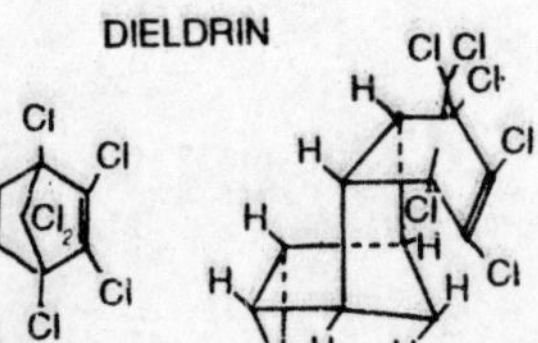

endo-exa

1,2,3,4,10 10-hexachloro-6,7,EPOXY 1,4,4a,5,6,7,8,8A-OCTAHYDRO-1,4 endo-end 5,8,-dimethanonaphalene

ENDRIN

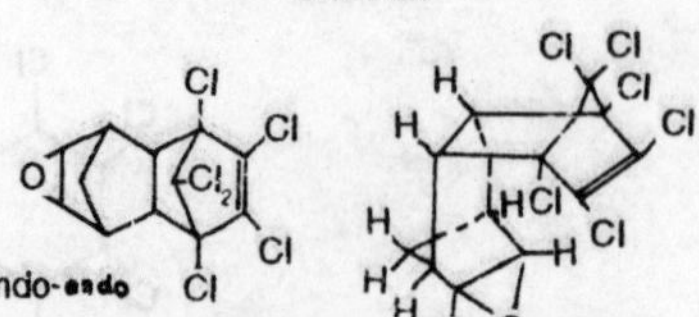

endo-endo

1,2,3,4,10,10-hexachloro-6,7,-EPOXY-1,4,4a,5,6,7,8,8a- octahydro-1,h-endo-endo 5,8-dimethanon aptnalene

of those atoms. HCH, for instance, has five isomers, named, after the Greek letters, *alpha, beta, gamma, delta* and *epsilon*. After much laboratory work in isolating and identifying these isomers, the chemists found to their great surprise that only the gamma isomer had insecticidal properties. In a normal mixture of HCH, the gamma isomer makes up only about 12 per cent of the total, leaving the other four isomer as inert material or insecticidally inactive ingredients.

Since the gamma isomer is the only active ingredient, methods were developed to manufacture lindane, a product containing 99 per cent gamma isomer, which is effective against most insect, because of its cest being high it is impractical for crop use.

Technical grade HCH has one highly undesirable characteristic, a prominent musty order and flavour. The order is form the inert isomers, which are more persistent than the orderless gamma isomer in animal and plant tissues as well as in soil. As a result, root and tuber crops planted in soils previously treated with HCH retain its order and are usually unsalable. The same problem is reported with leafy vegtables, poultry, eggs, and milk that directly or indirectly come in contact with HCH residues.

The effects of HCH on insects and mammals superficially resemble those of DDT. Lindane is a neurotoxicant whose effects are normally seen within hours and result in increased activity, tremors, and convulsions leading to prostration.

Lindane is orderless and has a high degree of volatility. It quickly became popular as a household fumigant sold as pellets to be attached to light bulbs or to small, decorative electric wall vaporizers. These were later found to be hazardous to humans and house pets and were removed from the market.

Cyclodienes

The cyclodienes, also known as the *diene-organochlorine insecticides*, are of more recent origin than DDT (1939) and HCH (1940) which were developed after World War II and are therefore of more recent origin than DDT (1939) and HCH (1940). The eight compounds listed here were first described in the scientific literature or patented in the year indicated : chlordane, 1945; aldrin and dieldrin, 1948; heptachlor, 1949; endrin, 1951; endosulfan, 1956; and chlordecone (Kepone®) 1958. Other cyclodienes developed in the

United States and Germany are of minor importance in a general survey. These include isodrin, adodan, bromodan and telodrin. Cyclodienes, which are stable in soil & relatively stable to ultraviolet action of sunlight are persistent insecticides. Consequently, they have been used in greatest quantity as soil insecticides (especially chlordane, heptachlor, aldrin, and dieldrin), for the control of termites and soil-borne insects whose immature stages (larvae) feed on the roots of plant. Because of their persistence, the use of cyclodienes on crops was restricted; undesirable residues remained beyond the time for harvest. To understand the effective- ness of cylodienes as termite control agents, consider in the year of their development are still protected from damage more than thirty years later. These insecticides are the most effective long-lasting, economical, and safest termite control agents known. However, several other soil insects became resistant to these materials in agriculture, resulting in a rapid decline in their use. In 1975 EPA banned most of the agricultural uses of cyclodienes.

The most valuable, and produced in the greatest quantity, were chlordane and dieldrin. Structures of the common cyclodienes are presented to illustrate their similarity and complexity.

The nomenclature and chemistry of the cyclodienes are quite complicated. The cyclodienes have three-dimensional structures and thus possess stereoisomers; that is, forms that have the same kinds and numbers of atoms, but their atoms differ in their spatial arrangement. For instance endrin is a stereoisomer of deildrin.

The toxic effects of not very clear. Cyclodienes on insects mammals and birds are of equal intensity but more on fishes perhaps because when the compound is introduced into water the fish continually respire and ingest any toxic compound contained in their aquatic environment.

The modes of action of the cyclodienes are not clearly understood. It is known that they are neurotoxicants, that have effects similar to those of DDT and HCH. They appear to affect all animals in generally the same way, first with nervous activity followed by termors, convulsions, and prostration. The cyclodienes undoubtedly disturb the delicate balance of sodium and potassium within the neuron but in a way differing from that of DDT and HCH.

Polychlorotepene Insecticides

Toxapliene (1947) and strobane (1951) are the only two polychloroterpene material. Neither have ever been considered urban insecticides. Toxaphene is manufactured by the chlorination of camphene, a pine tree derivative.

TOXAPHENE

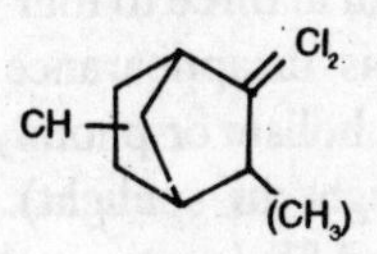

chlorinated camphene containing 67 to 69 percent chlorine

TOXAPHENE TOXICANT A

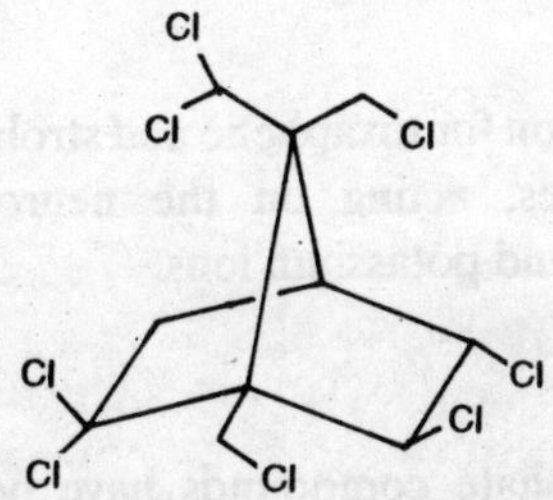

2,2,5-endo, 6-exo, 8,9,9,10-octachlorobornane
$C_{10}H_{10}Cl_8$.

Toxaphene had by far the greatest use of any single insecticide in agriculture. It was used on cotton, first in combination with DDT, for alone it has a low order of toxicity to insects. In 1965, after a number of cotton inesects became resistant of DDT, toxaphene was formulated in combination with methyl parathion, an organophosphate insecticide discussed later in this chapter. As late as 1976 some 11.8 million kg of toxaphene was used on cotton, or 41 per cent of all insecticides on cotton that year.

Toxaphene is an extraordinary mixture of more than 177 polychlorinated derivatives, which are 10-carbon compounds including Cl_6, Cl_7, Cl_8, Cl_9 and Cl_{10} constituents. Most are probably isomeric Cl_7-, Cl_8, and Cl_9- bornanes. No single component makes up more than a small percentage of the technical mixture. The toxaphene components of greater concern are those most toxic to mammals and fish. One of these is toxicant A, shown in its three-dimensional form (Saleh and Casida, 1978). Toxicant A makes up only 3 per cent of technical toxaphene, but it is 18 times more toxic to mice, 6 times more toxic to houseflies, and 36 times more toxic to goldfish than technical toxaphene.

These materials are persistent in the soil, though not as persistant as the cyclodienes, and narish in three to four weeks from the surfaces of mosts plant tissues. This disappearance is attributed more to volatility than to actual metabolism or photolysis (disintegration from the effects of ultraviolet light in sunlight). They are fairly easily metabolized by mammals and birds, are not stored in body fat to any great extent, as are DDT, HCH, or the cyclodienes. Despite low toxicity to insects, mammals, and brids fish the highly susceptible to toxaphene poisoning, in the same order of magnitude as to the cyclodienes.

The modes of action for toxaphene and strobane are similar to the cyclodiene insecticides, acting on the neurons and causing an imbalance in sodium and potassium ions.

Organophosphates

The organophosphate compounds have been replaced by the chemically unstable organophosphate (OP) insecticides, especially with regard to use around the home and garden. *Organophosphate* is

ORTHO PHOSPHORIC ACID

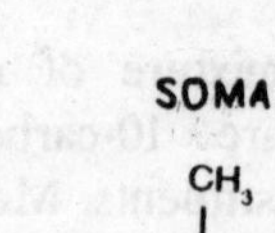

SHRIN

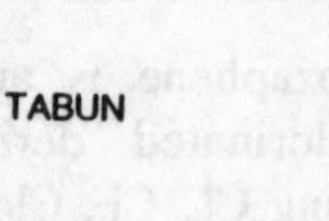

WORLD WAR II
GERMAN NARVE GASES

SOMAN

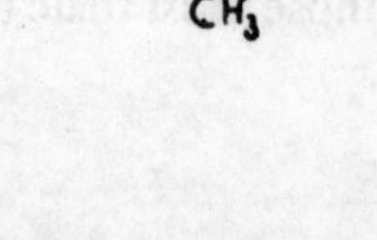

TABUN

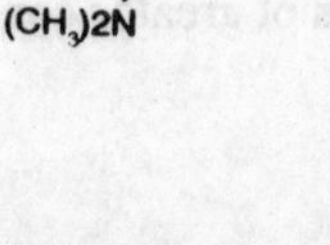

PHOSPHATE

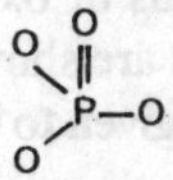

PHOSPHONATE

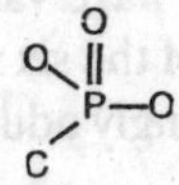

PHOSPHOROTHIOATE

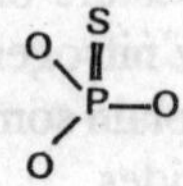

PHOSPHOROTHIOLATE

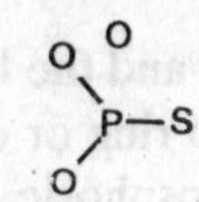

PHOSPHORODITHIOATE

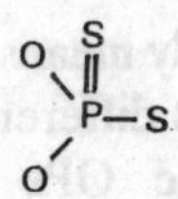

PHOSPHORAMIDATE

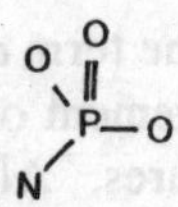

usually used as a generic term to include all the insecticides containing phosphorus.

The OPs have several other commonly used names, some of which are organic phosphates, phosphorus insecticides, nerve gas relatives, phosphates, phosphate insecticides, and phosphorus esters or phosphoric acid esters. They are all derived from phosphoric acid and are generally the most toxic of all pesticides to vertebrate animals. Because of their chemical structures and mode of action, they are related to the "nerve gases." Their insecticidal action was observed in Germany during World War II in the study of materials closely related to the nerve gases sarin, soman and tabun. Initially the discovery was made in search of substitutes for nicotine, which was in critically low supply in Germany.

The OPs have two distinctive features. First, they are generally much more toxic to vertebrates than are the organichlorine insecticides, and, second, they are chemically unstable or nonpersistent. It is this latter quality that brings them into agricultural use as substitutes for the persistent organochlorines, particularly DDT.

The OPs exert their toxic action by inhibiting certain important enzymes of the nervous system, cholinesterases (ChE). This leads to the accumulation of acetycholine (ACh), which interferes with the neuromuscular junction, producing rapid twitching of voluntary muscles and finally paralysis.

OPs that have attached to their phosphorus atoms combinations of different alcohols and different phosphorus acids are termed esters. Esters of phospherous have varying combinations of oxygen, sulfur & nitrogen. The nuclei of the six subclasses of OPs are shown to help explain some of the seemingly odd chemical names given to these insecticides.

The OPs are divided into three groups – the aliphatic, phenyl and heterocyclic derivatives.

Aliphatic Derivatives

The term *aliphatic* literally means "carbon chain," and the linear arrangement of carbon atoms differentiates them from ring or cyclic structures. All the aliphatic OPs are simple phosphoric acid derivatives bearing short carbon chains.

TEPP was the first OP introduced into agriculture is the only useful pyrophosphate and is probably the most toxic. It was never available for home use. Because TEPP is very unstable in water, it hydrolyzes (breaks down) quickly after spraying on crops and disappears with 12 to 24 hours.

PYROPHOSPHATE

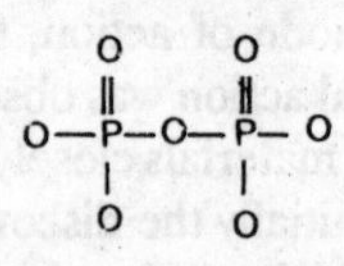

TEPP

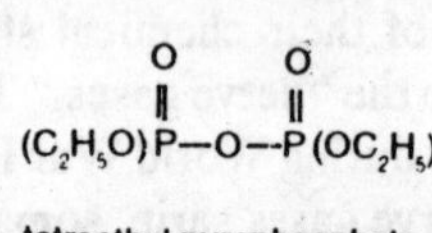

tetraethyl pyrophosphate

The oldest and most heavily used aliphatic OP is Malathion introduced it was quickly adopted by agriculture for use on most vegetables, fruits, and forage crops for control of an extensive range of insect pests. Malathion is so safe that it is suitable for home use aho since it is safe to use around pets, fast acting and can controlled

MALATHION

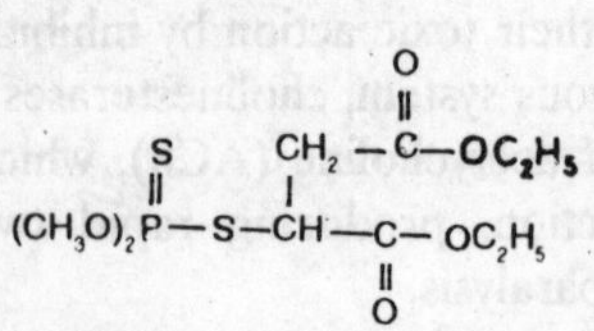

O,O-dimethyl-5-12-DI (carboethoxy)
ethyl phosphorodithiorate

DIMETHOATE (cygon)

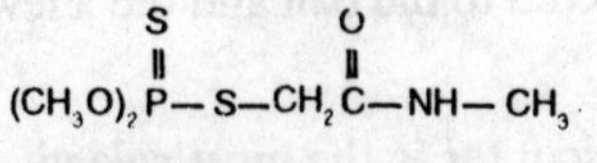

O,O dimethyl S-(2-(methylamino)
-2-oxoethyl) phosphorodithicate

OXYDEMETONMETHYL (Meta Systox R)

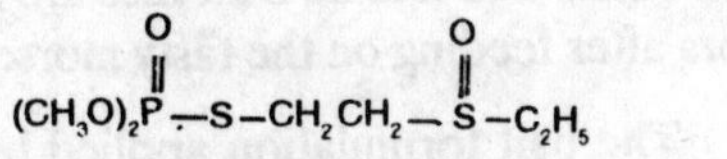

S(2-(ethylsulfinyl)ethyl]0,0-dimethyl
phosphorothioate

DICROTOPHOS (Bidrin)

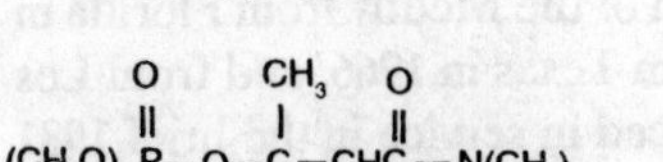

O,O-dimethyl 0-1 methylvinyl-
N,N-dimethyl CARBAMOYL PHOSPHATE

DISULFOTON (Di-syston)

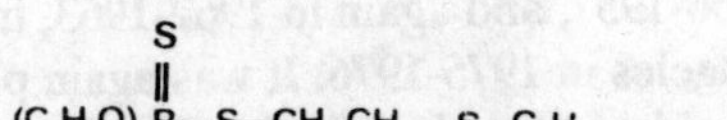

O,O-diethyl S--2-[(ethylthio)ethyl)
PHOSPHORODITHIOATE

practically every kind insect including garden and household aphids and cockroaches. It is so safe that it is perscribed by physicians for use on humans for the control of head, body, and crab lice. It commonly appears in flea powders for dogs, cats, and other domestic animals, and is used in dips for the control of mange mites.

TRICHLORFON (Dylox)

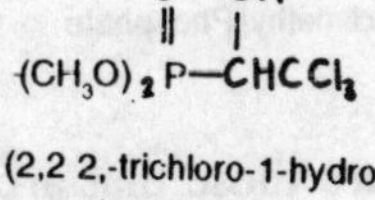

dimethyl (2,2 2,-trichloro-1-hydroxyethyl)
phosphonate

In 1981, and again in 1982, malathion became the insecticide of choice in the control of the Mediterranean fruit fly, which had invaded the rich fruit-growing areas of California. Malathion was mixed with a protein bait made of molasses and yeast and sprayed from ground equipment and by helicopter over the infested and surrounding areas. Both male and female fruit flies are attracted to the bait and die a few hors after feeding on the tasty morsels.

The bait formulation applied by helicopter is the most selective and inoffensive of all forms and methods of malathion use. With this technique malathion is appled at the astonishingly low rate of 171 g/ha mixed with 684 g of the bait. When applied by ground equipment it is used at the rate of 2 kg of active ingredient in 378 L of spray per hectare as a foliar or vegetation spray. This malathion-bait mixture was used successfully in the eradication of the Medfly from Florida in 1956-1957, and again in 1962-1963, from Texas in 1966, and from Los Angeles in 1975-1976. It was again placed in service in the brief 1981 Florida outbreak of the Medfly, because of its exceptionally low acute toxicity to humans and other warm-blooded animals.

Trichlorfon is a chlorinated OP, which has been useful for crop pest control and fly control around barns and other farm buildings.

Monocroptophos is an aliphatic OP containing nitrogen. It is a plant-systemic insecticide, but it has had limited use in agriculture because of its high mammalian toxicity. It is nto available to the home gardener.

MONOCROTOPHOS (Azodrin)

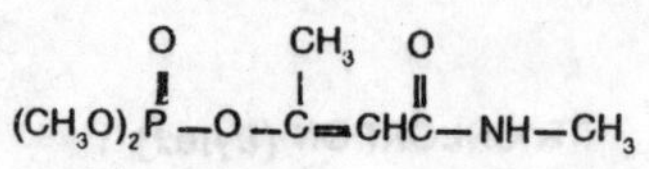

3-hydroxy-N-methyl-*cis*-crotonamide dimethyl Phosphate

Systemic insecticides are those that are taken into the roots of plants and translocated to the above-ground parts, where they are toxic to any sucking insects feeding on the plant juices. Normally caterpillars and other plan tissue-feeding insects are not controlled, because they do not ingest enough of the systemic-containing juices to be affected.

Contained among the alophatic derivatives are several plant systemics, dimethoate, dicrotophos, oxydemetonmethyl, and disulfoton, all of which can be used safely by the homeowner.

DICHLORVOS (Vapona)

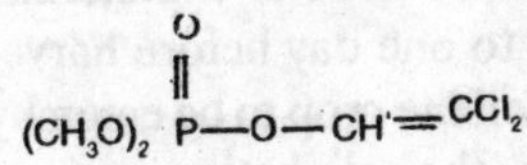

O,O- dimethyl- -2.2- dichloro vinyl phosphate

Dichlorvos is an aliphatic OP with a very high vapour pressure, giving it strong fumigant qualities. It has been incorporated into polychlorovinyl resin pet collars and pest strips, from which it is released slowly. It lasts several months and is useful for insect control in the home and other closed areas.

MEVINPHOS (phosdrin)

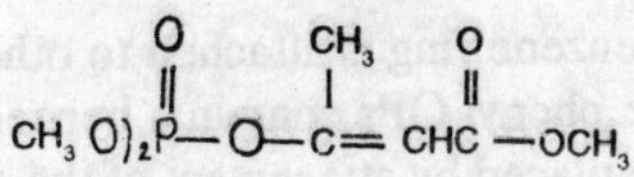

methyl (E) -3-hydroxycrotonate
dimethyl phosphate

METHAMIDOPHOS (Monitor)

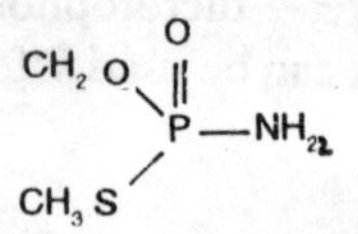

O, S-dimethylphosphoramidothioate

Mevinphos is a highly toxic OP. Because of its short insecticidal life, Mevinphos can be used in commercial vegetable production. It can be applied up to one day before harvest for insect control, yet it leaves no residues on the crop to be eaten by the consumer. Two of the recent arrivals in the aliphatic organophosphate structures are methamidophos and acephate. Both have proved highly useful in agriculture, especially for vegetable insect control.

In summary, the aliphatic organophosphate insecticides are the simplest in structure of the organophosphate molecules. They have wide range of toxicites, and several possess a relatively high water solubililty, giving them plant-systemic qualities several of which are important around the home.

ACEPHATE (Orthene)

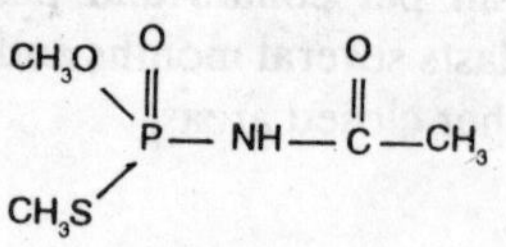

0-S-dimethyl acetylphosphoramidothioate

Phenyl Derivatives

When the Benzene ring is attached to other groups it is referred to as *phenyl*. The phenyl OPs contain a benzene ring with one of the ring hydrogens replaced by attachment of the phosphorus moiety and others frequently replaced by Cl, NO_2, CH_3, CN, or S. The phenyl OPs are generally more stable than the aliphatic OPs; consequently their residues are longer lasting.

Parathion is the most known of the phenyl OPs, being in 1947, the second phosphate insecticide introduced into agriculture. The first, TEPP, was introduced in 1946. As a result of its age and utility,

ETHYL PARATHION

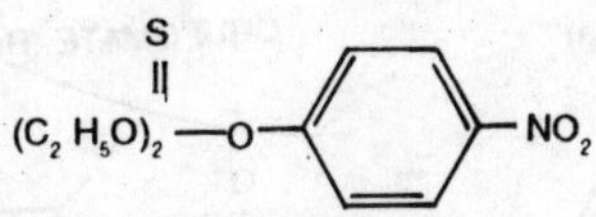

O.O-dimeth 0-p-nitrophenyl phosphorothioate

METHYL PARATHION

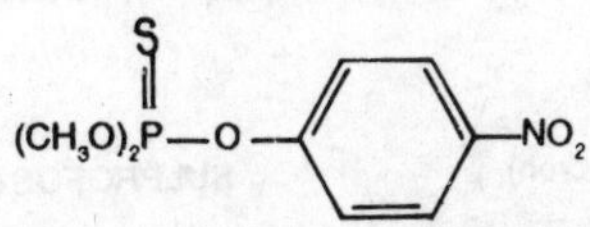

0,0-dimethyl 0-p-nitrophenyl phosphorothioate

parathion's total usage is greater than that of many of the less useful materials combined. Ethyl parathion was the first phenyl derivative used commercially and, because of its harmful effects has not been available for home use.

Methyl parathion became available in 1949 and proved to be more useful than (ethyl) parathion because of its lower toxicity to humans and domestic animals and broader range of insect control. Its shorter residual life also makes it more desirable in certain instances. This material is also not used by the layperson. Systemic insecticides are also found in the phenyl OPs. They are, however, usually animal systemics used for the control of the cattle grub; ronnel and crufomate are examples.

Stirofos is a home-safe OP much like malathion in its overall usefulness against home and livestock pests.

Profenophos and Sulprofos are two of the more recently registered phenyl derivatives that have a wide range of insecticidal activity and are used only on field crops today.

Isofenphos is used as a soil insecticide in field crops and vegetables, against corn rootworm and onion maggot, and also for white grubs, chinch bugs, and sod webworms in turf.

RONNEL (korlan)

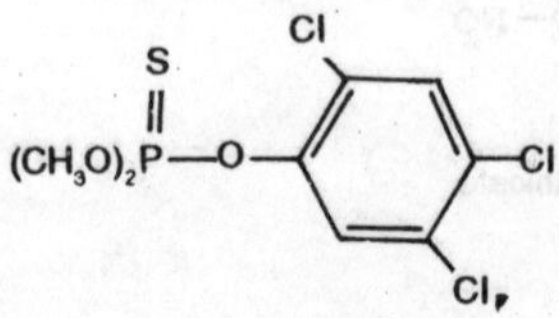

0,0-dimethyl 0-2,4,5,-trichlorophenyl phosphorothioate

CRUFOMATE (Ruelene)

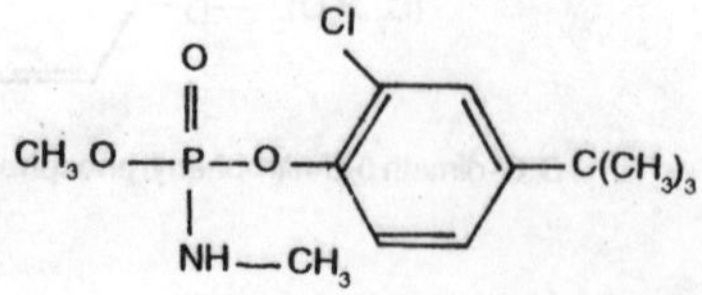

4-tert-butyl-2-chlorophonyl methyl methyl phosphoramidiate

PROFENOFOS (curacron)

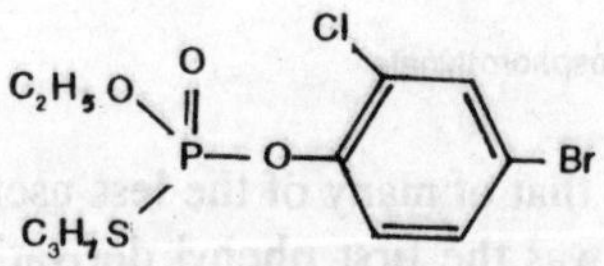

-(4-bromo-2cLOROPHENYLo-ethyl -S-prohyl PHOSPHOSOTHIOATE

SULPROFOS (Bolstar

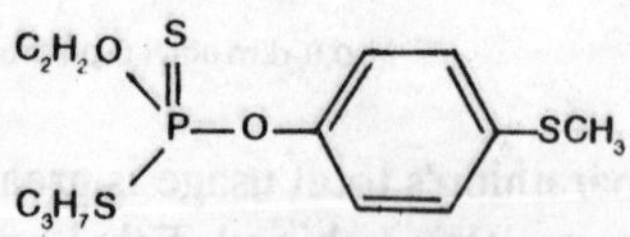

O-ethylS-PROPYl O (4-methylTHIO) phenyl phosphorodithioate

Heterocyclic Derivatives

The term *heterocyclic* means that the ring structures are composed of different or unlike atoms. In a heterocyclic compound, for example, one or more of the carbon atoms is replaced by oxygen, nitrogen or sulfur, and the ring may have three, five, or six atoms.

The first insecticide made available in this group was probably diazinon, in 1952. Note that the six-membered ring contains two

DIAZINON

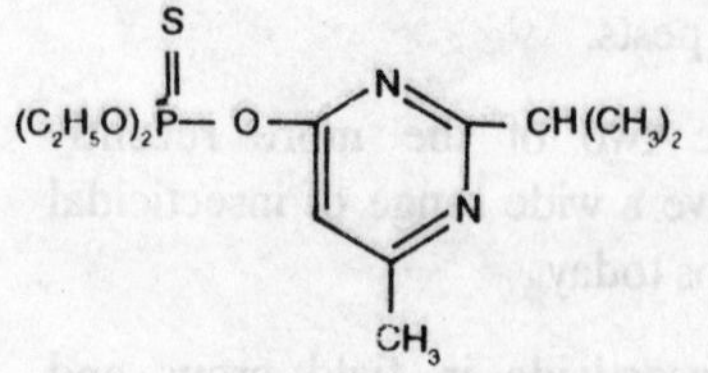

O, 0-diethyl 0-(2-isopropylh-meTHYL-6 pyrimidiny) phosphorothioate

AZINPHOSMETHYL (Guthion)

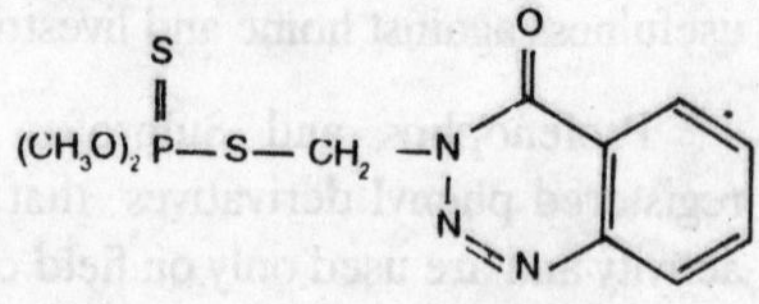

0,0- dimthyl S (4-OXO-1,2,3-benzotriazin 3(4H)-ymethyl) phosphoroclithioate.

nitrogen atoms, very likely the source of its proprietary name, since one of the constituents used in its manufacture is pyrimidine, a diazine.

Diazinon is a comparatively relatively safe OP that has an surprisingly good track record around the home. It has been effective for practically every conceivable use : insects in the home, lawn, garden, ornamentals, around pets, and for fly control in stables and pet quarters.

The second oldest member of this group is Azinphosmethyl (1954) and is used in U.S. agriculture. It serves both as an insecticide and acaricide in cotton production and is not available to the layperson.

Chlorpyrifos has become the most frequently used insecticided by pest control operators in homes and restaurants for controlling cockroaches and other household insects. Methidathion is not particularly new, but it has in the last few years acquired registrations for forage and field crops, true fruits, and nut crops for an extraordinay wide variety of insect and mite pests.

Phosmet has a set of registration credentials similar to methidathion, including the infamous boll weevil and the plum curculio, two closely related weevil pests.

FENSON

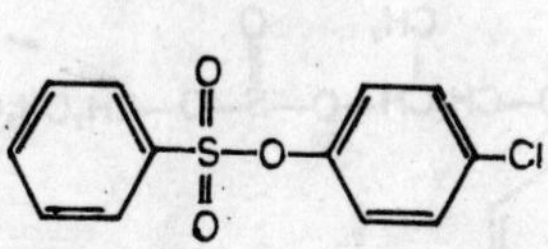

p-chlorophenyl benzenesulfonate

OVEX (Ovotran®)

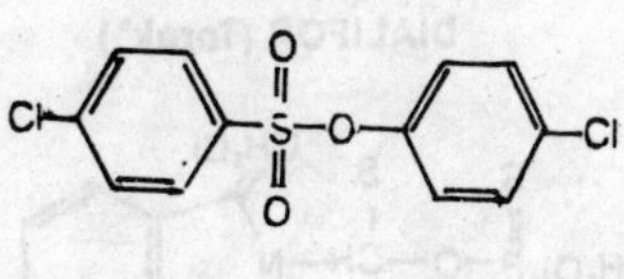

p-chlorophenyl p-chlorobenzenesulfonate

TETRASUL (Animert V-101®)

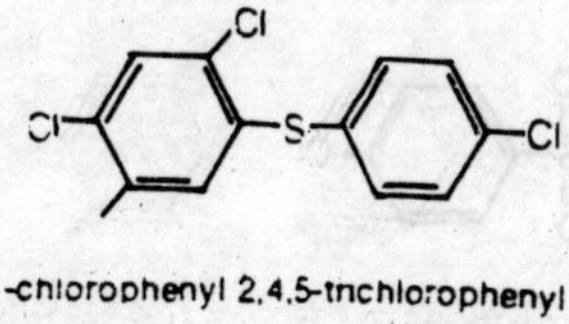

p-chlorophenyl 2,4,5-trichlorophenyl sulfide

PROPARGITE (Omite®)

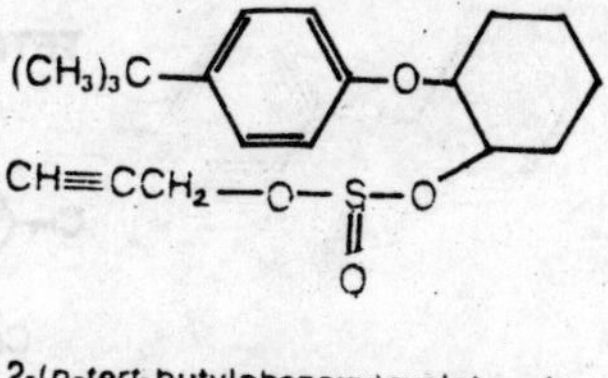

2-(p-tert-butylphenoxy)cyclohexyl 2-propynyl sulfite

Dialifor was first introduced in the mid-1960s as were methidathion and phosmet. Its uses are somewhat more limited, however, to apples, grapes, pecans and citrus.

The heterocyclic organophosphates have generally longer lasting residnes than many of the aliphatic not simple molecules. Also, because of the complexity of their molecular structure, their breakdown products (metabolites) are frequently many, making their residues sometimes difficult to measure in the laboratory. Consequently, their use by growers on food crops is some- what less than either of the other two groups of phosphorus– containing insecticides.

Organosulfurs

The organosulfurs, as the name suggests, have sulfur as their central atom. They resemble the DDT structure in that most have two phenyl rings.

METHIDATHION (Supracide®)

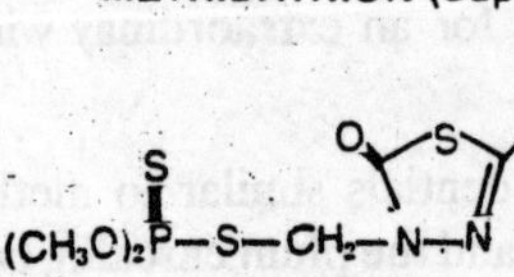

O,O-dimethyl phosphorodithioate *S*-ester with 4-(mercaptomethyl)-2-methoxy-Δ²-1,3,4-thiadiazolin-5-one

PHOSMET (Imidan®)

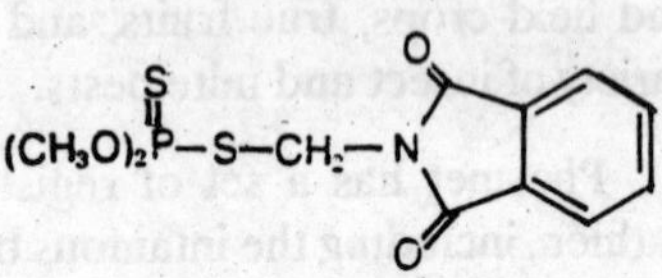

N-(mercaptomethyl)-phthalimide *S*-(*O,O*-dimethylphosphorodithioate)

DIALIFOR (Torak®)

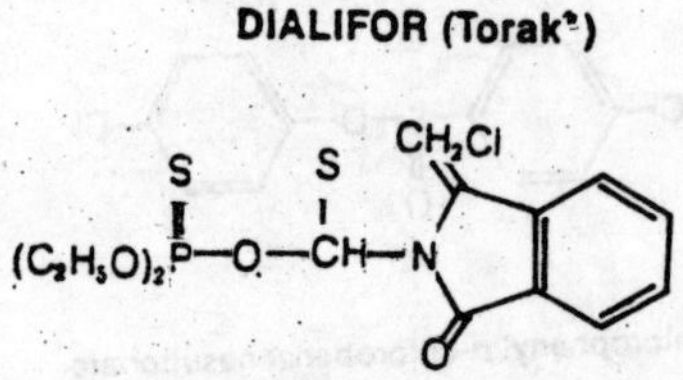

O,O-diethyl phosphorodithioate *S*-ester with *N*-(2-chloro-1-mercaptoethyl) phthalimide

Aramite®

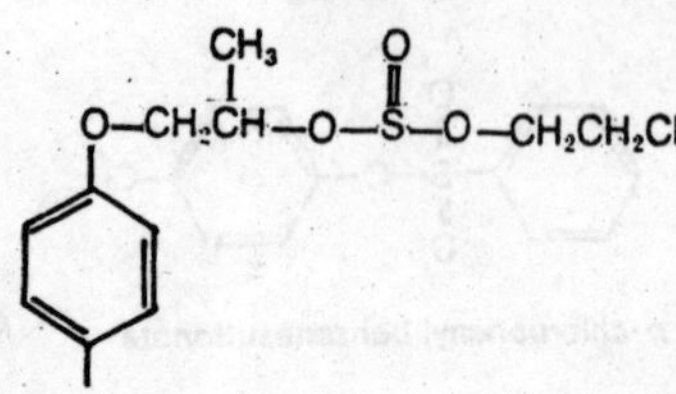

TETRADIFON (Tedion®)

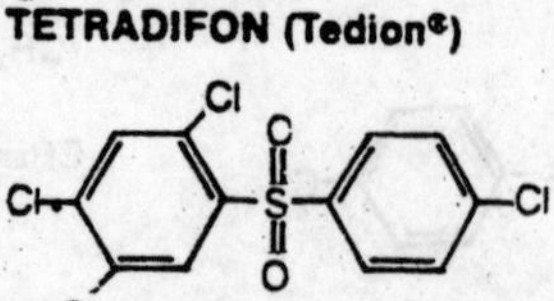

p-chlorophenyl 2,4,5-trichlorophenyl sulfone

In hot weather dusting sulfur alone is a good Acaricide (miticide). The organosulfurs, however, are far superior, requiring much less material to achieve control because sulfur in combination with phenyl rings is particularly toxic to mites. Of greater interest, however, is that the organosulfurs have very low toxicity to insects. As a result, they are used for selective mite control. This group has one other valuable property. They are usually ovicidal as well as being toxic to the young and adult mites.

Tetradifon is one of the older acaricides and typically bears the sulfur and twin phenyl rings, as do most of the organosulfurs.

No doubt the oldest of this group is Aramite®, introduced in 1951. Notice that Aramite has only one phenyl ring and is, therefore, an exception to the general rule that organosulfurs have two phenyl rings.

Carbamates

Since the organphsphate insectides are derivatives of phospohoric acid, the carbamates must be derivatives of carbamic acid $HO-\overset{\overset{\displaystyle O}{\|}}{C}-NH_2$. And like the organophosphates, the mode of action of the carbamates is that of inhibiting the vital enzyme, chloinesterase (ChE). In 1951 the carbamate insecticides were

ISOLAN

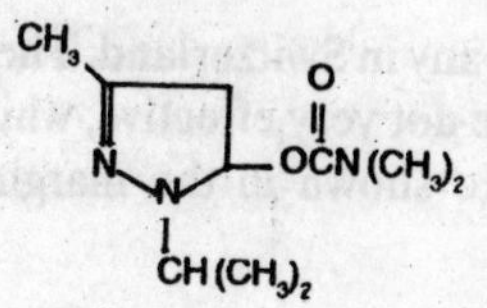

PYRAMAT

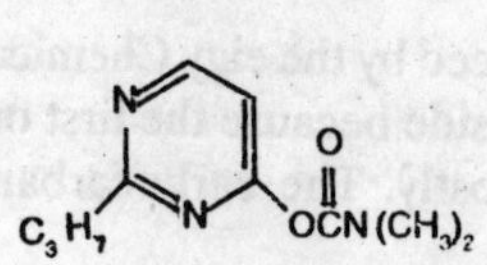

DIMETAN

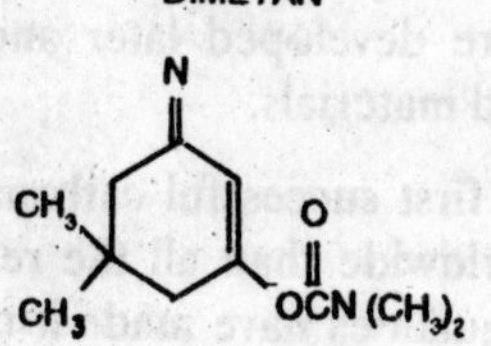

PYROLAN

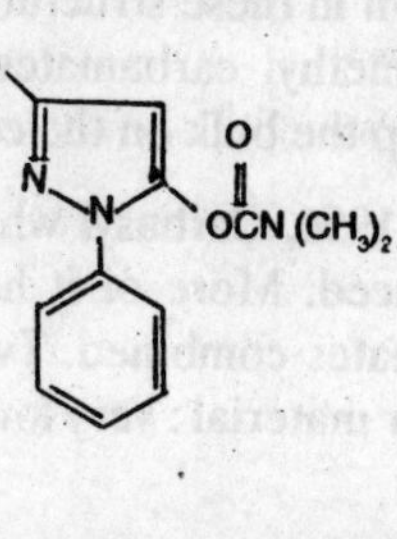

CARBARYL (Sevin)

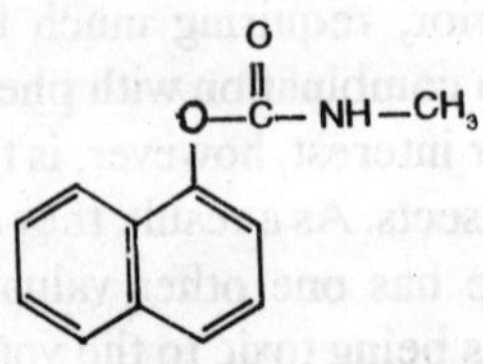

1-naphthl methylacarabamate

METHOMYL (Lannate, nudrin)

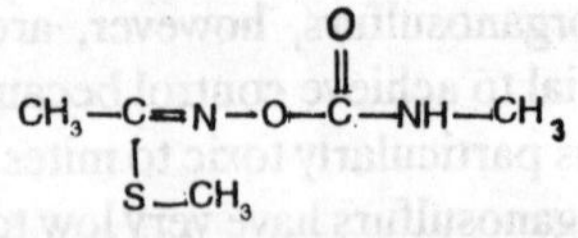

methyle N-(methylcarbamoyl) oxy) thioacetimidate

ALDICARB (Temik)

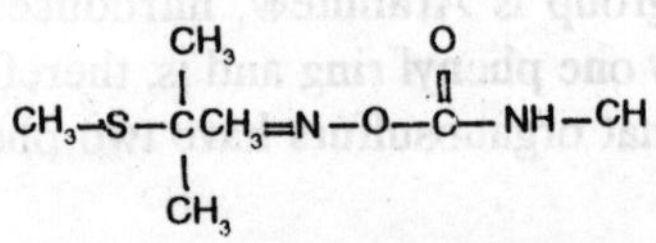

2-methyl-2-(methylthio)propionaldehyde O-(methylcarbamoyl) OXIME

OXAMYL (Vydate)

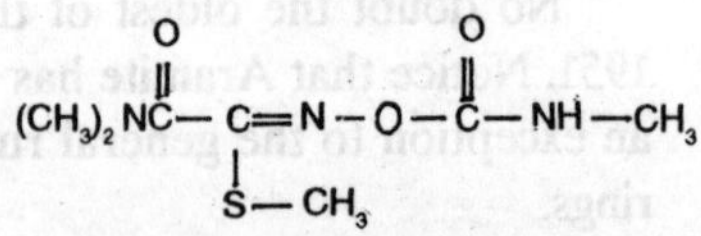

methyl N,N-d methyl N-(methyl carbamoyl) oxy)-1-thiooxamimidate

CARBOFURAN (Furadan)

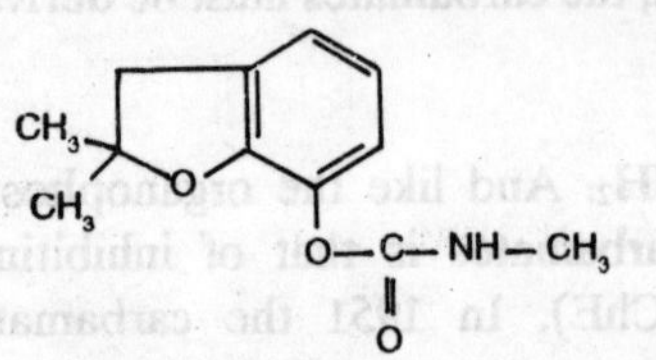

2,3-dihydro-2,2-dimethyl-7- benzofuranyl methylcarbamate

introduced by the eigy Chemical Company in Switzerland. They fell by the wayside because the first ones were not very effective, while being quite costly. The early carbamates are shown in the margin at the right.

At the time it was not known that the N,N-dimethyl carbamates, as shown in these structures, were generally less toxic to insects than the N-methyl carbamates, which were developed later and which make up the bulk on the currently used materials.

In 1956, Carbasyl which was the first successful carbanate, was introduced. More of it has been worldwide than all the remaining carbamates combined. Two distinct qualities have made it the most popular material : very low mammalian oral and dermal toxicity and a

BUFENCARB (Bux)

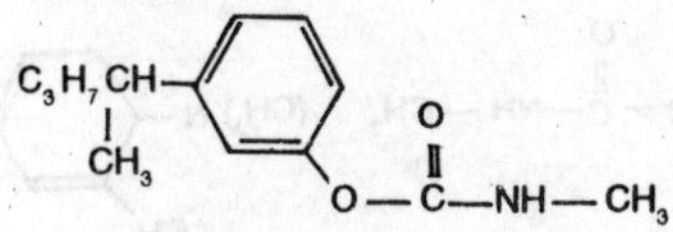

and

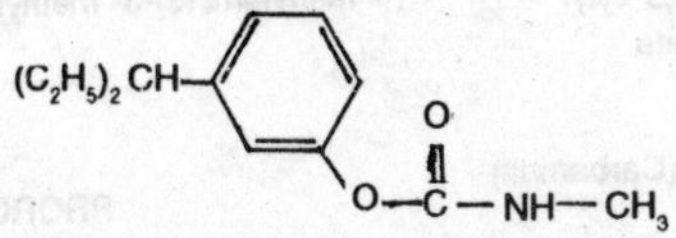

mexure of m-(ethylpropye) phenyl methylcarbomate and m-(1- methylbutyl) phenyl methylcarbamate (ratio 1:3)

rather broad spectrum of insect control. This has led to its wide use as a lawn and garden insecticide. Notice that carbarly is an *N-methyl carbamate.*

Most of the carbamates are plant systemics, indicating that they have a high water solubility, which allows them to be taken into the roots or leaves. They are also not readily metabolized by the plants. Methomyl, oxamyl, aldicarb, and carbofuran have distinct systemic characteristics, making them useful also as nematicidies. Of these, only aldicarb and carbofuram are used as soil insecticides and nematicides. Under rare circumstances, adlicarb has been detected in shallow groundwater following certain uses.

Methomyl has proved especially effective for worm control on vegetables. Bufencarb is used in agriculture exclusively as a soil insecticide, becoming a replacement for the long-residual organochlorine insecticides, aldrin, dieldrin and heptachlor. Methiocarb, aminocarb, and promecarb and effective against foliage-and fruit-eating insects. Methiocarb and aminocarb are both excellent molluscicides, used for slug and snail control in flower gardens and orgnamentals. Methiocarb is also registered as a bird repellent for cherries and blueberries and as a seed dressing.

For cockroaches and other household insects that develop resistance to the organochiorines & organophosphates, propoxure is highly effective. It is used by most structural pest control operators for cockroaches and other household insects in restaurants, kitchens and homes. For home use it is formulated in bottled sprays. Similarly,

METHIOCARB (Mesurol)

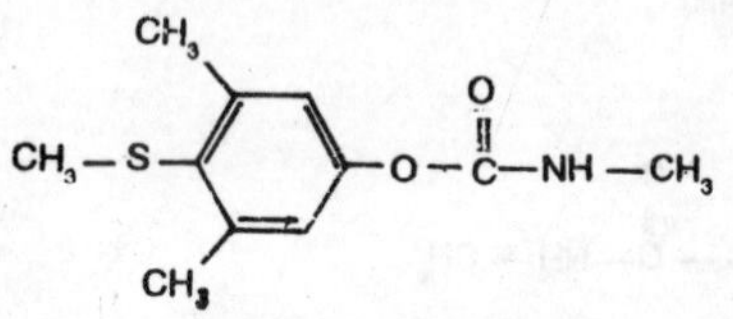

4 (methylthio)3,5-xylyl methylcarbamate

AMINOCARB (Matacil)

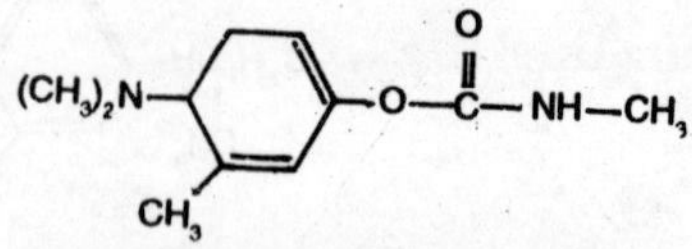

4-(dimethylamino)-3- methylphenol methylcarbamate

PROMECARB (Carbamult)

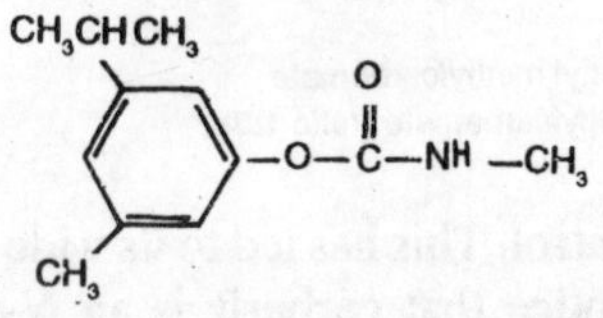

3-methyl-5-(1-methylethyl phenylmethylcarbamate

PROROXUR (Baygon)

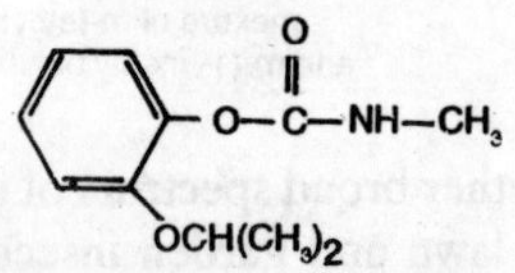

o-isopopoxyphenyl mathclarbamate

BENDIOCARB (Ficam)

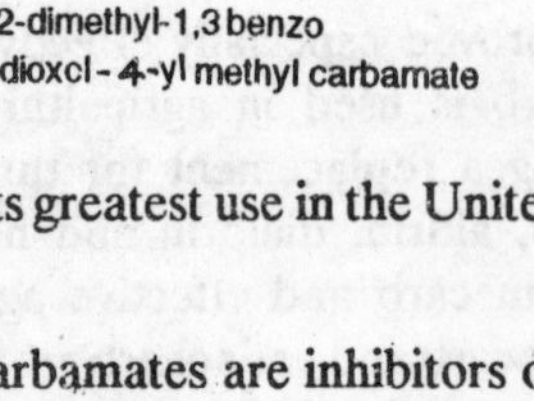

2,2-dimethyl-1,3 benzo dioxcl - 4-yl methyl carbamate

bendiocarb has found its greatest use in the United States as a residual household insecticide.

In summary, the carbamates are inhibitors of cholinesterase, are plant systemics in several instance, and are, for the most part, wide ranged in effectiveness, being used as insecticides, miticides, and molluscidides.

Formamidines

The formamidines consist a new, small, promising group of insecticides. Three examples are chloridimeform, formetanate, and amitraz. They are effective against the eggs and very young

CHLORDIMEFORM (Galecron, Fundal)

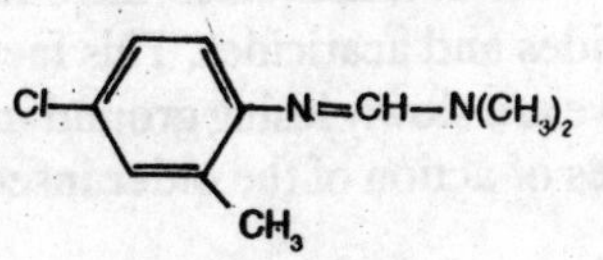

N-(4- Chloro-o - toly)-N,N-dimethylformamdine

FORMETANATE (Carzol)

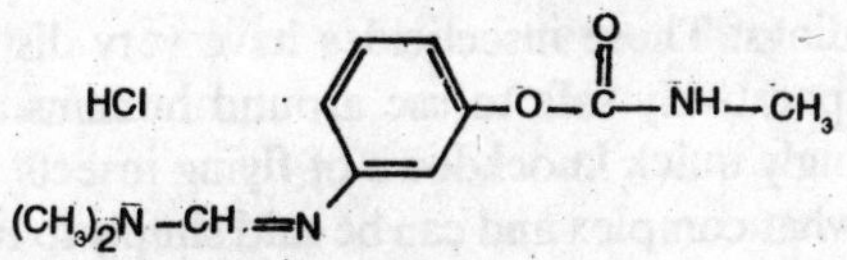

(3-dimethlamino-(methylene-iminophenyl)] -N methylcarbamate hydrochloride

AMITRAZ (Baam)

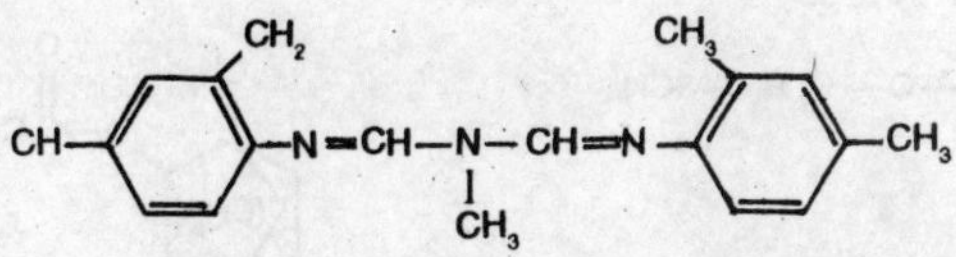

N-(2,4-dimethyLPHENYL)-N-(2,4-dimethyl-PHENYL) AMINO) METHYL)-N-methylmeth animidamide

caterpillars of several moths of agricultural importance and are also effective against most stages, of mites and ticks. Thus, they are classed as ovicides, insecticides and acaricides. Late in 1976, chlordimeform was removed from the market by its manufacturers, Ciba-Geigy Corporation, land Nor-Am Agricultural Products, Inc., because it proved to be carcinogenic to a cancer- prone strain of laboratory mice during high-level, lifetime feeding studies. In 1978 it was returned for use on cotton, but under very strict application restrictions.

Their present value lies in the control of organophosphate-and carbamates-resistant pests. Poisoning symptoms are distinctly deferent from other materials. It has been proposed that one possible machanism of action is the inhibition of the enzyme monoamine oxidase. This results in the accumulation of compounds termed *biogenic amines*. Thus, the formamidines introduce a new mode of action for the insecticides and acaricides. This fact alone makes them extremely useful, for we are slowly losing ground in the battle or insect resistance to the modes of action of the older insecticide groups.

Thiocyanates

Thicyanates have easily recognized structural formulas. Remembering the *theion* of the Greek word for "sulfur" and that the cyanides or cyanates end in -CN we have molecules that bear - SCN, or thiocynate endings. These insecticides have very distinct, creoste like odours, comparatively safe to use around humans and animals, and the astonishingly quick knockdown of flying insects. Their mode of action is somewhat complex and can be said simply to interfere with cellular respiration and metabolism. These materials may be found in aerosols to be used around horses and other farm animals. However, with the appearance of the synthetic pyrethroids, the thiocyanates have deceased demand and will probably sood disappear.

LETHANE 384

$CH_2\,CH_2 — O — C_2H_4 — SCN$
|
$O — C_4H$

2-(2-butoxyethoxy) ethylthiocyanate

THANITE

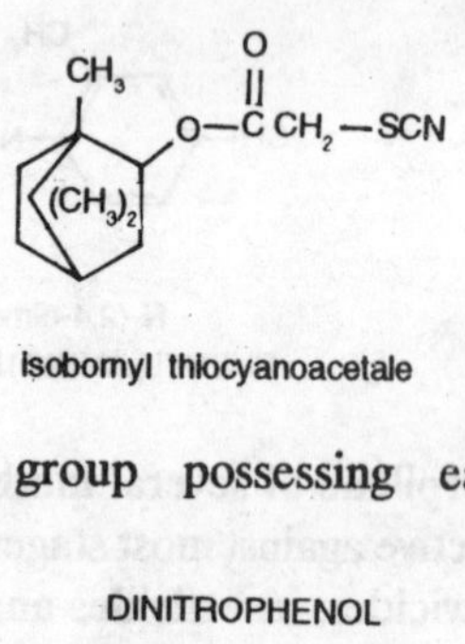

Isobornyl thiocyanoacetale

DINITROPHENOLS

The dinitrophenols are another group possessing easily recognized structural formulas :

PHENOL

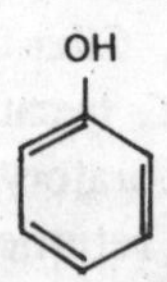

Di(two) nitro (NO_2) phenol

DINITROPHENOL

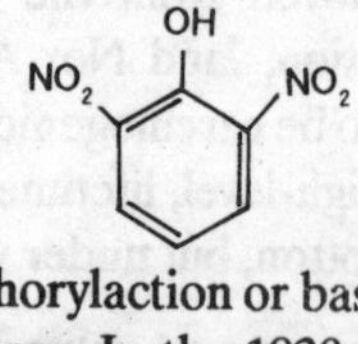

They act by uncoupling oxidative phosphorylaction or basically by preventing the utilization of nutritional energy. In the 1930s, certain dinitrophenols were given by ininformed physicians to their overweight patients to induce rapid weight loss. They were extremely effective, but quite toxic, and their use resulted in several widely publicized deaths.

The older of this group is DNOC (3,5-dinitro-o-cresol), introduced as an insecticixe in 1982. DNOC has also been used as an ovicide, herbicide, fungicide, and blossom-thinning agent. Its use has declined today to herbicidal applications in which all plants are to be killed. Dinoseb is used as a dormant fruit spray for control of many insects and mites.

Another a caricide which was introduced in 1960 is Binapacryl Dinocap in 1949) is still another acaricide as well as fungicide and is one of the rare materials made up of several related molecular structures, only one of which is shown. Dinocap is particularly effective against powdery mildew fungi. Owing to its safety to green plants, it has often replaced the phytotoxic sulfur that is so effective against powdery mildews.

In summary, the dinitrophenols have been used as pesticide in practically all classification : ovicides, insecticides, acaricides, herbicides, fungicides, and blossom-thinning agents.

DINITROCRESOL (DNOC)

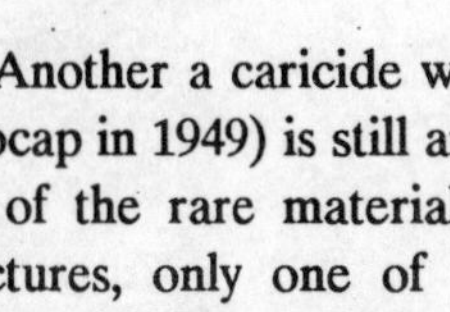

DINOSED

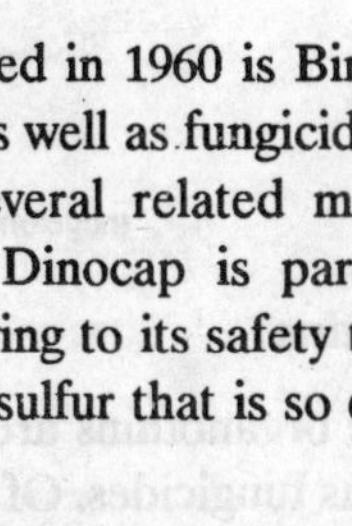

2-sec-butyl-4,6 dinitrophenol

BINAPA CRYL (Morocide)

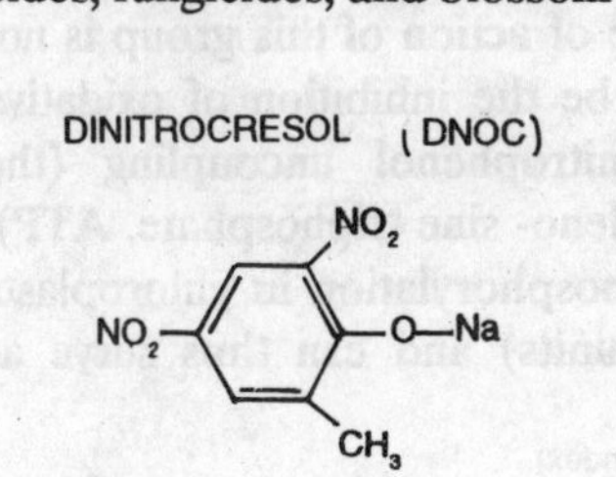

2 Sec-butyl-4,6-dinitrophenyl 3-methyl-2-butenoate

DINOCAP (karathane)

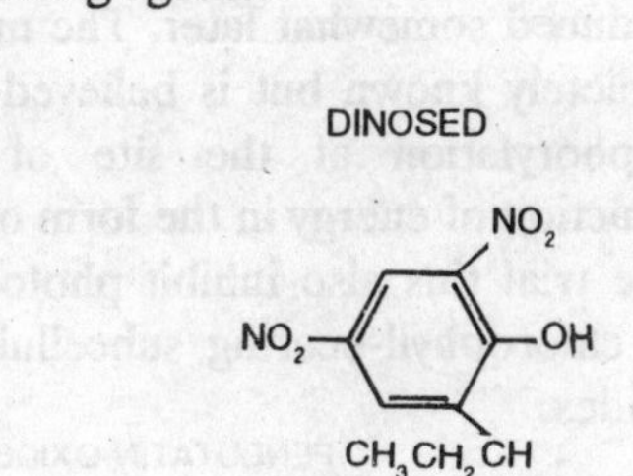

2-(1-methylepth)-4,6-dinitrophenyl crotonate

CYHEXATIN (plictran)

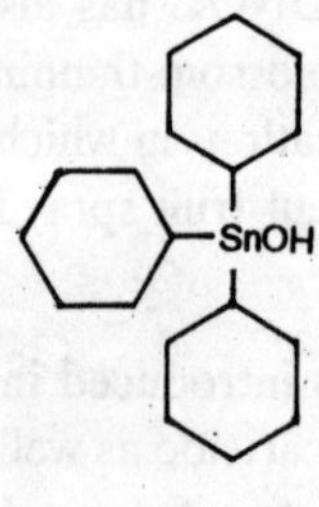

tricyclohexylhydruxytin

Organotins

The organothins are a relatively new group of acaricides, which double as fungicides. Of particular interest here in cyhenxatin, one of the most selective acaricides presently known introduced in 1967. Proved to be most effective against mites on deciduous fruits, citrus, greenhouse crops and ornamentals is fenbutatinoxide which was introduced somewhat later. The mode of action of this group is not completely known but is believed to be the inhibition of oxidative phosphorylation at the site of dinitrophenol uncoupling (the production of energy in the form of adeno- sine triphosphate, ATP). These trial tins also inhibit photo- phosphorylation in chloroplasts (the chlorophyll-bearing subcellular units) and can thus serve as algicides.

FENBUTATIN-OXIDE (Vendex)

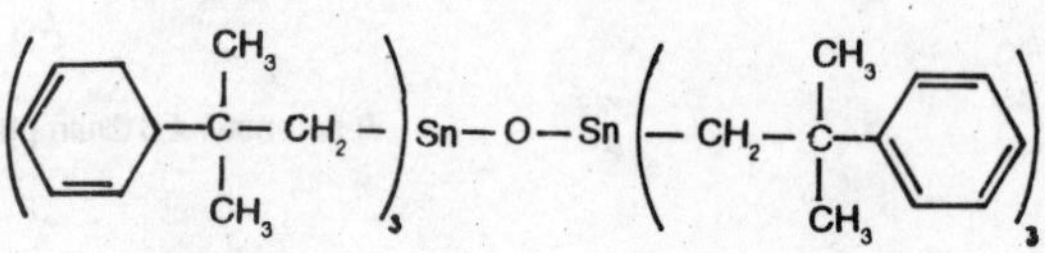

hexakis (2-methyle 2 phenylpropyl) distannoxane

Botanicals

Botanical insecticides are of great interest to many, because they are "natural" insecticides, toxicants derived from plants. Historically, the plant materials have been in use longer than any other group, with the possible exception of sulfur. Tobacco, pyrethrum, derris, hellebore, quassia, and turpentine were some of the more important plant products in use before the organized search for insecticides began.

Some of the widely used insecticides come from plants. The flowers leaves, and roots are finely ground and used in this form, or the toxic ingredients are extracted and used alone or in mixtures with other toxicants.

There are five natural or botanically derived insecticides that are of interest to gardeners in general, but especially to the organic gardener : pyrethrum, rotenone, sabadilla, ryania and nicotine. All except nicotine are exempt from the requirement of a tolerance when applied to growing fruit and vegetables. This is, they can be eaten anytime after application, but these batanical insecticides must be used according to label directions.

Batanical insecticide use reached its maximum in the United States in 1966 and has declined steadily since. Pyrethrum is now the only botanical of significance in use, typically in rapid knockdown sprays in combination with synergists and one or more synthetic organic insecticides formulated for use in the home and garden.

Botanical insecticides are naturally occurring chemicals, synthesized by plants. They are really no safer than most of the currently available synthetic insecticides, at least as compared to those available to the layman. These chemicals would be unaffordable as insecticides if they were synthesized in the laboratory, which is also possible, as Botanicals are expensive to extract from plant tissues.

Nicotine

Smoking tobacco was introduced to England in 1585 by Sir Walter Releigh. As early as 1690, water extracts of tobacco were reported as being used to kill sucking insects on garden plants. As early as about 1890, the active principle in tobacco extracts was known to be nicotine, and, from that time on, extracts were sold as

NICOTINE SULFATE

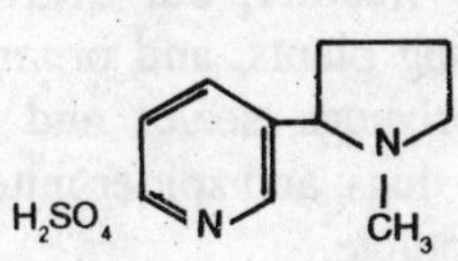

3-(1-methyl-2-pyrrolidyl) Pyridine sulfate

commercial insecticides, for home, farm and orchard. Today organic gardeners may soak a cigar or two in water overnight and spray insect-infested plants with the extract, achieving some success. "Black Leaf 40," which has long been a favourite garden spray, is a concentrate containing 40 per cent nicotine sulfate. Today steam distillation or solvent extraction are the two processes for the commercial extraction of Nicotine from Tobacco.

Nicotine which is a physiological compound is an alkaids & contains nitrogen. It shows prominent physiological properties. Other well-known alkaids, which are not insecticides, are caffeine (found in tea and coffee), quinine (from cinchona bark), morphine (from the opiumpoppy), concaine (from coca leaves), ricinine (a poison in castor oil beans), *Strychnos nux vomica*), confine (from spotted hemlock, the poison that killed Socrates), and, finally, LSD (a hallucinogenic derived from the ergot fungus attacking grain).

As its mode of action, nicotine mimics acetylcholine (ACh) at the neuromuscular (nerve-muscle) junction in mammals, and results in twitching, convulsions, and death, all in rapid order. In insects, the same action is observed, but only in the gangila of their central nervous systems.

Nicotine sulfur, as it is commonly marketed, is highly toxic to all warm-blooded animals, as well as insects; for example, the LD_{50} (dose that proves lethal for 50 per cent to the test population) for rats is 50-60 mig/kg. This makes it the most hazardous of the botanical insecticides of home gardeners. It has been used with great success since before the turn of the century.

Dusts are very much toxic to humans hence are not available for garden use. Nicotine sulfate us used primarily for piercing- sucking insects such as leafhoppers, aphids, scales, trips and whiteflies, but it can kill all insects and spider mites on which it is sprayed directly. Many caterpillar pests, however, are very resistant to nicotine. It is more effective during warm weather, but degrades quickly. It is registered for use on flowering plants, and ornamental shrubs and trees to control aphids, merlybugs, scales and thrips, lace bugs, leafminers, leafhoppers, rose slugs, and spider mites. Its use for most greenhouse pests is also acceptable.

Nicotine sulfate is registered for use on a variety of vegetables and fruit trees, but it cannot be used in most cases within seven days of harvest, as required by EPA. There are several ornamental plants that are sensitive to nicotine, such as roses. The label should identify those sensitive plants. Nicotine is also registered for out-of-door use as a dog and cat repellent as well as for furniture in the home. Additionally, tobacco dust is acceptable as a dog and rabbit repellent out-of- doors.

Rotenone

Rotenoids, the retenone-related materials, have been used as crop insecticides since 1848, when they were applied to plants to control leaf-eating caterpillars. However, they have been used for centuries (at least since 1649) in South America to paralyze fish, causing them to surface.

ROTENONE

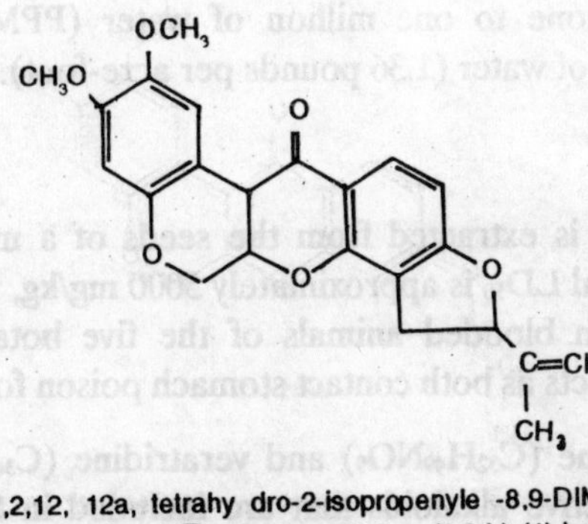

1,2,12, 12a, tetrahy dro-2-isopropenyle -8,9-DIMITHOXY-(1) BENZOPHYRANO-(3,4-b) furo-(2,3-b) (1) benzopyran-6(6Ah) one

Roteniods are produced in the roots of two genera of the legume (bean) family : *Derris*, grown in Malaya and the East Indies, and *Lonchcarpus* (also called cubed or cube), grown in south America.

Rotenone has an oral LD_{50} of approximately 350 mg/kg (in rats) and has been used for generations as the ideal general garden insecticide. It is harmless to plant, highly toxic to fish kind many insects, especially caterpillars, moderately toxic to warm- blooded animals, and leaves no harmful residues on vegetables. There is no waiting interval between application and harvest of a food crop.

Rotenone is marketed as spray concentrates and ready to use dust, it is both a contact & a stomach poison to insects. It kills insects slowly, but causes them to stop their feeding almost immediately. Like all the other botanical insecticides its life in the sun is short, one to three days. It is useful against caterpillars, aphids, beetles, true bugs, leafhoppers, thrips, spider mites, ants, rose slugs, whiteflies, sawflies, bagworms, armyworms, cutworms, leafrolles, midges and a host of other pests. Next to pyrethrum, rotenone is probably second in the number of approved use, exceeding 1000.

Rotenone is the most useful piscicide available for reclaiming lakes for game fishing. It eliminates all fish, closing the lake of reintroduction of rough species. After treatment, the lake can be restrocked with the desired species. Rotenone is a selective piscicide in that it kills all fish at dosages that are relatively nontoxic to fish foiod organisms. It also breaks down quickly leaving no residues harmful to the fish used for restocking. The recommended rate is 0.5 part of rotenone to one million of water (PPM), or 5.1 kg per hectarameter of water (1.36 pounds per acre-foot).

Sabadilla

Sabadilla is extracted from the seeds of a member of the lily lfamily. Its oral LD_{50} is approximately 5000 mg/kg, making it the least toxic to warm blooded animals of the five botanical insecticides discussed. It acts as both contact stomach poison for insect.

Cevatridine ($C_{32}H_{49}NO_9$) and veratridine ($C_{36}H_{51}NO_{11}$) are the two known active alkaloids that are included in Sabadilla. Neither chemical structure has been established. It is irritating to human eyes and causes violent sneezing in some sensitive individuals. It deteriorates rapidly in sunlight and can be used safely on food crops with no waiting interval required by the EPA. Sabadilla is probably the most difficult of the five botanical insecticides to purchase, simply because there was hardly any demand for it for about 15 years.

Sabadilla is registered for most commonly grown vegetables and will control caterpillars, grasshoppers, beetles, leafhoppers, thrips chinch bugs, stink bugs, larlequin and squash bugs, other true bugs, and patato psyllids. It is not very useful against and will not control spider mites.

Ryania

Ryania is a safe insecticide, quite safe for humans & domestic animals so safe that no waiting is required between the time of application to food crops and harvest, as there is for most other insecticides. Ryania is made from the ground roots of the ryania shrub grown in Trinidad and, like nicotine, belongs to the chemical class of alkaloids. It has an oral LD_{50} of approximately 750 mg/kg. It is a slow-actging insecticides, requiring as long as 24 hours to kill. Insects exposed to ryania usuallu stop their feeding almost immediately, making it particularly useful for caterpillars.

Ryanodine ($C_{25}H_{35}NO_9$),is the active principle of ryania, & its chemical structure is still not determined. Ryanodine affects insect muscles directly by preventing contraction, resembling the effects of strychnine in mammals.

The preferred uses for ryania are against fruit-and foliage-eating caterpillars on fruit trees, especially the codling moth on apple trees. HOwever, it is useful against almost all plant- feeding insects, making it an ideal material lfor small orchards of deciduous fruits. It is not effective against spider mites.

Ryania is exempt from a waiting period between application and harvest and is registered by the EPA for the control of a host of insect pests on a wide variety of plants, shrubs and trees. Vegetable garden pests include aphids, cabbage loopers, Colorado patato beetles, corn borers, cucumber beetles, diamond back moths, flea beeles, leafhoppers, Mexican bean beetle, spittle bugs, and tomato hornworms.

Ryania is registered for deciduous fruit trees of control aphids (except the woodly aphid), codling moth, Japanese beetle, and cherry fruit fly. It has long been used to control citrus thrips on all citrus. Though not very effective, ryania can be used in the home to control ants, silverfish, cockroaches, spiders and crickets. On ornamentals it is registered for aphids and lace bugs. It can used on roses against aphids, Japanese beetles, thrips, and whiteflies; on brambles, for aphids, raspberry fruitworms, and sawflies; and on grapes for aphids (except the woolly aphid) kand the Japanese beetles.

Ryania is difficult to obtain since its importation into the United States has been stopped by its major distributor.

Pyrethrum

Pyrethrum is extracted from the flowers of a chrysanthemum grown in Kenya, Africa and Ecuador, South America. It has an oral

LD_{50} of approximately 1500 mg/kg and is one of the oldest household insecticides available. In the early 19th century during the Napoleonic wars, ground dried flower heads were used to control body lice. Pyrethrum acts on insects with phenomenal speed causing immediate paralysis, thus its popularity in fast knockdown household aerosol

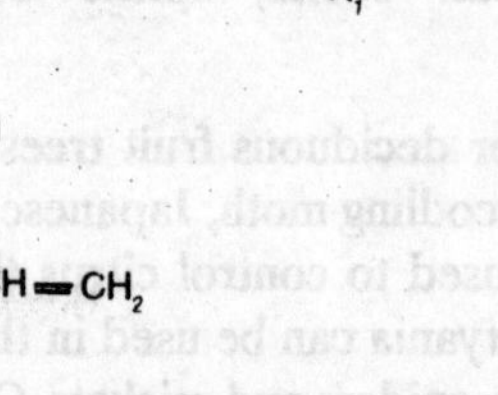

PYRETHRIN 1

$R_1 = -CH_3$

$R_2 = -CH_2CH{=}CHCH{=}CH_2$

CINERIN I

$R_1 = -CH_3$

$R_2 = -CH_2CH{=}CHCH_3$

PYRETHRIN II

$R_1 = -\underset{\underset{\displaystyle O}{\|}}{C}-OCH_3$

$R_2 = -CH_2CH{=}CHCH{=}CH_2$

CINERIN II

$R_1 = -\underset{\underset{\displaystyle O}{\|}}{C}-OCH_3$

$R_2 = -CH_2CH{=}CHCH_3$

FENVALERATE (Pydrin®)

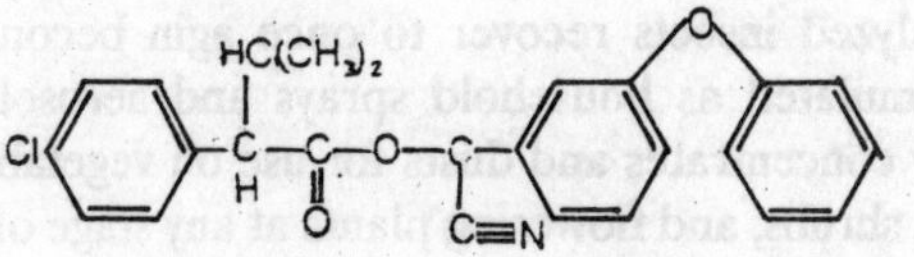

cyano (3-phenoxyphenyl) methyl 4-chloro-α-(1-methylethyl) benzeneacetate

PERMETHRIN (Ambush®, Pounce®)

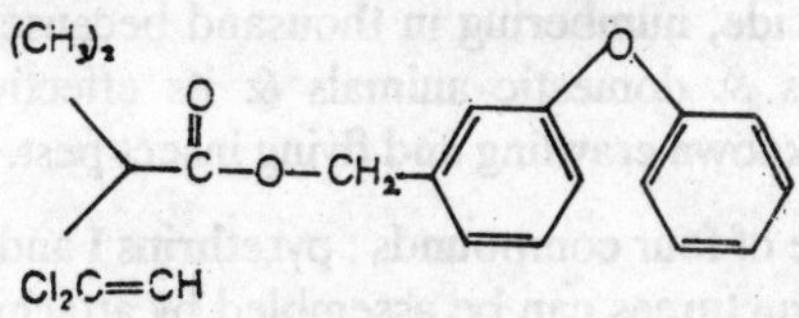

m-phenoxybenzyl (±)-*cis, trans*-3-(2,2-dichlorovinyl)-2,2-dimethylcyclopropanecarboxylate

ALLETHRIN (Pynamin®)

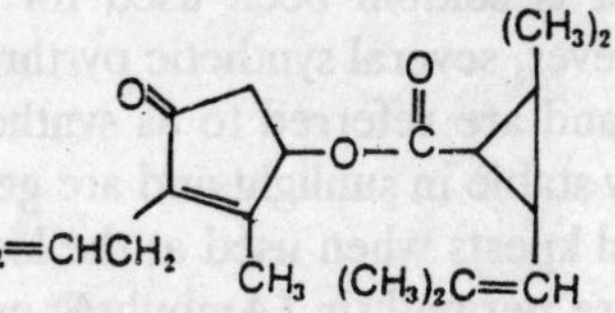

2-methyl-4-oxo-3-(2-propenyl)-2-cyclopenten-1-yl 2,2-dimethyl-3-(2-methyl-1-propenyl)cyclopropanecarboxylate

TETRAMETHRIN (Neo-Pynamin®)

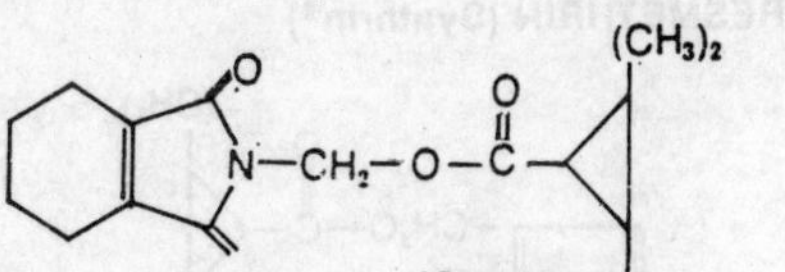

1,3,4,5,6,7-hexahydro-1,3-dioxo-2*H*-isoindol-2-yl)methyl 2,2-dimethyl-3-(2-methyl-1-propenyl)cyclopropanecarboxylate

spray. However, unless it is formulated with one of the synergists, most of the paralyzed insects recover to once agin become pests. Pyrethrum is formulated as household sprays and aerosols and is available as spray concentrates and dusts for use on vegetables, fruit trees, ornamental shrubs, and flowering plants at any stage of growth. Vegetables and fruit sprayed of dusted with pyrethrum may be harvested or eaten immediately; there is no waiting interval required between application and harvest of the food crop.

Pyrethrum has been approved by EPA, and has more uses than any other insecticide, numbering in thousand because of its general sagety to humans & domestic animals & its effectiveness against practically every known crawling and flying insect pest.

It is a mixture of four compounds : pyrethrins I and II and cinerin I and II. Their structurees can be assembled by attaching the R_1 and R_2 in their proper positions on the large ester structure to the left.

Synthetic Pyrethroids

The natural insecticide is quite unstable in sunlight and aho it is very costly, therefore, it is seldom been used for the agricultural purposes. Recently however, several synthetic pyrthrin-like materials have become available and are referred to as synthetic pyrethroids. These meterials are very stable in sunlight and are generally effective against most agricultural kpests when used at the low rate of 0.11 to 0.32 kig/ha. Examples are permethrin (Ambuhs® or Pounce®) and fenvalerate (Pydrin®).

The synthetic pyrethroids, or more correctly *pyrethroids*, have a rather long and successful history. For ease of classification, they are placed in four categories, or generations.

RESMETHRIN (Synthrin®)

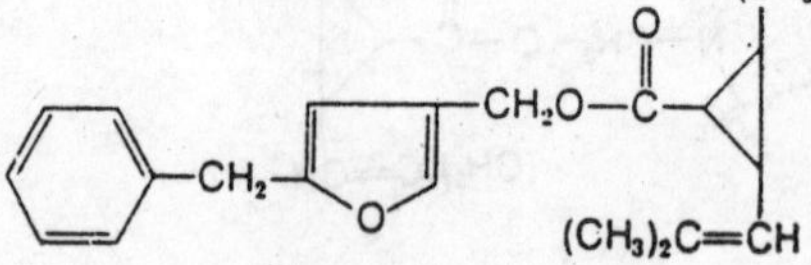

(5-phenylmethyl-3-furanyl)methyl
2.2-dimethyl-3-(2-methyl-1-propenyl)
cyclopropanecarboxylate

BIORESMETHRIN

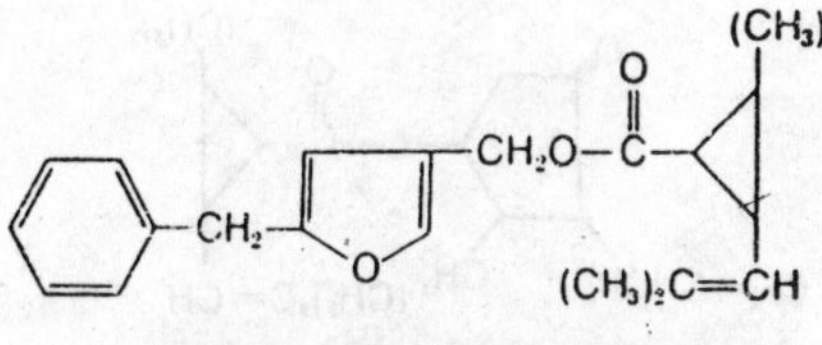

(5-phenylmethyl-3-furanyl)methyl
(1R-*trans*)-2,2-dimethyl-3-(2-methyl-
1-propenyl)cyclopropanecarboxylate
Stereoisomer of RESMETHRIN

The first generation contains but one pyrethroid, allethrin (Pynamin®). Allethrin was commercially available in 1949. It consisted of 22 chemical ieactions to produce the final insecticide, in this sense it marked the begining of an era of complex syntheses. Allethrin is merely a synthetic duplicate of cinerin I (a component of pyrethrum), with a slightly more stable side chain, and it is more persistent than pyrethrum. Equally effective against houseflies and mosquitoes, but less so against cockroaches and other insects it was readily synergized by the common pyrethrum synergists.

The second generation includes tetramethrin (Neo- Pynamin®), which appeared in 1965. It gives stronger knockdown of flying insects than allethrin and is readily synergized. Resemthrin (NRDC-104, SBP-1382, and FMC-17370) appeared in 1967, is approximately 20 times omre effective than pyrthrum in housesefly knockdown, and is not synergized to any appreciable extent with pyrethrum synergists. Bioresmethrin (NRDC-107, FMC-18739, and RU-11484), also described in 1967, is 50 times more effective than pyrethrum against normal (suceptible to insecticides) houseseflies, and also not synergized with pyrthrum synergists. Both resmethrin and bioresme-thrin and more stable than pyrethrum, and decompose fairly rapidly on exposure to air and sunlight, which explains why they were never developed for agricultural use. (Resemthrim has become the most used kof the second generation pyrethroids for sprays and aerosols to control flying and crawling insects indoors.)

In 1969 Bioallethrin (d-trans-allethrin) was introduced & it is more potent than allethrin and readily synergized, but it is not as effective as resmelthrin. The last of this period was phenothrin

d-trans-ALLETHRIN (Bioallethrin®)

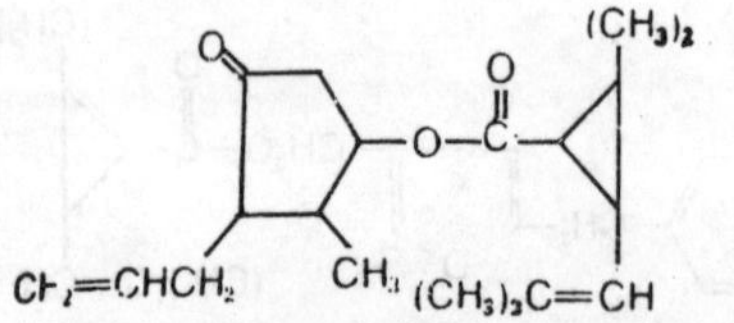

2-methyl-4-oxo-3-(2propenyl)-2-cyclopenten-1-yl 2,2-dimethyl-3-(2-methyl-1-propenyl), cyclopropanecarboxylate

PHENOTHRIN (Sumithrin®)

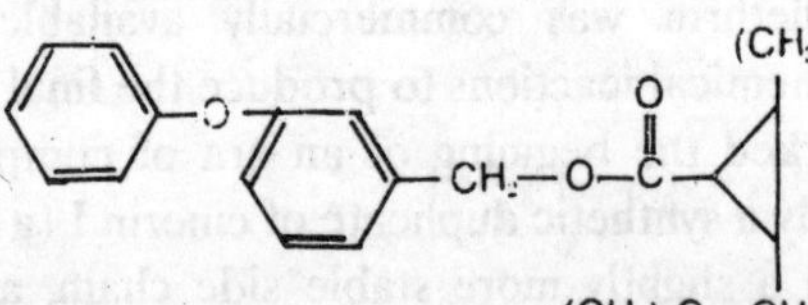

(3-phenoxyphenyl)methyl 2,2-dimethyl-3-(2-methyl-1-propenyl)cyclopropane-carboxylate

(Sumithrin ®), introduced in 1973. It, too, is intermediate in quality and slightly enhanced by synergists.

The third generation includes fenvalerate (Pydrin®) and permethrin (Ambush®, Pounce®, and Pramex®), which appeared in 1972 and 1973 respectively. These became the first agricultural pyrethroids because of their exceptional insecticidal activity (0.11 kgAI/ha) and their *photostability*. Photostability means they are unaffected even by ultraviolet in sunlight, lasting four to seven days on crp foliage as effective residues.

DECAMETHRIN (Decis®)

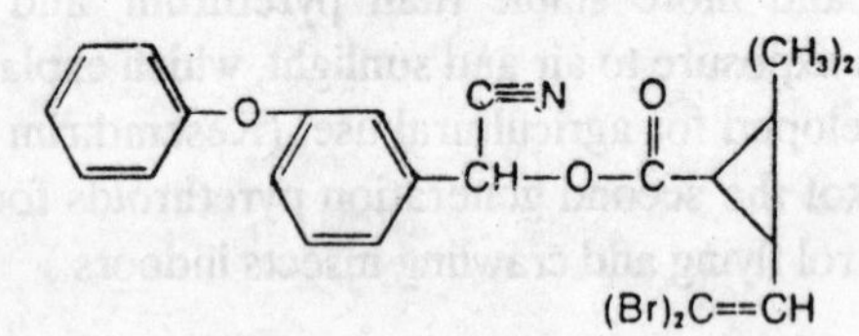

(1R(1 ;S*),3))-cyano(3-phenoxyphenyl) methyl 3-(2,2-dibromoethenyl)-2,2-dimethyl-cyclopropanecarboxylate

The contemporary forth generation, still being developed and regsitered, is truly exciting, for their rates of application are again reduced to one-tenth of the previous generation, or to the order of 0.11 to 0.06 kg AI/ha. This is truly phenomenal compared to the rate of 1.1 to 2.3 kg) AI/ha required of the organophosphate, carbamate, and organochlorine insecticides. Emerging in this truly revolutionary form of chemical insect control are the pyrthroids; cypermethrin (Ammo®, Cymbush®, and Ripcord®), fenpropathrin, flucythrinate (Pay- Off®), fluvalinate (marvil®), and decamethrin (Decis®). All these insecticides are photostable, providing long residual effectiveness in the field, and are not significantly imporved with the addition kof pyrethrin synergists.

Synergists or Activators

Activators or Synergists as the name indicates are not in themselves considered toxic or insecticidal, but are materials used with insecticides to synergize or enhance the activity of the insecticides. Synergists are added to certrain insecti- cides in the ratio of 8 : 1 or 10 : 1. The first synergist was introduced in 1940 to increase the effectiveness of pyrethrum. Since then many materials have been introduced, but only a few are still marketed beccause of cost and ineffectiveness. Synergists are found in practi- cally all the "bug-bomb"aerosols to enchance the action of the fast knockdown insecticides pyrethrum, allethrin, and sometimes resemthrin against flying insects.

Although initially developed for use with pyrethrum, they have since been observed to synergize some, but not all, organophosphates,

PIPERONYL BUTOXIDE

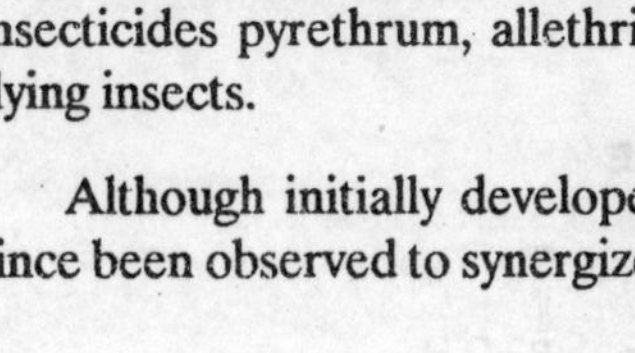

α-[2-(2-butoxyethoxy)ethoxy]-4,5-methylenedioxy 2-propyltoluene

MGK 264

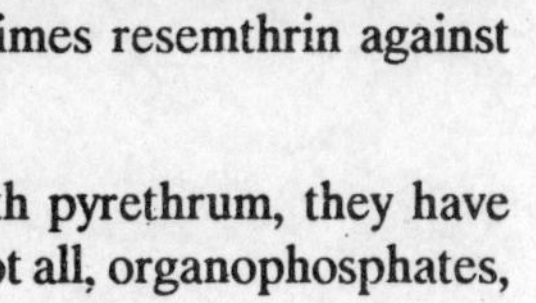

N-(2-ethylhexyl)-5-norbornene-2,3-dicarboximide

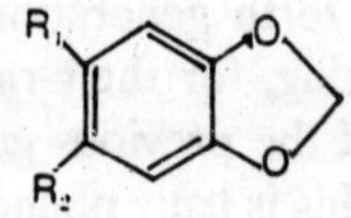

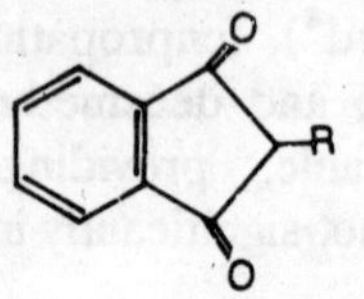

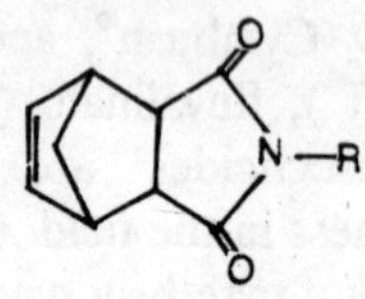

SESAMEX

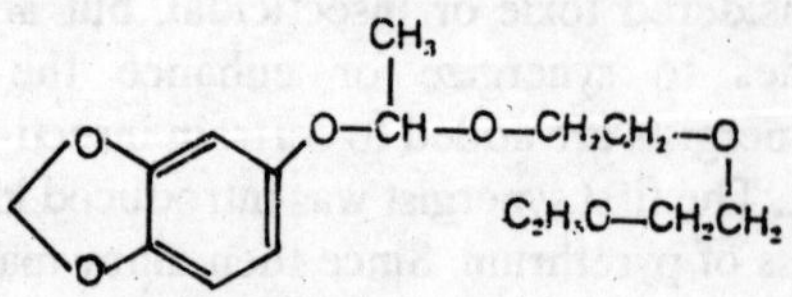

2-(2-ethoxyethoxy)ethyl-3,4-(methylenedioxy) phenyl acetal of acetaldehyde

SULFOXIDE

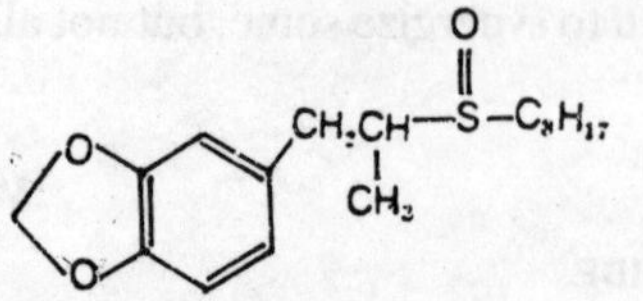

1,2-methylenedioxy-4-[2-(octyl sulfinyl)pro benzene

organochlorines, carbamates, as well as a few of the botanical, or plant- derived, insecticides. The mode of action of the synergists is to inhibit mixed-function oxidases, enzymes that metabolize foreign com- pounds, which in this instance would be the pyrethrum.

The most popular synergists belong to the only two molecular groups, or moieties. The first is the methylendioxyphenyl moeity. The R_1 and R_2 are simple or oxygenated dcarbon chains or other groups of varying combinations.

The second synergistic moiety does not have a single name but is characterized by either of the following sturctures.

Notice that all three moieties involve a five-membered ring associated with two oxygens. Because their mode of action kis the inhibition of insecticide-metabolizing enzymes, it is likely that this steric three-dimensional structure is generally the most effective in enzyme binding. The synergists are usually used in sprays prepared for the home and garden, stored grain, and on livestock, particularly in dairy kbarns. Synergists are quite expensive, thus seldom if ever used on crops.

Sesamin was the name given to a material containing the methyl endioxyphenyl groups, first found is sesame oil. As mentioned earlier, many compounds having this moeity are synergistic, but only the structures of piperonyl butoxide, mostly in livestock and animal shelter sprays.

In summary, the synergists are used in many of the insecticide mixtures for home, garden and barn. Their mode of action of their binding to oxidative enzymes the would otherwise degrade the insecticide.

Inorganics

These insecticides which do not contain carbon are called Inorganic insecticides . Usually they are white and crystalline, resembling the salts. They are stable chemicals, do not evaporate, and are frequently soluble in water. They are mentioned here for their historical siignificance.

Sulfur, 1, is very likely the oldest known *effective insecticide*. Sulfur and sulfur candles were burned by our great-grandparents for every conceivable purpose, from bedbug fumigation to the cleansing of a house just removed from medical quarntine of smallpox. Sulfur is a highly useful pesticide in integrated pest management programmes where target pest specificity is important. Sulfur dusts are especially toxic to mites of every variety, such chiggers and spider mites, thrips, newly-hatched scale insects, and as a stomach poison for some caterpillars. Sulfur dusts and sprays are also fungicidal particularly against powdery mildews.

Several other inorganic materials have been used as insecticides. These include compounds of mercury, boron, thallium, arsenic, antimony, selenium and fluoride. The only one of these used extensively today is arsenic, which is used in two forms, the arsenites (salts of arsenious acid) and the arsenates (salts of arsenic acid). The first commonly used arsenical was Paris green (green because of its copper content). It's a water soluble assenite. Next was lead arsenate. Finally, the third and last of these arsenicals was calcium arsenate, which was used for a time on vegetables in the 1930s and on cotton in the 1930s and 1940s. *Arsenicals* are truly stomach poisons, exerting their toxic action following ingestion by the insects. Their action is attributed to the arsenite or arsenate ion.

The arsenicals have a rather complex mechanism of action. First, they uncouple oxidative phossphorylation (bysubstitution of the arsenite ion for phosphorus), a major enter-producing step of the cell. Second, the arsenat ion inhibits certain enzymes that contain enzymes that contain sulfhydryl (–SH) groups. And , finally both the arrsenite and arsenate ions coagulate protein by causing the shape or configuration of proteins to change.

Arsenical insecticides were very useful agricultural tools form 1930 until 1956, as we were making the transition form the simple to the coimplex synthetic kmolecules. They were, in fact, responsible for the initiation of large-scale insecticide applications eventually leading to the intensive use of fungicides and herbicides in modern agriculture.

Fluorine insecticides also included organic fluorine compounds, but these were of little importance and seldom used. The inorganic fluorides were sodium fluoride, used for cockroach and ant control around the home, and barium fluosilicate, sodium silicofluoride, and cryolite (NaF, $BaSiF_6$, and Na_3ALF_6, respectively). The last three were used for a time in plant protection. The fluoride ion inhibits many enzymes that contain iron, calcium and magnesium. Several of these enzymes are involved in energy production in cells, as in the case of phosphatases and phosphorylases.

Boric acid (H_3BO_4), used as an insecticide against cockroaches and other crawling household pests in the 1930s and 1940s, has returned in the 1980s. Although manufacturers claim it is "safe, does not evaporate, and continues to kill for years," in fact, it is only moderately effective, acting as both a stomach poision and adsorber of insect cuticle wax.

The last group of inorganics are the silica gels or silica aerogels. These are light, white, fluffy silicates used for household insect control. The silica aerogels kill insects by adsorbing waxes from the insect cuticle, permitting the continuous loss of water from the insect body. The insects then gradually become desiccated and die from dehydration. These include Dir-Die®, Drianone®, and Drione®. The latter two are fortified with pyrethrum and synergists, which enchance their effectiveness.

Fumigants

The fumigants are small, volatile, organic molecules that become gases at temperatures above 5°C. They are usually heavier than air and commonly contain one or more of the halogens (CL, Br, or F). Most are highly penetrating, reaching through large masses of materials. They are used to kill insects, insect eggs, and certain microorganims in buildings, warehouses, grain elevators, soils, and green-houses and in packaged products such as dried fruits, lbeans, grain, land breakfast cereals.

Fumigants, as a group, are narcotics. The mechanism of action of the fumigants are more physical than chemical. The fumigants are liposoluble (fat soluble); they have common symptomogoly; their effects are reversi- ble; and their activity is altered very little by structural changes in their molecules. Narcotics induce narosis, sleep, or unconsciousness, which is effect in their action on insects. Liposolubility appears to be an important factor in the action of fumigants, since these narcotics lodge in lipid-containing tissues, which are found in the nervous system. Some of the more common fumigants' names and structures are presented in the list of the left.

Microbials

Certain microbes or micro-organisms are and used for controlling the insects, hence are called microbial insecticides like mammals insects are susceptible to diseases caused by fungi, bacteria,

Methyl bromide	CH_3Br
Ethylene dibromide	$BrCH_2CH_2Br$
Ethylene dichloride	$ClCH_2CH_2Cl$
Hydrogen cyanide	HCN
Chloropicrin	Cl_3CNO_2
Sulfuryl fluoride (Vikane®)	SO_2F_2
Vapam	$CH_3NHC(=S)—S—Na$
Telone	$ClCH{=}CH—CH_2Cl$
D-D®	$CHCl{=}CHCH_2Cl$ plus $ClCH_2CHClCH_3$
Chlorothene	CH_3CCl_3
Nemagon® (DBCP)*	$BrCH_2CHBrCH_2Cl$*
Ethylene oxide	
Naphthalene (crystals)	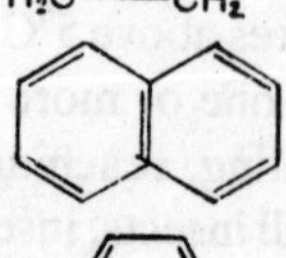
p-dichlorobenzene (PDB crystals)	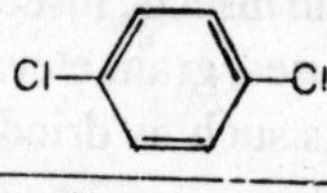

* Registered uses were canceled by the EPA in 1980.

and viruses. In several instaces, these have been isolated, cultured, and mass-produced for use as pesticiedes.

The insect disease-causing microorganisms do not harm other animals or plants. The reverse of this is also true. This method of insect control is ideal in that the diseases are usually rather specific. Undoubtedly the future holds many such materials in the arsenal of insecticides, since several new insect pathogens are identified each year. However, at the persent only a few are produced commercially and approved by the EPA for use on food and feed crops. There is still some concern regarding the every remote chance of human susceptibility to these diseases of invertebrate animals, hence the slow advances into this relatively new field exceptional precautionary testing.

The microorganizm *Bacillus thuringiensis* is a disease-causing bnacterium whose spore are necessary induction. These spores produce compounds that injure the gut of the insect larve in such a way that invasion of the body cavity follows. This organism produces four substances toxic to insect. The first, and most important, is a crytalline protein whose ingestion by caterpillars results in paralysis of their gut. The second is a toxin, a water-soluble nucleotide derivative, that passes unchanged through the gut of a cow to kill certain fly maggots breeding in the manure. The commercial preparation of Bacillus caused the loss of remaining two substance which are enymes.

Only one insect virus has been registered for agricultural use by the EPA. Two of the most destructive pest caterpillars in agriculture, knows as the *Heliothis* nuclear polyhedrosis virus, it is specific for *Heliothis zea* (corn earworm, cotton bollworm) and *Heliothis virescens* (tobacco budworm). Recently registrtion has been extended not only for cotton but also for all crops attacked by *Heliothis* species, including beans, corn, lettuce, okra, peppers, sorghum, soybeans, tobacco and tomatoes. Proprietary names for this knaturally occuring viral pathogen are Elcar® and Biotrol VHZ®.

Several other insect viruses are in the development stage, all of which are aimed at caterpillars and are only experimentally available. Viruses are highly specific and have modes of action that may not be identical throughout. Generally, the viruses result in crystalline protenis that are eaten by the larva and begin their activity in the gut. The virus units then pass through the gut wall and into the blood. There are units multiply rapidly and take over complete genetc control of the cells, causing thier death.

A recent and innovative development in the agricultural use of microbial insecticides is the addition of feeding or gustatory stimulants. The feeding stimulants apparently attract the caterpillars to treated foliage, which increases their consumption of the microbial. Two successfully marketed products are Coax® and Gustol®, both of which are wettable powders and used at 1.1 to 2.3 kg/ha.

Abbott Lboratories developed a mycoacaricide, mycar miticide which is a new biological cosntrol. The microorganism is *Hirsutella thompsonii*, a parasitic fungus that infects and kills the citrus rust mite. Under optimum conditons *H.thompsonni* can infect spider mites and other nontarget mites. It is, however, consistenty effective only against the citurs rust mite; this it is a selective miticide. When sprayed on plants the spores grow into colonies that attach to the mite. In the presence of ample free moisture the spore germinates and infects the mite. The mite dies in about three days, and the fungus spreead, continuing to propagate itself. In this manner it offers potentially long residual effectiveness. Because the active agent is a fungus, all commercial fungicides, copper chemicals, and certain other metal salts such as zinc, lead, and manganese are detrimental to *H. thompsonii*. Because of its specificity *H. thompsonii* should be highly compatible with other suppression techniques used in the integrated past managemnt of citrus crops.

For the control is giasshoppers, Sandoz Inc. developed a new biological control, *Nosema Locustae*. Marketed under the name NOLOC® and Trojan 10®, the microorganism is a protozoan. Depending on the method of application, climatic conditions, and grasshopper density, the portozoa kill up 50 per cent of the insects and sterilize up to 30 per cent of the survivors. Maximum effect occurs over a two-to four-week period. Through subsequent generation by transmission through the eggs upto three or four years the residual effect of a single treatment continues to control grasshoppers.

Insect Growth Regulators

The first generation insecticides were the stomach poisons, such as the arsenicals, heavy metals, and fluorine compounds. The second

alpha-ECDYSONE

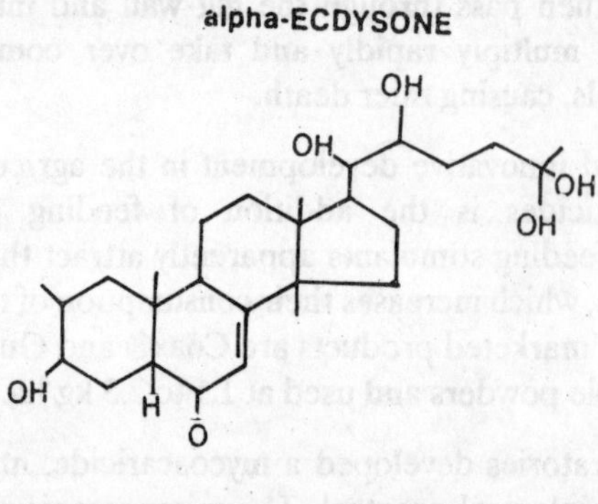

TYPICAL JUVENILE HORMONE

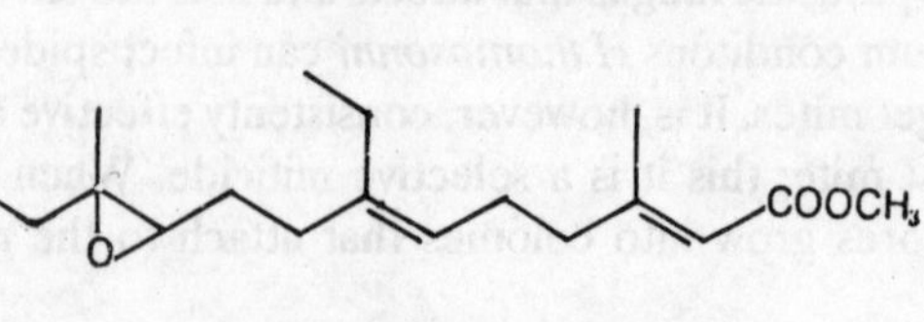

CHROMENE NUCLEUS

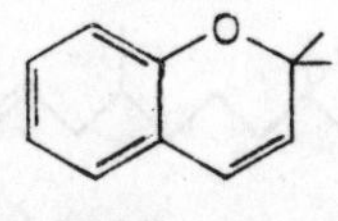

generation includes the familiar contact insecticides roganochlorines, organophosphates, carbamates and formamidines.

The third generation of insecticides are the *biorationals*. These chemicals are environmentally sound, closely resembling or identical to chemicals produced by insects and plants. Among these are the insect growth regulators (IGRs), a relatively new group of chemical compounds that alter growth and developments in insects. Their effects have been observed on embryonic and larval and nymphal development on metamorphosis, on reproduction in both males and females, on behaviour, and on several forms of diapause. They include ecdysone (the molting hormone), juvenile sghormone (JH), JH mimic, JH analog (JHA), and their broader synonyms, juvenoids and juvegens. More recently another growth effect, chitin inhibition, has been identified and is discussed later, with the EPA-registered IGRs.

The IGRs are effective when applied in very minute quatities and apparently have no undersirable effects on humans and widlife. They are, however, nonspecific, since they affect not only the target species, but most other arthropods as well. IGRs may play an important role in future insect control, consequently, when used with precision.

Several hormone producing glands in insects perform the principal functions of which are the control of reproductive processes, motting land metamorphosis. Here we are interested only in the hormone ecdysone, which is responsible for molting, and JH, which inhibits or prevents metamorphosis.

When insects are treated with ecdysones, they usually die in all stages of growth, making ecdysones smiliar to second-generation insecticides. One attractive feature of ecdysones as potential tools is their widespread distirbution in plants,, and these may play as yet unrecognized roles in insect-plant relationships.

Keen interest has been directed toward JHs. These are not in the usual sense, toxic to insects. Instead of killing directly, they interfere in the normal mechanisms of development and cause the insects to die

METHOPRENE (Altosid®)

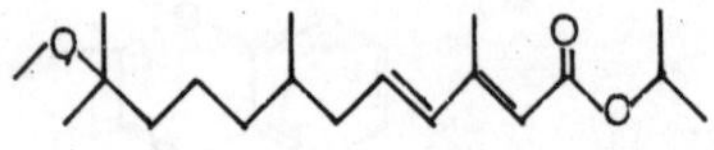

isopropyl (2*E*-4*E*)-11-methoxy-3, 7, 11-trimethyl-2, 4-dodecadienoate

DIFLUBENZURON (Dimilin®)

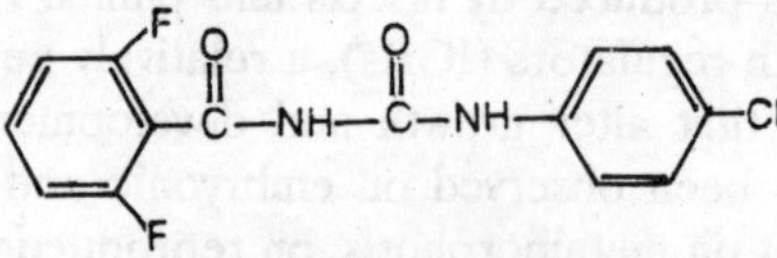

1-(4-chlorophenyl) 3-(2,6-difluorobenzoyl)urea

before reaching the adult stage. One JH is the classical juvabione, found in the wood of balsam fir. Its effect was discovered quite by accident when paper towels made from this source were used to line insect-rearing containers, and the insects' development was suppressed.

Some of these plant-derived substances actually serve to inhibit the development of insects feeding thereon, thus protecting the host plant. These are referred to broadly as antijuvenile hormones, or more accurately, *antiallatotropins*. Recently, antillatotropins, called *precocenes*, have been discovered. These are considered a potential fourth generation of insect control agents. Incidentally, the name precocene results from the precocious metamorphosis stimulated by compounds having the chromene nucleus. It is known that precocenes depress the level of juvenile hormone below that normally found in immature insects, although the mode of action is still unclear.

Dramatic results with JHs have been obtained in the laboratory, the most promising effects being on mosquito larvae, caterpillars, and hemipterans (bugs), although effects have been observed on practically all insect orders. Most insect species respond to treatment with JHs by producing extra larval, nymphal, or pupal forms that vary from giant, almost perfect, forms to intermediates of all sorts between

KINOPRENE (Enstar®)

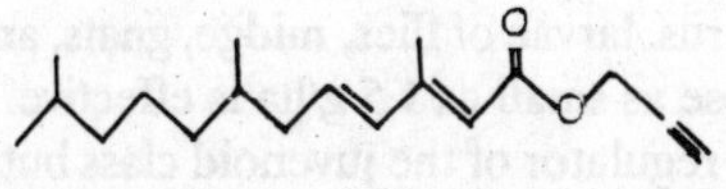

2-propynyl (*E,E*)-3,7 11-trimethyl-2 4-dodecadienoate

the immatures and the adults. For the most part, the periods of greatest sentitivity for metamorphic inhibition are the last larval or nymphal stages and the pupa, in those having complete metamorphosis. One recognizable problem is the precision of timing applications to achieve meximum damaging effect on the uncoming life stage of a particular insect.

For practical purposes. IGRs could be used on crops to suppress demaging insect numbers. They would be applied with the purpose lof preventing pupal development or adult emergence, thus keeping the insects in the immature stages, resulting eventally in their deaths.

To date, only three IGRs are registred by the EPA. The first, methoprena (Altosid®), manufactured by the Zoecon Corporation, was registered early in 1975 as a mosquito growth regulator, for use against second – through fourth – larval-stage floodwater mosquitoes at 0.11 to 0.14 kg/ha to prevent adult emergence. Larvae exposed to methoprene continue their development to the pupal stage, when they die. When applied to pupae or adult mosquitoes. Additionally methoprene has been formulated as PrecorØ for indoor control of dog and cats fleas. It acts by interrupting the flea's life cycle at the larval stage, preventing emergence of adult fleas for lup to 75 days. Its use in combination with a conventional insecti- cides is usually necessary to control adult fleas not affected by the growth regulator.

Methoprene is also registered for use on tobacco to control cigarette beetles (KabatØ), on cattle feed to control horn flies, and on mushroom-growing media to control hungus gnats (ApexØ). In order to control mosquitoes in drinking water it has been approved by the World Health Organization for its use.

The second IGR is diflubenzuron (Dimilin®), manufactured by Thompson-Hayward Chemical Company. Currently registered for gypsy moth caterpillars in forest and for the boll weevil in cotton, it is awaiting registration for a wide variety of pests. Experimentally, it has controlled insect pests in forest, woody oramentals, fruit, vegetables, cotton, soybeans, citrus, larvae of flies, midge, gnats, and mosquitoes : on mosquitoes, a dose as small as 1.5 g/ha is effective. Diflubenzuron is not truly a growth regulator of the juvenoid class but rather another insecticide with a different mode of action. However, it is tentatively classed with the IGRs. It acts on the larval stages of most insects by inhibiting or blocking the synthesis of chitin, a vital and almost indestructible part of the hard outer covering of insects, the exoskeleton.

DIMETHYL PHTHALATE

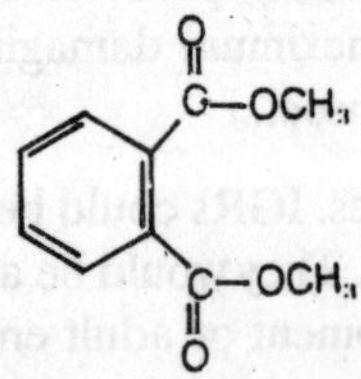

dimethyl 1,2-benzenedicarboxylate

Indalone®

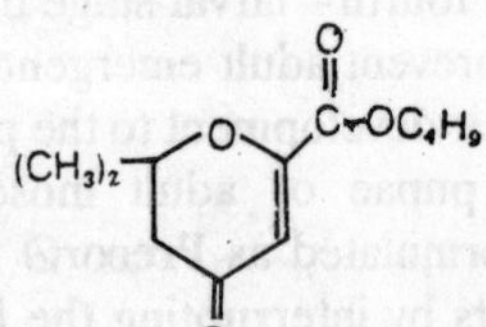

butyl 3,4-dihydro-2,2-dimethyl-4-oxo-2H-pyran-6-carboxylate

RUTGERS 612

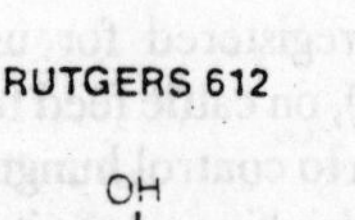

2-ethyl-1,3-hexanediol

The third, and most recent, IGR is kinoprene (EnstarØ), also developed the Zoecon Corporation. It is effective agsinst aphids, whiteflies, mealybugs, and scales (both soft and armored) on ornamental plants and vegetable seed crops grown in greenhouses and shadehouses. It is specific for insects in the order Homoptera, and results in a gradual reduction rathar than an immediate kill, by inhibiting development, reduction egg-laying, killing eggs already laid, and sterilizing mature whiteflies and aphids. It is virtually nontoxic to humans and other warm blodded animals because it is specific for insects like all other IGRs.

Cans IGRs become successful pest control agents ? Certainly they can in time. They will, however, have to meet general criteria for other pests control agents; thus, they must be effective in reducing insect populations below economic damage levels, be competitive with second-generation insecticides in cost, and have no undersirable side effects. In summary, IGRs hold intriguing possibilities for future use in practical insect control. It should be kept in mind that IGRs are insect-controlling chemicals and thus fall within the same legal confines as other insecticides. Being Toxic to populations of insects rather than to individuals they are however distinguished from others. Likewise the ability to control the fecundity (reproductive capacity) of the pest to the cause of the ultimate success of any pest control agent.

Insect Repllents

Historically, repellents have included smokes, plants huge in dwellings or rubbed on the skin as the fresh plant or brews, oils, pitches, tars, and various earths applied to the body. Camel urine sprinkled on the clothing has been of value, although questionable, in certain locales. Camphor crystals sprinkled among woolens have been used for decades to repel clothes moths. Before humans developed a more edified approach to insect olfaction and behaviour, it was assumed that if a substance was repugnant to humans it would likewise be repellent to annoying insects.

Prior to World War II, there were only four principal repellents : (1) oil of citronella, discovered in 1901, used also as a hair- dressing fragrance by certain Eastern cultures, probably to control lice and other head ectoparasites; (2) dimethyl, phthalate, discovered in 1929;

DIBUTYL PHTHALATE

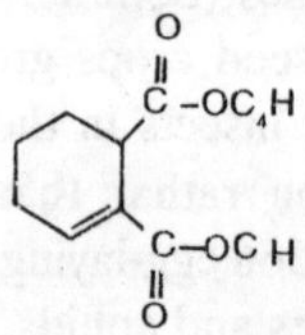

di-n-butyl phthalate

(3) Indalone®, introduced in 1937, and (4) Rutgers 612, which became available in 1939.

It became essential to find new repellents that would survive both time and dilution by perspiration, with the onslaught of World War II and the introduction of American military personnel into new environments, particularly the tropics. The ideal repellent would be nontoxic and nonirritating to humans, nonplasticizing, and longlasting (12 hours) aginst mosquitoes, biting flies, ticks, fleas and chiggers. Unfortunately, the ideal repellent has still not been found. Some repellents have unpleasant odours, require massive dosages, are oily or effective only for a short time, irritate the skin, or soften paint and plastics.

Insect repellents come in every conceivable formulation– undiluted, diluted in a cosmetic solvent with added fragrance, aerosols, creams, lotions, treated cloths to be rubbed on the skin, grease sticks, powders, suntan oils, and clothes- impregnating laundry emulsions. Regardless of the formulation, the periods of protection they offer vary with the chemical, the individual, the general environment, insect species, and avidity of the insect.

What happens to repellents once applied ? Why aren't they effective longer than one or two hours ? No single answer is satisfactory–but generally it is because they evaporate, are absorbed by the skin, are lost by abrasion of clothing or other surfaces, and are diluted by perspiration.

The following chemical structures are those of the most commonly found in today's repellents. Of these, diethyl toluamide

STA- WAY

$CH_3\overset{\displaystyle O}{\overset{\|}{C}}OCH_2CH_2OCH_2CH_2OC_4H$

2(2-butoxyethoxy) ethyl acelate

MGK REPELL ENT 874 (ants, cockroaches)

$CH_3(CH_2)_7SCH_2CH_2OH_2$

2 -(octylthio) ethanol

MGK REPELLENT II

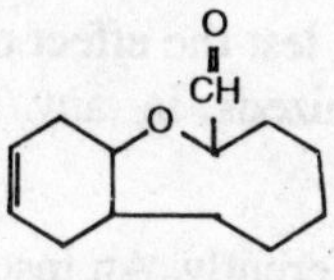

hexahydro -4a (4H)-
dibenxofuran carboxal

MGK REPELENT 326

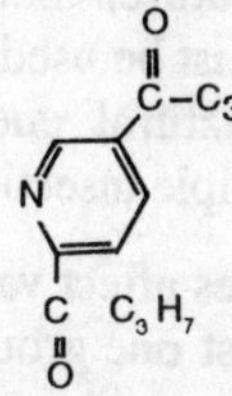

di-n- propyl 2 5- pyridinedicarboxylate

CLOTHING IMPREGNANT (ticks, chiggers)

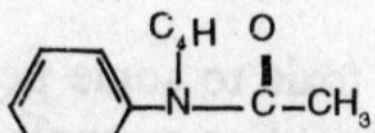

N-butyl acetanilide

DIMETHYL CARBATE (Dimelone)

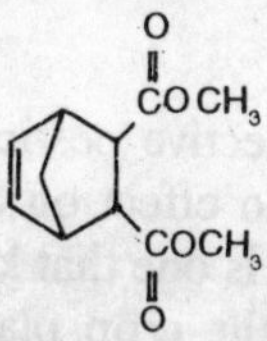

dimethylcis- bicyclo (2,2,1)- 5- heptane
2,3-dicaboxylate

DEET (Delphene)

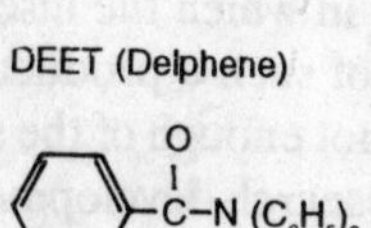

N,N-diethyl-m-toluamide

BENXYL BENZOATE (ticks chiggers)

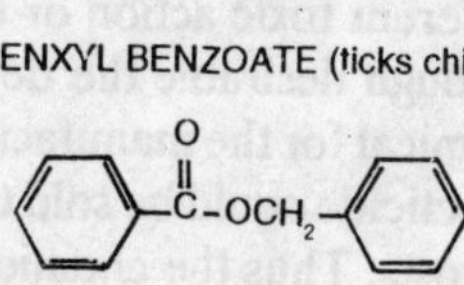

benzyl benzoate

TABATREX (buildings only: ants,
cockroaches and filies)

$$\begin{array}{l} \quad\;\; O \\ \quad\;\; \| \\ CH_2COC_4H \\ | \\ CH_2COC_4H \\ \quad\;\; \| \\ \quad\;\; O \end{array}$$

di-n-butylsuccinate

(Delphene®) is by far superior to all others against biting and mosquitoes.

Insecticide Selectivity

We must supplement insecticides with other forms of insect control as their are on main tool or insect control at least or the foreseeable future. Selective insecticides must be used and insecticides must be used in selective way, lest the effect of beneficial insects and natural enemies be minimizeds in any programme requiring multiple insecticide applications.

Insecticides affect various insects differently. An insecticide that is lethal against one group may have little, if any, effect on another. Generally, however, most insecticides kill many kinds of insects; broad-spectrum insecticides are those that kill more kinds than others.

A selective pesticide is one that is toxic to some pests, but has little or no effect on other similar species. For example, a selective herbicide is one that kills certain undersirable species of weeds while harming the crop plants little or not at all. Certain fungicides are selective to the point that they control only powdery mildews and no other fungi. A selective insecticide kills selected insects, but sprares many or most of the other insects, including beneficial species, either through different toxic action or the manner in which the insecticide us used. Though desirable the development of such a product would be uneconomical for the manufactures since not enough of the spectes specific insecticide could be sold to recoup research development and marketing costs. Thus the chemical industry seek rather to develop a general purpose, broad-specturum chemical that will control not only elm leaf beetles, but boll weevils on cotton, caterpillars on vegetables, the gypsy moth in deciduous forest, green bugs in small grains, alfalfa weevis, the codling moth in apples, and scales on citrus. General-purpose insecticides offer the best possible incentives in the economic market : satisfied growers and contented stock-holders.

A selective insecticide, in the narrowest use of term, simply means that the active ingredient is inherently more toxic to one group of insects than to others; in other words most groups of insects are physiologically more tolerant to the chemical than are the few readily killed by it. More commonly, selective is used to designate an insecticide that does not harm a beneficial species of insect while killing pest species.

This physiological selectivity results from differences between target and nontarget species in (1) penetration rate of the toxicant, (2) tissue binding and loss of the toxicant, (3) the speed of toxicant and metabolite excretion, (4) metabolic alteration or detoxification, (5) target sites of action or biochemical lesion, or (6) polyfactorial selectivity where more than one of the above processes is involved. (For a clear, detailed discussion of physiological selectivity, see Hollingworth, 1976.)

Physiologically nenselective insecticides may achieve use selectivity if they are applied in such a way that certain insect groups are more adversely affected than others. Use selectivity may result from the timing of the application, the formulation used, the dosage level, or numerous ther techniques. From a practical standpoint this approach offers the most immediate hope of being included in precision integrated pest management programmes. The selective use of an insecticide to kill a particular pest while permitting other insects to escape, particularly beneficials, achieves the same selective effect as the use of physiologically selective insecticides.

The existence of both the pest and the desirable species, and utilize some difference in their habits, distribution, or biology to devise a discriminatory method of application, all these must be taken into a count by the planner if they use selectivity. Although some insecticides are intrinsically more toxic physiologically to certain insects, and some are equally toxic to different species, they can be used selectively to affect pests more than beneficial insects.

Insecticide selectivity thus can be achieved not only by restircted toxicity but also by precision in timing, the calculation of effective rates of application, the use of materials with short residual lives, methods of application such as seed treatment, spot treatments restricted to areas in which pests vastly outnumber beneficials, or use of those formulations that tend to enhance the survivial of beneficials. In general, selectivity is obtained by one of the following paractices : (1) use of physiologically selective insecticides that are more toxic to pest species than to others; (2) timing of applications such that detrimental effects on the pest's natural enemies are minimized; and (3) reduction of dosages to levels that adequately control the target pest while sparing relatively large numbers of natural enemies of the target pest and other potential pests (Watson, 1975).

The selective use of insecticides is not simple. In most instances it requires an intimate knowledge of not only the target pest but also the secondary pests beneficial species. Researchers are directing their efforts towards integrating the use of insecticides and beneficial insects. Effective insect control will require both toxicological research in the development of physiologically selective insecticides and applied ecological research in the use of available insecticides to achieve selective action. The selective use of insecticides requires an intimate knowledge of not only the target pest but also the secondary pests beneficial species.

8

Insecticides & Their Applications

Insecticides are the chemical substances used in controlling the pests. The usefulness of any insecticide depends in a very large measure upon its proper application, and this is determined by the properties of the insecticide, the nature of the pest or pest complex to be controlled, and the site to which the application is to be made. The three general methods of applying insecticides are as *sprays*, in which water or oil is used as the carrier for the toxicant, as *dusts*, in which a fine dry powder is the carrier, and as *fumigants*, in which the insecticide in applied as a gas. The equipment for the application of insecticides ranges from such simple devices as the puff duster and "flit gun" to complex machines such as the mist blower and spraying helicopter, and this development is entirely a product of the last 100 years. Before that time, liquid insecticides were applied by a bundle of twigs, a feather, or a brush broom, and dusts by a bellows or blowing tube. The progress which has been made is a product of detailed knowledge of physics and engineering coupled with a vast amount of trial and error. Much remains to be learned, and relatively simple devices may yet be produced which will effect veritable revolutios, as for example, the development of the aerosol "bomb," which from its humble begining in 1942 has become a standard household article, of which 65,900,000 units were sold in 1959 for insect control.

Cautions in Insecticide Applications

Usually insecticides are applied when populations of a pest species reach the economic threshold. In this way economic damage is prevented and needless treatments are avoided. Other factor must often be con- sidered as well, such as the timing of applications so as to coincide with the stage of development of the insect pest which is most easily killed or with the stage in the seasonal development of the plant when it well best withstand the treatment. Timing to observe the safe interval before harvest which is required to attenuate insecticide residues to legal levels and to avoid periods of bloom for protection of bees and other pollinating insects is also extremely important. Since proper timing varies widely with different crops, different insecticides, and in various areas of the country, the grower should follow only approved spray programmes of state and federal agencies, observe all label precautions and directions, and whenever in doubt should secure the advice of a trained entomologist who knows local conditions regarding spray programmes for different crops and insects. In general, the earlier the insecticide is applied after the insects appear, the easier it is to destroy them, but in most cases applications should be made only when populations reach economic thresholds.

Most of insecticides are violent poisons to human beings and livestock, as well as to insects. Therefore, they should be plainly labeled and, together with mixing vessels, kept out of reach of chidren. Animals must be kept away from liquid sprays and not allowed to pasture in treated areas. All discarded insecticide containers should be burned or buried. The spray applicator should familiarize himself with the hazards of various insecticides before use and determine the proper protective measures necessary for safe application, as furnished by the manufacturer. In general, these will include the use of protective clothing, rubber gloves, and goggles when handling spray concentrates and a protective mask if there is any possibility of inhaling toxic dust, mist, or vapour. Prolonged wetting by sprays or other contamination should be avoided, and clothes changed at least twice daily, followed by the liberal application of soap and water to remove any skin contamination. In rainy weather the spraying and dusting should be avoided although many sprays will adhere satisfactorily to plants if the spray has time to dry before rain falls. Winter sprays should not be applied when the temperature is below freezing or when the trees are wet with snow or rain. When possible , spraying should also be avoided in very hot weather, above 90° F., because of increased susceptibility of plants to insecticide damage and because of the increased hazard to the operator from skin and inhalation absorption of poisons.

Sprays and Spraying

Historically, dusts have been the simplest formulations of pesticides to manufacture and the easiest to apply. In recent years, because of the increased effectiveness of the new organic insecticides, concentrate sprays have largely replaced dusts on many field and vegetable crops. Sprays are made up of solutions, emulsions, or suspensions of the toxicant. The liquid phase is usually water, but light oils are also employed. Insecticidal sprays are con- veniently described as *space sprays* or *aerosols,* directed against flying insects, and *residual sprays,* applied to surfaces of plants, animals, or structures frequented by insects.

Importance of Droplet Size.

The size property of spray droplet is very significant and conveniently measured in microns (μ a or 0.001 millimeter). In every spray cloud there is a considerable spectrum of droplet sizes and the cloud is most conveniently characterized by a mean value which expresses the normal frequency distrubution of the droplets. The most commonly used characteristic is the mass median diameter MMD or D_m, which is that throretical droplet diameter which divides the volume of the spray into equal parts. The surface median diameter SMD or D_o is the theoretical droplet diameter which has the same surface to volume ratio as that of the total spray. D_m is a useful measure of the effectiveness of the spray in terms of the amount of active material deposited per unit area, while D_o defines the amount of exposed liquid surface and thus the coverage which may be obtained. D_m ranges from 1.1 to 1.6 times D_o, depending upon the nature of the atomizing device. However, with the common fan, swirl, and twin-fluid atomizers used in the application of insecticides. The average ratio D_m/D_o is relatively constant with a value of 1.28, as studied by Fraser.

Droplet-size Requirements.

The droplet-size of the sprays as per requirements for various insecticidal spraying operations cover a 300-fold range of droplet diameters and a 27,000,000 range of droplet volumes and extend from the 1-to 30- micron size of true aerosols to the 100- to 500-micron size of hydraulic sprays. With contact sprays, the zone of efficiency for air-blast sprays is about 30 to 80 microns and for hydraulically driven sprays, between 100 and 300 microns. Special problems are encountered in aircraft spraying, and in general the most efficient range is from 100 to 300 microns.

Space sprays or aerosols. The principle of space spraying is to suspend a cloud of droplets in the space through which the insect pests are flying and thus to force the insect to accumulate a lethal deposit by colliding with the droplets. The successful application of this method is dependent upon physical laws governing the behaviour of aerosols, which are defined as colloidal suspensions of matter in air and involve an understanding of the insect flight behaviour. It is clear that the insecticide will be effective only as long as the droplets remain suspended in the free-air space.

SPRAYING EQUIPMENT

There are Narious types of spraying equipments. The size and type of sprayer to be used will depend upon the amount and kind of work to be done. However, spraying is at best a disagreeable task, and every effort should be made to have equipment of sufficient capacity so that the work will be done thoroughly and efficiently. The best spraying equipment can easily be ruined by failure to give it proper care. Clean water should always be pumped through the sprayer after use, to flush out all corrosive spray materials, and the metal parts should be oiled to prevent rust. It is not advisable to apply insecticides with spraying equipment which has previously been used for the application of herbicides of the plant-growth-regulator type such as 2,4-dichlorophenoxyacetic acid (2,4-D), because very small traces of such materials may seriously injure certain plants.

There are many types of sprayers, but the most common sprays are *hydraulic sprayers,* which employ only pressure nozzles for the formation and distribution of the spray, and *air-blast sprayers,* which utilize an air stream for distribution and/or breakup of the spray.

Hydraulic sprayers

The various types of hydraulic sprayers include (*a*) the compression sprayer, which is actually a special type in which compressed air is used to force the liquid through a pressure atomizing nozzle. The most familiar type is a 4-gallon cylindrical tank with a hand piston pump which is operated at 40 to 80 pounds per square inch and is very popular with the home gardener. The bucket or stirrup pump sprayer (*b*) has a single-or double-action plunger pump clamped into a 2-gallon bucket and operates at up to 150 pounds per square inch; (*c*) the napsack sprayer is a 2- to 5-gallon tank carried as a napsack and containing a diaphragm pump and agitator operations at 50 to 80 pounds per square inch by a lever carried under the operator's shoulder. These are especially suitable for small garden opertions or treatments in inaccessible spots, as in mosquito larviciding or residual house spraying. The barrel pump (*d*) is a larger plunger-type spray pump which operates at a pressure of 200 to 300 pounds per square inch and uses a 15- to 45-gallon container as a reservoir. The wheelbarrow sprayer (*e*) is essentially a portable barrel pump which is suitable for spraying small fields, orchards, or farm buildings. Most modern wheelbarrow sprayers have rubber tires and contain a two-cycle gasoline-engine-driven pump of 1.5 to 3.0 gallons per minute at 200 to 250 pounds per square inch and have a capacity of 15 to 50 gallons. Hydraulic power sprayers (*f*) range from the power whellbarrow sprayer through trailer-type field and row-crop sprayers to truck-mounted orchard sprayers. These power sprayers have 50- to 500-gallon capacities, operate up to 400 to 800 pounds per square inch, and have discharge capacities ranging from 5 to 80 gallons per minute. Discharge may be carried through one or more spray hoses or through a horizontal or vertical spray boon equipped with multiple swirl spray nozzles.

Parts of a sprayer. The important components of a spray outfit are similar for the various types, some of which are as follows: The *cylinder* is the portion of the pump in which the pressure is developed. It must be of noncorrosife material, and in large power sprayers, which have two to four cylinders, the lining is generally of acid-resistant porcelain. The *plunger* or *piston* forces the spray liquid through the cylinder and is made to fit tightly inside the cylinder by means of *packing*. Plungers may be fitted with molded rubber and fabric *plunger cups* which act as inside packing, expanding against the cylinder wall on the pressure stroke, or outside packing may be used which is compressed against stainless-steel plungers. *Valves,* usually of the ball type, and *valve seats* made of hardened stainless steel, direct the flow of spray liquid. They must be readily accessible for cleaning in case of clogging by particles of dirt or spray materials. An *air chamber* is used to equalize the pressure and remove excessive strain on the pump. The discharge opening leading to the nozzle is located at the bottom of the air chamber. The *tank*, in which the spray ingredients are mixed and held, is usually constructed of metal with an enamel lining or of stainless steel.

Since many spray ingredients, such as wettable powders, are not soluble in water, an *agitator* in necessary to keep the finely divided particles in suspension or the emulsion evenly distributed. In power sprayers this is commonly effected by two or more flat paddles or propellers on a rotating shaft near the bottom of the tank. A *pressure gage* indicates the pressure being maintained, and the *pressure regulator* maintains uniform pressure on the spray nozzles and allows the pump to operate at a greatly reduced load when spray is not being discharged. This is accomplished by a spring-and-ball valve, which can be set to lift at the desired pressure and permit the excess liquid to bypass to the tank. *Strainers,* over the intake to the tank and over the suction or intake pipe leading to the cylinders, keep the spray liquid free of troublesome foreign matter.

The *discharge pipe* should be free from abrupt angles that cut down the pressure between pump and nozzles. The *hose* for orchard spraying should be at least 25 feet long, of 1/2-to 3/4-inch inside diameter and four-to seven-ply strength. The friction of the spray

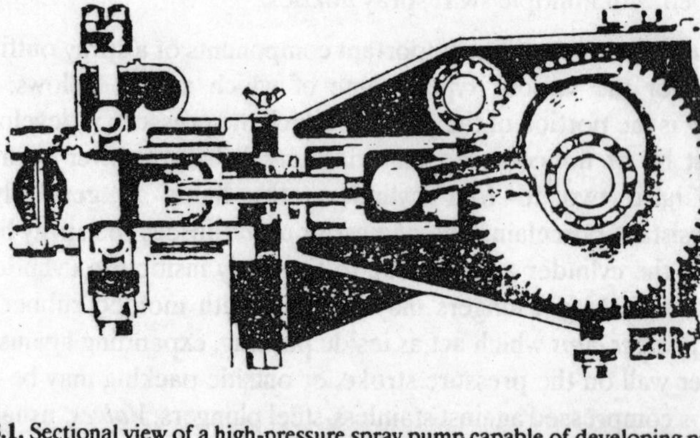

Fig. 8.1. Sectional view of a high-pressure spray pump capable of developing pressures of 000 to 1.000 punds per square inch. When the plunger-moves right, spray solution is drawn through the lower ball valve. Upon the return stroke of the plunger, the lower bal valve is forced shut and the solution under pressure passes through the upper valve into the compression chamber. (*From Food Machinery and Chemical Corp., Bean Division.*)

liquid against the inside surfaces of the spray hose results in a considerable loss of pressure at the nozzle. For example, with a 50-foot hose 1/2-inch in diameter and a flow rate fo 10 gallons per minute, the friction loos in 70 pounds per square inch. The *pump* of most power sprayers is of the direct- displacement- plunger type. The recent development of high-capacity, high-pressure boom sprayers for orchard spraying has been possible through the utilization of centrifugal pumps with capacities of up to 125 gallons per minute at pressures of 800 to 1,000 pounds per square inch. Such pumps, however, are subject to severe wear by abrasie wettable powders. *Spray booms.* The boom is a light hollow tube or pipe which is used to carry the spray liquid under pressure from the pump to one or more nozzles. The simplest type is the spray rod of lance, which is a tube from 3 to 8 or even 12 feet in length, with a cut off valve at one end and a simple swirl spray or flat fan nozzle at the other. With power sprayers, a broom or fog-drive gun is sometimes used for spraying fruit trees. This consists of a spray rod from 3 to 5 feet long, with a cutoff at the base and a group of three to eight nozzles at the tip.

Residual spraying indoors. The spraying of houses and barns with a heavy deposit of a long-lasting residual insecticide is widely practiced for the control of flies, mosquitoes, bed bugs, cookroaches, silverfish, and ectoparasites of animals and is one of the principal weapons for the eradication of malaria. This type of application is most efficiently performed with a small- capacity, portable spray pump such as the napsack, compression, or bucket sprayer or the wheelbarrow power sprayer. The World Health Organization has specified that residual-spraying operations be conducted with a spray pump having a regulator to produce a constant pressure of 40 pounds per square inch and that a flast fan nozzle discharging at 0.2 gallon perminute with a spray angle of 60 to 65 degrees be used. During the spraying operation, the nozzle should be held approximately 18 inches away from the surface being streated, and about 150 to 250 square feet of wall surface should be covered per minute. Adherence to the suggested conditions will avoid wasteful and unsightly discharge and produce a spray of maximum impingement and deposition.

Air-blast Sprayers

In the air-blast sprayers a relatively large volume of high-velocity air is used to break up the spray droplets and to carry them to the target, as in the twin-fluid atomizers. For convenience, however, we shall also discuss air-blast equipment in which the spray may be produced by other means but dispersed by means of an air blast. The common hand atomizer or "flit gun" is perhaps the simplest familiar device of this type. Here a piston-type air pump furnishes the air blast, which also forces the spray liquid from a reservior up a small-diameter delivery tube, and atomization results from the appo- sition of the air and liquid streams, according to equation (7). The compressd air-paint spray gun is a larger example of the type.

In the use of air-blast equipment for the treatment of field or orchard crops the 30- to 80-micron droplet range seems most suitable because of the questions of coverage and dynamic catch, and thus the matter of air volume and velocity is of great importance. This is particularly true in orchard spraying, where it is necessary to have sufficient air-moving capacity to agitate the air in all parts of the tree and to displace most of it. It has already been stated that volumes of air required with various-sized delivery tubes to produce adequate velocities at various distances from the blower. It is evident that to secure adequate dynamic catch of 30-micron droplets at distances of 25 to 50 feet from the bolwer, very large volumes of air, ranging from 7,500 to 28,000 cubic feet per minute are required. It has also been shown that to spray a moderate-sized orchard tree 20 by 20 by 20 feet (8,000 cubic feet) with a mist blower travelling at 0.5 mile per hour, 17,600 cubic feet per minute of spray are required, and of course this requirement is directly proportional to the volume of the tree and the speed of travel of the sprayer.

TABLE 8.6

Air Velocities at Various Distances

Diameter of round outlet, in.	*Volume of air, cu. ft./min.*	*Velocity at indicated distance from blower, m.p.h.*				
		Outlet	*10 ft.*	*25 ft.*	*50 ft.*	*100 ft.*
4	1,300	170	40	8	2	0
8	3,800	125	40	10	3	0
12	7,500	125	60	30	7	2.5
24	28,00	120	75	40	17	7
54	27,000	50	35	30	8	3

Fig. 8.2. Speed Sprayer in operation. (*From Food Machinery and Chemical Corp., Bean Division*.)

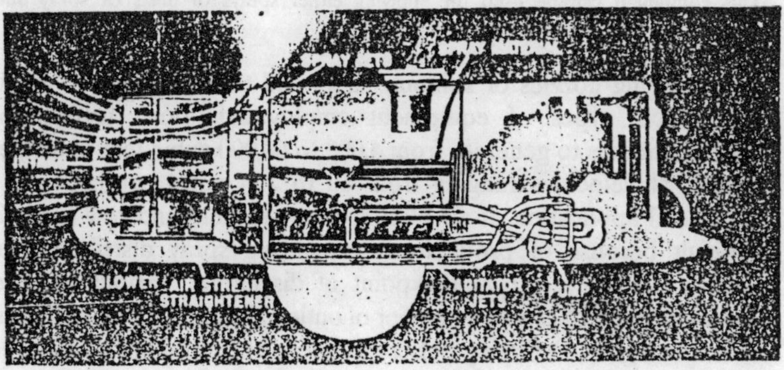

Fig. 8.3. Sectional view of Speed Sprayer. (*From Food Machinery and Chemical Crop., Bean Division.*)

Two types of air-blast sprayers have been constructed the *spray blower* and the *mist blower*. In the spray blower, the liquid is broken up by hydraulic nozzles and carried to the target by the air blast, while in the mist blower the spray liquid is finely dispersed by means of

Fig. 8.4. Large-spray duster showing the spray nossoes and dust orifice on one side. This equipment can be used for applying either sprays or dusts or spray dust combinations. (*u. isc. Calif. Agr. Exp. Sta.*)

air-atomizing nozzles or spinning disks and then carried by the air blast. Both types of equipment employ axial-flow turbines or centrifugal fans to generate from 4,000 to 60,000 cubic feet of air per minute at velocities ranging from 60 to 250 miles per hour. The successful application of air-blast sprayers to field and row crops is difficult because of problems in equalizing spray distribution at varying dis- tances from the point of discharge. Various devices, including manifolds with a number of outlets, fishtails, elongated slots, and rotating or oscillating outlets to sweep the area, have improved the evenness of coverage. Air blasts of 3,500 to 4,000 cubic feet per minute at 60 miles per hour have given effective results over swaths of 15 to 35 feet. The commercial untis available, however, are generally modifications of orchard sprayers and use air blasts of 20,000 to 40,000 cubic feet per minute at moderate velocities of 70 to 90 miles per hour to avoid crop damage, damage, directed at right angles to the path of travel from either one or both sides. The rows over which the machine passes are simultaneously sprayed with a low-volume boom. Such equipment will treat a swath 40 to 60 feet wide suing 50 to 100 gallons per acre.

A number of large spray blowers are in extensive commercial use for orchard spraying, and the Speed Sprayer is perhaps the best known. This equipment will apply from 50 to 500 or more gallons per acre and delivers up to 45,000 cubic feet per minute of air, with nozzle pressures of 50 to 70 pounds per square inch and pump capacities of 55 to 140 gallons per minute. The spray is delivered into the air blast from as many as 58 to 264 swirl spray nozzles. The spray duster (Fig. 8.9) employs 40,000 cubic feet per minute of air passing through two vertical fishtail outlets to distribute high-pressure spray at 100 to 1,000 gallons per acre from vertical banks of nozzles. The air flow is periodically changed in direction by the oscillation of a hinged section of the outlet to aid the spray penetration by moving leaves and twigs. The most difficult problem with air-blast sprayers in orchard spraying is equalizing the coverage between the top and bottom of the tree.

Aerosol Generators

In the aerosol culture, we have raised being bombs, hair sprays, under arm deodorants, adhesives, car wax, fabnc, finishers etc. Aerosols for insecticidal use have been produced by a number of methods: (*a*) by burning the insecticide; (*b*) by spraying a solution of insecticide onto a heated surface; (*c*) by forcing the insecticide, dissolved in a liquefied, low-boiling gas through a capillary tube; (*d*) by spraying at high pressure through a swirl spray nozzle of low capacity; (*e*) by forcing a solution of insecticide between two closely apposed and rapidly spinning disks; and (*f*) by the use of a high-velocity stream of gas, air, or steam acting upon a stream of liquid in some type of twinfluid atomizer.

The exhaust aerosol generator uses the hot exhaust gases from an internal-combustion engine to produce atomization. The droplet size is directly proportional to the ratio of flow of the liquid to the flow of air and can be predicted from Eq. (7). The use of the hot exhaust gases favours the production of droplets of aerosol dimensions because of the very high velocities, 600 to 1,400 feet per second, obtainable with the hot gases and aerosols are not, however, produced at pressures below 25 pounds per square inch gage. These low-or moderate-pressure aerosols require a more critical nozzle design than the simple capillary tube of the high-pressure aerosol. In order to introduce turbulence and the formation of small bubbles of gas which serve as nuclei for boiling, a mixing chamber is provided into which the liquefied gas flows through a 0.015-inch-diameter constriction and from which the mixture issues through a 0.020-inch orifice. The dispenser valve consists of a ball of metal or nylon held by a spring against a valve seat of synthetic rubber.

The use of low-pressure aerosols requires higher boiling propellants or mixtures such as 54 per cent dichlorodifluoromethane and 46 per cent trichlorofiuoromethane, which develops 37 pounds per square inch gage at 70°F. Other sueful combinations include the addition of up to 25 per cent by volume of isobutane of *n*-pentane to the above mixture, II and III, I and IV, and I and V from Table. Mextures such as these form bubbles in the nozzle more readily than single- component propellants and produce better aerosol dispersion at low pressures. The insecticids most commonly dispersed by the liquefied-gas method are DDT, methoxychlor synergized pyrethrins or allethrin, lindane, chlordane and dieldrin. High purification is necessary to prevent clogging of the capillary nozzle and decomposition of the insecticide and consequent corrosion of the container, although this can be inhibited by including 0.1 per cent propylene oxide. Auxiliary solvents such as petroleum distillates, polymethylnaphthalenes, or cyclohexanone have been used to solubilize the insecticides in the propellant gas. A small percentage of nonvolatile material such as 10-W lubricationg oil has been found effective in preventing too rapid volatilization of the aeroso droplets. The ultimate droplet size is determined by the concentration of

High-pressure Aerosol	*Per Cent*
DDT	3
Pyrethrins, 20% extract	1
Alkylated naphthalenes	16
Dichlorodifluoromethane	80

Low-pressure Aerosol	*Per Cent*
Methoxychlor	2.0
Pyrethrins	0.25
Piperonyl butoxide	1.0
Alkylated naphthalenes	11.75
Dichlorodifluoromethane	30.0
Trichloromonofluoromethane	55.0

Fig. 8.9. Applying aerosol for the control of greenhouse insects and mites. Note the nrotective mask, which should be worn when applying highly toxic materials such as parathion and tetraethyl pyrophosphate. (*From Eston Chemicals, Inc.*)

nonvolatile materials, and the optimum for flying insects of 5 to 20 microns (D_m *10 to 15 microns) is obtained with about 20 per cent nonvolatile material in a high pressure formulation and about 15 per cent in a low-pressure formulation. Typical formulations of both types are:*

Such aerosols should be used at the rate of 2 to 4 grams per 1,000 cubic feet of space and the ischarge rate is approximately 1 gram per second. The most popular containers are the 1-pound size are available for larger enclosures and for greenhouse work. For the latter purpose a formulation of 5 to 10 per cent parathion or TEPP in methyl chloride, 90 to 95 per cent, is widely used in a 10-pound bomb with a 2-foot rod and nozzle. This should be used only when wearing a protective mask. The liquefied-gas method has been used also to produce bombs delivering residual- type sprays for the control of cockroaches, ants and clothes moths and carpet beetles, at a low

Roach and Ant Spray	*Per Cent*	*Mothproofing Spray*	*Per Cent*
Dieldrin	0.5	DDT	3.0
Pyrethrins	0.04	Perthane	3.0
Piperonyl butoxide	0.1	Petroleum distillate	59.0
Petroleum distillate	69.36	Propellant	35.0
Propellant	30.0		

pressure of 20 pounds per square inch gage. For this purpose, the amount of propellant is reduced to 30 to 35 per cent of the total, and this produces a relatively coarse discharge of about 30 to 50 microns, which settles rapidly and has good wetting properties. Typical formulations are :

Dusts and Dusting

Dusting is the oldest and the simplest method of poisoning plants for insect control It has been much more extensively practices during the past 50 years. This is largely due to the following undesirable features of dusting: (*a*) decreased efficiency due to less efficient deposition on the plant, (*b*) increased drift problems with poisonous materials, (*c*) increased cost of dust diluents compared with water, (*d*) tendency of carrier and toxicant to separate in the air unless the diluent particles are coated with the toxicant, (*e*) difficulties in incorporating several insecticides or other agricultural chemicals into a single dust, and (*f*) increased hazard to operator from inhalation of toxic dusts. Despite these disadvantages, dusting is in many cases much easier, lighter, and several times faster than thorough coverage spraying. The equipment is considerably simpler and lighter than the spray rig and can be better used in hilly territory and under muddy conditions. Dusting is obviously the favoured means of insect control in regions with a limited water supply and is a useful means of achieving insect control just prior to harvest without exceeding insecticide-residue tolerances.

Most insecticidal dusts are very finely divided, and typical surface mean diameters are talc 1.8 microns, pyrophyllite 2.2 microns, lead arsenate 6 to 10 microns, calcium arsenate 1 to microns, and ground rotenone roots 6 microns (Table 7.3. Such dusts pass nearly completely through a 325-mesh screen of 44 micron aperture. Thus when failling free in air these dust particles should behave similarly to aerosol droplets. However, dust particles not only very greatly in size and shape but also tend to agglomerate during the dusting operation, so that many particles adhere tightly together, and it is very difficult to predict their settling and drifting pattern. Since the settling velocity of these small particles is directly proportional to particle density, dusts of such materials as lead and calcium arsenates settle considerable faster than dusts of botanicals such as pyrethrum or rotenone. The density of the dust diluent and the presence of dust conditioners such as stabilizers and fluffing agents also affect the dusting behaviour.

The electrostatic charge on dust particles has often been suggested as a means of importance in causing the dust to adhere to plant surfaces which are negatively charged. A number of devices have been designed to increase the charge on the insecticidal dusts, either by friction or by passing the dust particles through a flow of positive ions from a highpotential electrical discharge. Inorganic dusts such as arsenicals, fluorides, copper, lead, and sulfur assume positive charges, while botanicals assume negative charges. Diluents such as pyrophyllite and gypsum are also positively charged, and diatomite, clays, and talcs are negatively charged. The use of electrostatically charged dusts has been claimed to prevent agglomeration, to increase adherence and even distribution on both sides of leaves, and to increase the dust deposition as much as 4 to 10 times. Insecticidal dusts are commonly applied to crop plants at rates of 10 to 50 pounds per acre. However, organic insecticides such as DDT give somewhat similar insect control as long as the amount of active ingredient and plant coverage are constant, regardless of whether applied as 3 to 10 pre cent dust. The use of dusts impregnated with oily materials such as mineral oil or polymethyl naphtalenes has sometimes been of advantage in increasing deposits and decreasing drift.

For some insecticidal applications, such as in the control of the European corn borer in the whorls of the corn plant, ants and other soil-inhabiting insects, and mosquito larvae in water under dense vegetation, the use of relatively coars granular dusts of 30- to 60-mesh (250 to 590 microns) serves to prevent the insecticide from adhering to plant foliage and enables it to reach the soil or water surface. The use of these granular materials also greatly decreases drift problems.

Dusting Equipment

A good duster, like a good sprayer, is one that will spread the insecticide – in this case in dry form – in such a manner as to give the most uniform coating possible to the plants being treated. The simplest dusters are the small *hand dusters* adapted to the home garden in which a plunger discharges an air blast through a chamber containing from 0.5 to 2.0 pounds of dust and the dust cloud is emitted through a small flared nozzle or tip. Next in size are the *blower dusters,* which consist of an enclosed fan rotated by a hand crank which sends a continuous blast of air through a small chamber into which the dust is fed by an adjustable gate. An agitator stirs the dust in the hopper and provides for an even feed of the dust into the discharge chamber. The dust cloud is forced out through a delivery tube extending nearly to the ground, which may end in Y or fishtailed nozzle. Such a duster will contain 5 to 10 pounds of dust and can be used to protect 1 to 2 acres of truck crops. The most recent innovation in blower dusters is a back pack unit containing a small gasoline engine which drives a fan providing a steady stream of dust. The *bellous* or *knapsack duster* of similar capacity uses the extension and compression of a bellows attached to the back of the duster to force air through the discharge chamber. This gives a discharge of the dust in puffs instead of the continuous stream as in the blower type. This design is best adapted to somewhat isolated plants such as small trees and shrubbery.

Traction dusters have been widely used for vegetable-and field-crop dusting, especially for small acreages of cotton. In this type the fan is driven by the wheel or wheels on which the duster runs. Such maçhines, containing 50 to 100 pounds of dust, may develop enough power to turn a 12- to 16-inch fan at 2,500 to 3,000 revolutions per minute and may operate with a boom containing up to eight nozzles for a four-row operation. Because the speed of the machines will vary with the rate at which they are drawn through the field and the wheels may slip in soft ground, it is always difficult to maintain an even discharge from traction-type dusters.

Power dusters operated by small gasoline engines are the most practical for orchard and field-crop work. The fan is operated at about 3,000 revolutions per minute. For field and row crops, booms up to 30 feet in length with from 8 to 18 delivery nozzles are used. These may be individually connected by flexible tubing to a peripheral manifold surrounding the fan, or they may be attached to a tapering hollow-boom manifold to equalize the discharge rates at each nozzle.

The *orchard dusters* commonly have a single discharge outlet or a double fishtail arrangement for discharging dust from both sides and a centrifugal or squirrel-cage fan to discharge large volumes of air, up to 20,000 to 40,000 cubic feet per minute. This equipment will project dusts for considerable distances even to the tops of forest trees. In these machines the dust is mechanically fed by an agitator or metered by a worm or helix arrangement directly into the fan or blower, assuring a fairly even rate of discharge. However, in practice, the uniformity of discharge of power dusters may vary over a severalfold range because of the changes in the head of dust above the hopper opening and the tendency of dusts to cake and is greatly inferior in the respect to comparable spray equipment. The most recent developments to remedy this deficiency include equipping the hopper with an elevator which lifts the dust to the top, where it is force-fed into the fan, or the use of a rotating hopper which scoops up the same measure of dust at each revolution and feeds it into the fan, and the employment of a conveyer-belt system at the bottom of the hopper which transports an even layer of dust through a variable shutter into the fan. Such desings have reduced the variability of discharge to a few per cent.

The *spray duster* was developed to apply dusts simultaneously with a mist spray from nozzles along the edge of the fishtail orifice, which served to wet the foliage and increase the initial deposit of dust. Certain types of mist blowers are also equipped for simul- taneous spray-dusting, and results have been secured by the simultaneous emission of 1 gallon of aqueous oil emulsion to each pound of dust, which were equal to or better than results of high- pressure spraying. However, this type of application has been superseded by the use of concentrate sprays. Many other innovations have been developed for dust appli- cations. For use with volatile materials such as nicotine an hopper with a mixing attachment in which the dust is mixed immediately before application has been used. The performance of nicotine dusts has also been improved by partially confining them under a gasp of canvas trailer 10 to 30 feet wide dragged behind the duster for a distance of 10 to as much as 100 feet. The trailler serves to confine the vapour of the insecticide given oft by the dus and to thus expose the insects to it for a longer period.

Spraying of Insecticides by Aircraft

Practically the fust use of the airplane for insect control was in the application of lead arsenate dusts to control the catalpa sphinx, *Ceratomia catalpae,* in Ohio in 1921. The advantages of aircraft applications in rapidity, cheapness, and ease of treatment were readily apparent, and the uses in creased rapidly. Calcium arsenate dust was applied by air in 1923 to control the cotton boll weevil, *Anthonomus grandis,* and the cotton leafworm, *Alabama argillacea,* and paris green dust was applied by air to control anopheline mosquito larvae in the same year. By 1925, forest insects were being dusted by air in Germany, the United States and Canade. During the nest 20 years, most aircraft operations employed dusts because of the difficulties in formulating suitable sprays with the water- insoluble arsenicals.

During the Second World War, malaria control became an essential adjunct to military operations in many areas of the globe and DDT was found to be exceptionally effective for mosquito control. However, its poor dusting qualities resulted in the development of aircraft spraying, and literally hundreds of devices were employed using a range of aircraft such as the Piper Cub L-4 and Stearman PT-17 equipped with booms and nozzles, fighter planes with wing tanks, and heavy multiengined ships such as the B-25 bomber and C-47 transport with large tanks emptying through vertical discharge pipes. Both larviciding and adulticiding operations were proved practicable on the very largest scale and have since been extended to pest mosquito control in Alaska and Canada and to the control of the tsetse flies, *Glossina* spp., in Central Africa and to black flies, *Simulium* spp., in the Adirondack mountains.

Aircraft have been employed extensively for the control of locusts and grasshoppers by a veriety of means. Oil solutions of dinitro-*o*-cresol and BHC have been sprayed on swarms of *Locusta migratoria migratoriodes* both in flight and at rest on the ground. Aldrin, chlordane, and toxaphene have also been used both as sprays and baits for various species of *Melanoplus,* while arsenical-molasses baits have been applied on a very wide scale. Aircraft applications have been used very successfully for the control of forest insects where other means of application are completely impractical. Notable campaigns have involved the use of calcium arsenate dust and, more recently. of DDT dusts and sprays for the control of the gypsy moth,

Fig. 8.10. Dusting forest trees with high-velocity blower.

Porthetria dispar, and of DDT sprays of the spruce budworm, *Choristoneura fumiferana*. It has been estimated that in the gypsy moth operations, the payload of a single C-47 airplane treats an area as large as that which could be covered by a truck-mounted spray rig in 4 years.

Today, in the United States, the airplane plays a major role in the appication of pesticides or example, in 1955, 45,316,000 acres were treated by air for insect control and 265,808,000 pounds of dust and 51,274,000 gallons of spray were applied. On California farms in the same year, of a total of 5,756,941 acres treated by commercial pest-control operators, 4,853,462 were treated by air craft.

The aircraft has brought an astonishing increase in efficiency to many insect-control operations (Fig.8.11) An ordinary dusting plane flying at 100 miles per hour can treat an area at the rate of about 10 to 40 acres per minute, depending upon the effective swath width, varying from 50 to 200 feet; allowing for an operating period of only about 3 hours per day and the time required for loading, from 500 to 2,000 acres are commonly treated. Even greater retes of treatment are obtained in specific operations with larger aircraft. Thus in mosquito-larviciding operations, several square miles may be treated in an hour, and in grasshopper-bait spreading operations with a C-47, 10,000 acres have been treated per day. In tussock moth, *Hemerocampa leucosigma,* control, 450,000 acres of forest were sprayed within a few weeks. The aircraft operations have also made insecticidal applications possible in many areas which were previously inaccessible. These included swamps and marshes where mosquitoes and other biting insects breed, jungles and forests, vast plains areas where grasshoppers and locusts are found, and certain cultivated crops such as rice, sugarcane, mature corn, and cotton, where ground treatment is impractical or destructive.

In assessing the place of aircraft in any insect-control operation, it is well to keep in mind that the advantages of aircraft are (*a*) speed

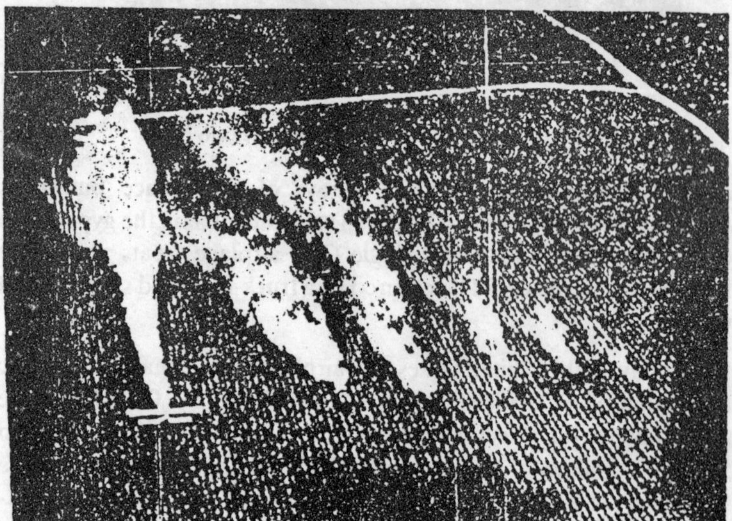

Fig. 8.11. Aerial view of airplane duster, power duster, traction dustr, one-mule duster, saddle gun, and hand-dusting cotton.

and timeliness of application, (*b*) freedom from crop injury and soil compaction, (*c*) no need for farmer preparation of area to be treated, and (*d*) accessibility of all types of areas. Against these must be balanced the disadvantages: (a) dependency upon optimum weather conditions, (b) lack of uniform coverage, especially under leaves, (*c*) difficulty in confining application to treatment area which results in severe drift problems, (*d*) very hazardous operation, with equipment and sometimes human lives lost through accidents, (*e*) expense of operation and inefficency on small acreages, and (*f*) inflexibility after application has begun due to difficulty in communication with pilot.

The use of the airplane or helicopter for the application of insecticides introduces and additional dimension of complexity into the operation, that of the turbulences imparted to the air by the rotation of the propeller and the downdraft resulting from the flight of the aircraft. The influence of these factors on sprays and dusts has been given a great deal of study, and the principles involved are well understood although they are often ignored in practice. A thorough appreciation of these factors is very important in designing and operating aircraft application equipment which will have the maximum effectiveness on the insects to be controlled are reduce hazardous or objectionable drift to a minimum.

Airflow The lift of an aircraft in flight is obtained by imparting a downward motion to the air. This downwash or downdraft may amount to about 600 feet per minute for the ordinary biplane flying at 80 to 100 miles per hour and as much as 1,100 feet per minute for the helicopter. The downdraft helps to carry the insecticidal discharge toward the ground and also moves foliage, which aids in penetration and distribution of the insecticide. The lift is secured by the airfoil of the wing and results in a decreased air pressure above the wing and an increased pressure below it. As a result, the high-pressure air under the wing flows out and around the wing tips to the low- pressure area above the wing, producing a rotary movement at each wing tip, the trailing wing-tip vortex, which may persist for several seconds after the passage of the aircraft. Superi- mposed on these forces is the rotary slipstresm of the propeller, which displaces the airflow in a counterclockwise direction. The velocities of the downdraft and the wing-tip vortices increase as the speed of the aircraft decreases, since the downward displacement of the air must equal the lift required to support the aircraft. With the helicopter, the forces are very similar.

Pronounced vortices with outward and upward components occur at the ends of the rotor, while the downdraft is most pronounced under the central section of the rotor. The strength of these forces is greatest when the helicopter is hovering, and they are materially decreased under conditions of normal forward flight.

9

Effects of Pesticides on Life

Almost all agro-chemicals have their direct and indirect effects on living organisms. How they effect on various physiological needs the great attentional studies. These studies are essential not only for understanding the pesticide's biological behaviour out also for understanding how to treat human poisonings, desing antidotes, and establish safety criteria for humans and animals. From this creative, imaginative research into chemicals with rather specific modes of action also comes the knowledge needed to design new and better pesticides. Such knowledge can also be used in the exploration of biology, ecology, and even aspects of human health.

In terms of their modes of action, insecticices fall into six classes: physical toxicants, protoplasmic poisons, nerve poisons. meta- bolic inhibitors, muscle poisons and alkylating agents.

Physical Toxicants

There are certain Chemicals causing the use chemical obstructions terminate life are the physical toxicants. Thus *physical toxicants* are those materials that are to block a metabolic process, not be a biochemical or neurological reaction, but mechanically, such as oils used to control mosquito larvae by blocking or clogging the respiratory openings or the gill. Heavier oils applied to fruit trees during the dormant season control scales by clogging their spiracles. Other physical toxicants are the inert, abrasive dusts such as boric acid, diatomaceous earth, silica gel, and aerosilica gel. These kill insects by adsorbing waxes from the insect cuticle, effecting the continuous loss of water from the insct body. The insects become desiccated and die from dehydration. Thus the physical toxicants help in killing the pests.

Protoplasmic Poisons

These are the chemical substances destroying the metabolic activities physiologically. *Protoplasmic poisons* attack all enzymes in the insects, system, seemingly leading to their precipitation. These include mercury and its salts, all strong acids, and several of the heavey metals, including cadmium and lead. Phenomenal quantities are required to kill in this manner as compared to present-day insecticides.

NERVE POISONS

This category includes the pesticides affecting the nervous functioning of the organisms. It includes:

Narcotics

The fumigants, particularly the hologenated [containing chlorine, bromine, or fluorine] fumigants, as a group, are narcotics; that is, their mode of action is more physical than chemical. The fumigants are liposoluble (fat soluble); they have common symptomology; their effects are reversible; and their activity is altered very little by structural changes in their molecular architecture. Pharmacological narcotics induce *narcosis* sleep., or unconsciousness, which in effect is their action on insects. These narcotics lodge in lipid-containing tissues, including the nerve sheaths and the lipoproteins of the brain. There are a few fumigants whose mode of action goes beyond the narcotic; for example, methyl bromide, ethylene dibromide, hydrogen cyanide, and chloropicrin. Many of the fumigants are odourless, and chloropicrin is commonly added in trace quantities to them as an olfactory warning agent to users.

Axonic Poisons

The axon is the longest part of a neuron or nerve cell is an elongated extension of the cell body and is especially important in the transmission of nerve impulses from the region of the cell body to other cells. Virtually all axonic transmission of impulses is electrical.

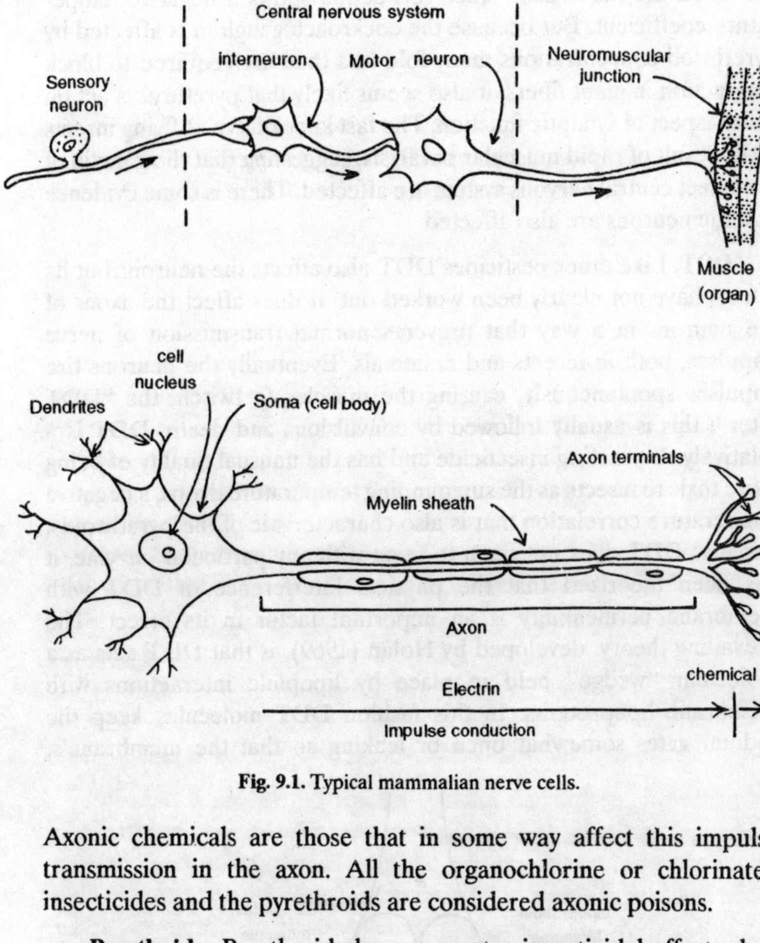

Fig. 9.1. Typical mammalian nerve cells.

Axonic chemicals are those that in some way affect this impulse transmission in the axon. All the organochlorine or chlorinated insecticides and the pyrethroids are considered axonic poisons.

Pyrethoids. Pyrethroids have a greater insecticidal effect when the temperature is lowered : that is, they have a negative temperature coefficient, as does DDT. The affect both the peripheral and central nervous system of the insect. Pyrethroids initially stimulate nerve cells to produce repetitive discharges and eventually cause paralysis. These effects are produced in insect nerve cord, which contains ganglia and synapses, as well as in giant nerve fiber axons. The stimulating effect of pyrethroids is much more pronounced than that of DDT. The exact sites of action of pyrethroids at synapses are not known. It is probable that the toxic action of pyrethroids is primarily due to its blocking action on the nerve axon since this action shows a negative temperature coefficient. But because the cockroach ganglion is affected by pyrethriod concentrations many fold less than are required to block conduction in giant fibers, it also seems likely that pyrethroids act on some aspect of synaptic function. The fast knockdown of flying insects is the result of rapid muscular paralysis, suggesting that the ganglia of the insect central nervous system are affected. There is some evidence that the neurons are also affected.

DDT. Like other pesticides DDT also affects the neurons but its effects have not clearly been worked out. It does affect the axons of the neurons in a way that prevents normal transmission of nerve impulses, both in insects and mammals. Eventually the neurons fire impulses spontaneously, causing the muscles to twitch, the "DDT jitter"; this is usually followed by convulsions and death. DDT is a relatively slow-acting insecticide and has the unusual quality of being more toxic to insects as the surrounding temperature drops, a negative temperature correlation that is also characteristic of the pyrethroids. Because DDT does not seem to react with any particular enzyme, it has been theorized that the physical interference of DDT with membrane permeability is an important factor in its errect. The prevailing theory, developed by Holan (1969), is that DDT acts as a molecular "wedge" held in place by lipophile interactions with membrane lipoproteins. In this fashion DDT molecules keep the sodium gates somewhat open or leaking so that the membrane's,

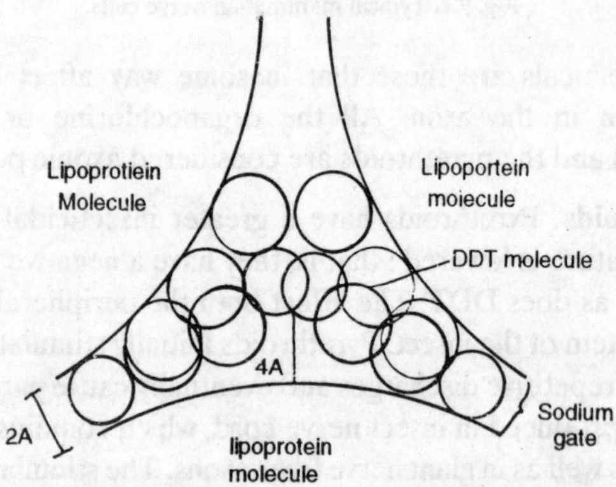

Fig. 9.2. Intermolecular fit of DDT: Each large are represents a lipoprotein molecule.

resting potential cannot be restored. This repeated discharge causes the train of spikes that results in tetanus. Thus DDT and its chemical relatives in some complex manner destroy the delicate balance of sodium and potassium around the exon, thereby preventing it from normally conducting electric impulses. Thus the neurons are functionally desturbed.

Cyclodienes. Unlike DDT and the pyrethroids, the cyclodienes have positive temperature correlation; that is, their toxicity increases with increases in the surrounding temperature. Beyond this large difference, the similarity of dieldrin poisoning and that of DDT is striking. Trains of repetitive discharges can be observed for the nerve cords of cockroaches treated with cyclodienes. There are always time lags or latent periods between application and the appearance of symptoms, varying from 2 to 8 hours, depending on the insecticide. It is quite likely that the mechanism of cyclodiene poisoning also involves changes in ion permeability at axonic membrane levels. Because all the cyclodienes have two electron-rich sites positioned oppsite each other along the line of symmetry, it is theorized that they should fit into a particular biological site in the nervous system in such a way as to block its normal physiological function. These compounds do become bound with the outer covering or sheath of the nerve; however, this bonding does not completely explain the mode of action.

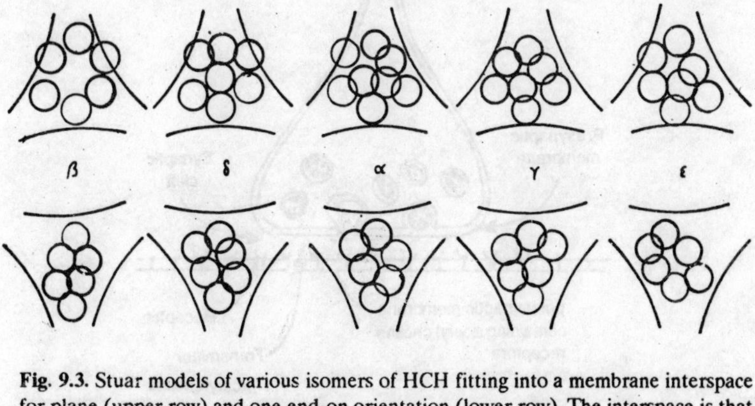

Fig. 9.3. Stuar models of various isomers of HCH fitting into a membrane interspace for plane (upper row) and one end-on orientation (lower row). The interspace is that employed for DDT (See Fig. 9.2). While all end-on orientations are possible. Only the *Y*-isomer (lindane) fits into the interspce in plane orientation.

HCH (lindane). The cohien is the gamma isomer. Lindane active isomer of HCH (hexachlorocyclohexane) acts much faster than DDT or the cyclodienes and causes a much higher rate of respiration in insects than DDT. Lindane toxicity also exhibits a negative temperature correlation, but not as pronounced as that of DDT. The repetitive spikes seen in treated nerve cord occur in groups of two to four in series, a phenomenon not yet understood but definitely different from the repetitive discharges or trains observed in DDT poisoning. Lindane may be classified as belonging to a group of neurotoxicants similar to DDT, though its mode of action is noticeably different from that of DDT. There are, however, more similarities between lindane and DDT poisonings than differences. Fascinating in the chemistry of HCH are the tremendous differences in biological activity found among the closely related isomers. These differences suggest that a rigid spatial arangement of the molecule is necessary to

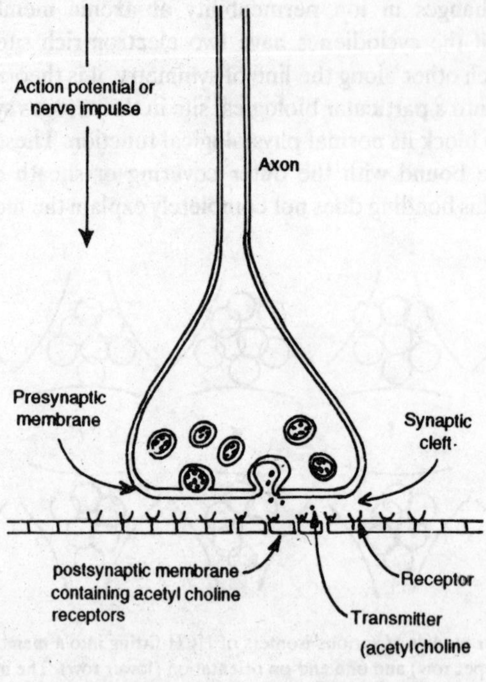

Fig. 9.4. Diagram of a nerve synapse.

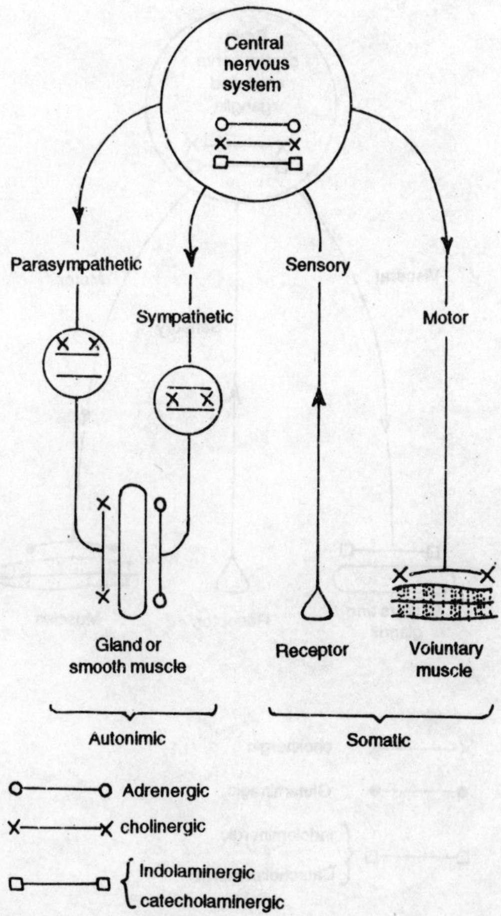

Fig. 9.5. Diagram of the nervous system of a vertebrate showing locations of adrenergic and cholinergic junctions.

secure its insecticidal activity. One widety accepted theory is that when lindane enters the hypothetical intorspace, the nerve membrane is thrown out of equilibium by the attractive force of chlorine atoms applied against the membrane components. The membrane then becomes excited due to untimely ion leaks caused by distortion of the lipoprotein molecules.

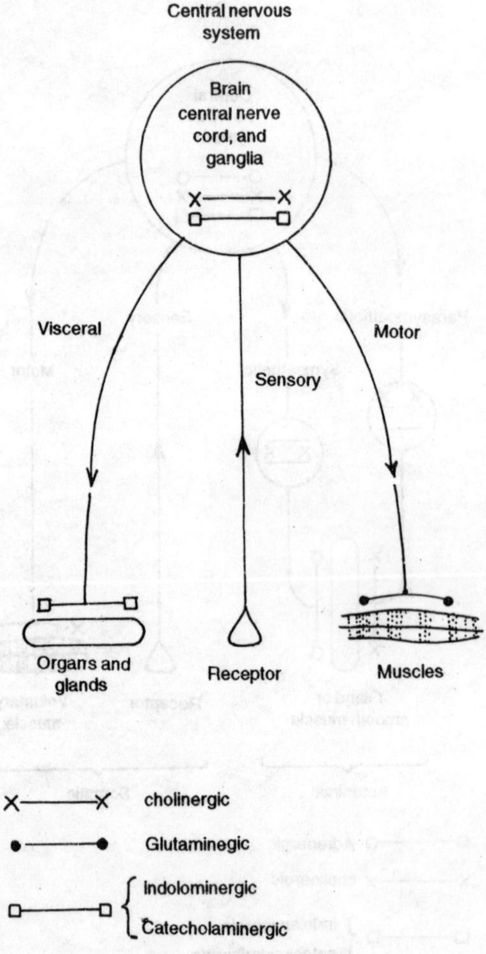

Fig. 9.6. Diagram of the nervous system of an insect showing the (proposed) locations of cholinergic, glutaminergic, and indolaminergic/catecholaminergic junctions.

In summary, it is known that lindane attacks the nervous system and requires a complete reflex arc to be effective, like DDT. Since lindane does not have any special functional group, it appears that physical interaction involving such electonegative centers takes place between lindane and some biological site in the nervous system that normally carries out an important role.

Synaptic Poisons

This category includes the pesticides affecting the synaptic transmission of impulses. As stated already there are two different methods of nerve impulse transmission in the nervous system; Axonic transmission, conveys an impulse from its arrival point along the axon to a neuron or a muscle, gland, or sensory receptor cell. Across the junction between cells, synaptic transmission occurs. A synapse is the junction of a neuron with other cells, including the junction between neuron and muscle, or neuromuscular junction. Virtually all synaptic transmission is chemical. When an impulse, travelling along an axon, reaches a synapse, the impulse dies out, while causing to be released from the end of the axon a small charge of a chemical transmitter substance. This substance moves across the synapse (gap) and sets off another impulse if the synapse is between neurons, or an appropriate response if the synapse is between a neuron and a muscle or gland. There are two well-known chemica transmitters: *acetylcholine* and norepinephrine; however, there are probably others. Synapses that utilize acetylcholine are referred to as cholinergic. Those that utilize norepinephrine are called adrenergic.

Anticholinesterase insecticides. Cholinesterase is the enzymatic substance found in the nervous tissue to help in the transmission of nervous impulses. The organophosphates (OPs) exert their toxic action by tying up or inhibiting certain important enzymes of the nervous system, namely cholinesterases (ChE). At the synapse an impulse is transmitted by acetylcholine (ACh), which is then destroyed by the ChE enzyme so the synapse will be cleared for another transmission. These chemical reactions happen within micro-seconds and continue constantly, as needed, under normal conditions. However, the OPs attach to the enzyme ChE in a way that prevents it from clearing away the ACh transmitter; in effect, the electric trans-mission circuits jam because of the accumulation of ACh. In mammals the accumulation of ACh interferes with the neuromuscular junction, producing rapid twitching of voluntary muscles, finally resulting in paralysis and death due to respiratory failure. Symptoms in insects follow the pattern of nerve poisoning: restlessness, hyperexcitability, tremours and convulsions and paralysis. A unique feature of all organophosphate (and carbamate) insecticides is their structural "fit" with the ChE enzyme. These insecticides mimic the molecular shape of ACh, the natural substrate of cholinesterase (AChE. AChE has two active sites, the esteratic site and anionic site.

The inhibition of AChE by an insecticide is very similar to the early stage of the hydrolysis (splitting) of AGh. The acetyl carbon $(CH_3-\overset{O}{\overset{\|}{C}}-O-)$ of ACh. or phsphorus moiety $(RO_2\,\overset{O}{\overset{\|}{P}}-OR')$ of the organophosphate, or the carbamyl moiety $(R_2-N-\overset{O}{\overset{\|}{C}}-O-)$ of, the carbamate binds to the esteratic site on the amino acid serine, while the remainder of the molecule attaches to the anionic site. Through a series of very rapid steps, ACh is hydrolyzed or split into one molecule of acetic acid and one of choline. The organophosphate is hydrolyzed so that the alcohol fragment (HOR') leaves the reaction and the phosphorus remains attached to the serine, phosphorylating the enzyme, thus inhibiting it for an extended period of days. This phosphorylation prevents the enzyme from acting further in its role of clearing away ACh, and is a strong and difficult bond to break. Phosphorylated ChE is commonly referred to as being irreversibly inhibted, because of this strong bonding. It is not truly irreversible, but rater slowly reversible.

When the carbamate is hydrolyzed, the alcohol fragment (HOR) leaves the reaction and the $(CH_3-\overset{H}{N}-\overset{O}{\overset{\|}{C}}-O-)$ remains attached to the serine, thus carbamylating the enzyme for a relatively short period of time, usually hours. Carbamylated ChE is reversibly inhibited; the relatively short attachment period is usually several hours.

Carbamates inhibit cholinesterase as organophosphates do, and they behave in an almost identical manner in biological systems, but with two main differences. Some carbamates are potent inhibitors of aliesterase (miscellaneous aliphatic esterases whose exact functions are not known), and their selectivity is sometimes more pronounced against the ChE of different species. Second, ChE inhibition by carbamates is apparently reversible. When ChE is inhibited by a

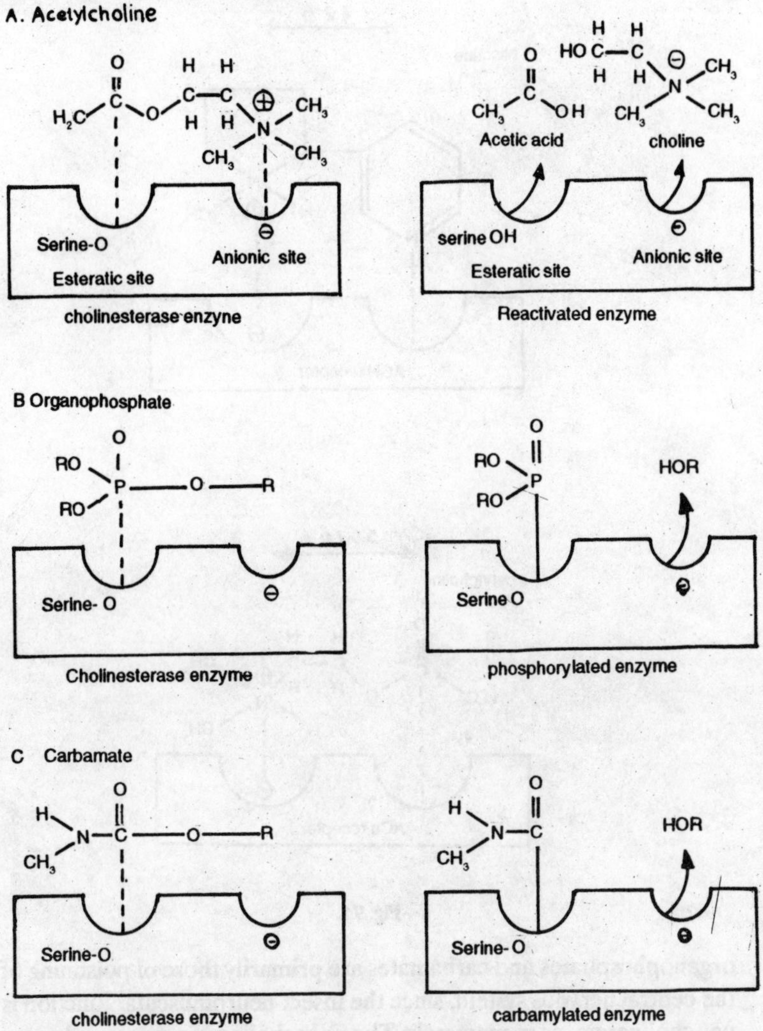

Fig. 9.7. Simplified reactions of acetylcholine, organophosphate, and carbamate with the enzyme cholinesterase. (A) Acetylcholine is hydrolyzed to acetic acid and cholire, resulting in reactivated enzyme. (B) Organophosphate is hydrolyzed and phosphorylates the enzyme by attaching to serine at the esteratic site. (C) Carbamate is hydrolyzed and carbamylates the enzyme by also attaching to serine at the esteratic site.

carbamate, it is said to be carbamylated, as when an organophosphate results in the enzyme being phosphorylated. In insects, the effects of

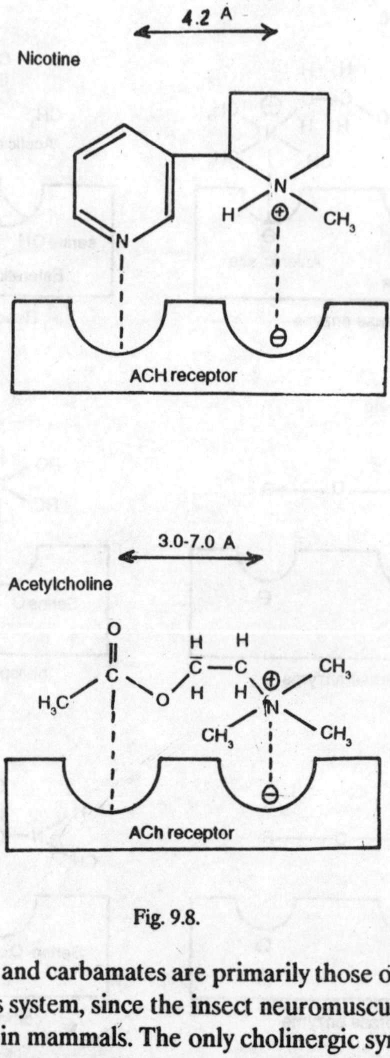

Fig. 9.8.

organophosphates and carbamates are primarily those of poisoning of the central nervous system, since the insect neuromuscular junction is not cholinergic, as in mammals. The only cholinergic synapses known in insects are in the central nervous system. The chemical neuromuscular junction transmitter in insects is thought to be glutamic acid, but that has not been proved.

Postsynaptic Insecticides. Nicotine and nicotine sulfate are the insecticides produce symp toms that are described as *nicotinic effects* because they resemble the tremors seen in nicotine poisoning. Symptoms of nicotine poisoning in insects progress in the follwing sequence: excitation, convulsions, paralysis and death. In low concentrations, nicotine stimulates the heartbeat; at high concentrations, it decreases heartbeat. The outstanding feature of nicotine effects in insects is the selective inhibitory action on nerve ganglia and synapses in the voluntary nervous system. When nicotine is applied to the ganglion controlling an isolated leg, the leg muscles tremour violently. When the leg is severed, the tremours stop. Further, nicotine has no effect on the isolated (severed) leg.

In warm-blooded animals, nicotine acts like the transmitter acetylcholine (ACh) and binds to its receptor at the neuromuscular junction and at the ganglion. As a resul, it causes stimulation of voluntary muscles and of the ganglia, glands, and smooth muscles. Nocotine thus attacks some of the cholinergic junctions and all neuromuscular junctions of mammals by ACh-like action. Nicotine does not inhibit or bind cholinesterase, but rather acts by mimicking ACh. It is very selective in choosing the binding sites in both insects and mammals: the neuromuscular junction of mammals and the synaptic ganglion of insects. It seems that the receptors cannot distinguish between ACh and nicotine. A comparison of the structures of the two molecules reveals similarities both in dimension and in relative distribution of charge and polarity.

Adrenergic insecticides. The degrading enzyme of the transmitter *norepinephrine* is *monoamine oxidase.* Its precise localization in insects is not known; however, it has been identified in the insec nervous system, as has *octopamine* in the firefly. Formamidines induce abnormal behaviour in pests: reduced feeding, dispersal from plants, erratic mating behavior, and detachment of ticks from their host. There are now several formamidine insecticides, ovicides, and acaricides whose mode of action resembles that of the first of these compounds, chlordimeform. In rats, chlordimeform apparently induces accumulation in the brain of serotonin (5-hydroxytryptamine) and norepinephrine, two important biogenic amines. Studies with rat liver slices indicate that chlordimeform inhibits the activity of the enzyme monoamine oxidase, which removes serotonin and norepinephrine from their sites of action. Chlordimeform also inhibits oxidative phosphorylation while stimulating mitochondrial ATPase activity in high concentrations; ATPase is an enzyme that breaks down ATP. In insects, chlordime- form is toxic only to eggs and the youngest larvae of caterpillars. The same is true for plant-feeding mites. Recently it has been discovered that formamidines inhibit octopamine, the neurotransmitter in the light organ of a firefly. It has been suggested that octopamine has multiple actions as a neurohormone and neurotransmitter, actions similar to those of epinephrine in vertebrates. Octopamine is also found in considerable amounts in the central nervous system of several insect species.

METABOLIC INHIBITORS

Inhibitors of Mitochondrial Electron Transport

This category includes the pesticides inhibiting the metabolism by affecting into chandrial electron transport.

The electron transport chain is the series of cytochromes in the mitochondria involved in the production of energy from the oxidation of carbohydrate, lipid, and protein molecules. Pesticides for which there is good evidence that their mode of action is on the electron transport chain are rotenone, fumigants that work through the cyanide ion (CN), dinitrophenols, organotin acaricides and fungicides, and the oxathiin fungicides.

Rotenone. In insects rotenone poisoning results in slowing of the heartbeat, depression of respiratory movement, reduction in oxygen consumption, and finally limp paralysis. Rotenone is a respiratory enzyme inhibitor, acting between NAD^+ (a coenzyme involved in oxidation and reduction in metabolic pathways) and coenzyme Q (a respiratory enzyme responsible for carrying electrons in some electron transport chains), resulting in failure of the respi- ratory functions. Rotenone is highly toxic to most insects, and its mode of action is interference in energy production. Its extreme toxicity to fish and insects and its low toxicity to mammals are explained by the way each metabolizes rotenone. Insects and fish convert rotenone to highly toxic metabolites in large quantity, while mammals almost exclusively produce nontoxic metabolites.

HCN. HCN is an inhibitor of the electron transport system. It reacts with the terminal cytochrome oxidase, an enzyme in the respiratory chain, by attaching to the heme iron, the same attachment site as for oxygen. Other poisons that also chemically attach to this location and act in the same specific way are CO (carbon monoxide), NO (nitrous oxide), N_3^- (azideion), H_2S (hydrogen sulfide), and CN^-ion.

Dinitrophenols. The dinitrophenols are "uncouplers" or inhibitors of oxidative phosphorylation. In the biochemical reactions of a cell incontact with a dinitrophenol, the close coupling between the respiratory chain and phosphorylation is broken, respiratory control is lost, and electron transport along the chain occurs at full speed without producing ATP, the high- energy molecule.

Organotins. The organotin compounds used as acaricides inhibit oxidative phosphorylation by blocking rather than uncoupling the formation of ATP, much as in the action of dinitrophenols.

Inhibitors of Mixed-Function Oxidase

Certain chemicals like pyrethrins and some of the synthetic pyrethroids are the only insecticides that are used commercially with the addition of a synergist, usually piperonyl butoxide. The synergist, though expensive, is worthwhile because pyrethrins are quite expensive and not persistent; the synergist maximizes the effect of the pyrethrins during their short life. Several years of mystey surrounded the mode of action of synergists, because, to further confuse matters, these compounds produce no effect with some insecticide classes and even antagonize the action of others. It has since been well established that the synergist mode of action is the inhibition of mixed-function oxidases (MFO), enzymes produced by microsomes, subcellular units gound in the liver of mammals and in some insect tissues (e.g., fat bodies). Microsomes were once thought to have no specific function other than to oxidize miscellaneous fat-soluble foreign compounds intro- duced into the system, such as insecticides. Oxidation in the micro- somes leads either to detoxification or activation of an insecticide. If the inhibited enzyme normally detoxifies the insecticide, the insecti- cide is left intact to exert its effectiveness, and appears to

---

Fig. 9.9 The respiratory electron transport chain, showing a likely arrangement of electron carriers, the three phosphorylation sites (ATP), and the four points (double lines) where pesticides inhibit this process Arrows indicate the direction of electron transport.

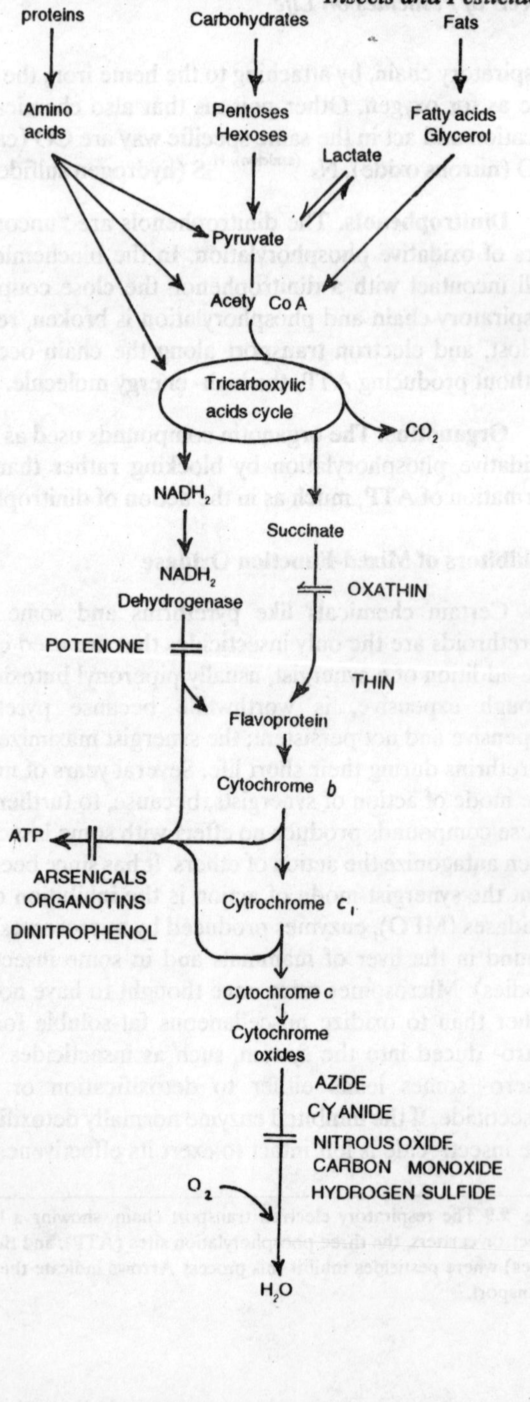
proteins
Carbohydrates
Fats
Amino acids
Pentoses Hexoses
Fatty acids Glycerol
Lactate
Pyruvate
Acety Co A
Tricarboxylic acids cycle
CO_2
$NADH_2$
Succinate
$NADH_2$ Dehydrogenase
OXATHIN
POTENONE
THIN
Flavoprotein
Cytochrome *b*
ATP
ARSENICALS
ORGANOTINS
DINITROPHENOL
Cytrochrome c_1
Cytochrome c
Cytochrome oxides
AZIDE
CYANIDE
NITROUS OXIDE
CARBON MONOXIDE
HYDROGEN SULFIDE
O_2
H_2O

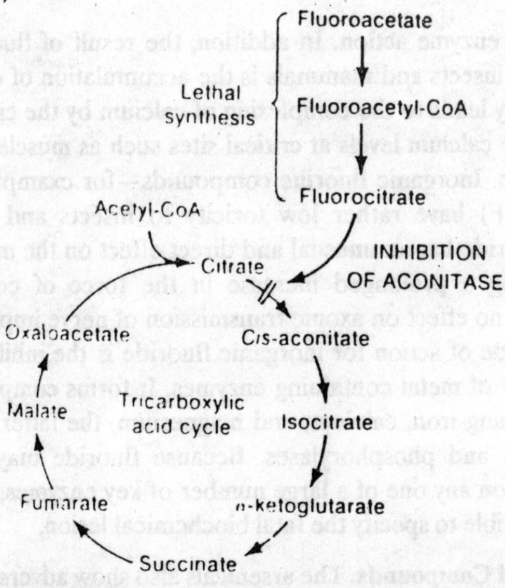

Fig. 9.10. Ihibition of the tricarboxylic acid cycle enzyme aconitase by lethal synthesis of fluorocitrate from fluoroacetate.

be synergized. If, however, the inhibited enzyme normally activates the insecticide, as with some phosphorothioates, the insecticide is not activated, and appears to be inhibited or antagonized in its effectiveness. It is now known that most of the biodegradability of current insecticides depends on their successful biotransformation by micro- somal oxidases of various types, such as MFO. These enzymes are also responsible for synerginsm.

Glycolysis – Citric Acid Cycle

Fluorine compounds. There are certain chemicals like fluorine compounds having a classic mode of action. In mammals the organofluorine compounds (fluoroacetate) cause symptoms only after a 20- to 60-minute delay, when convulsions begin. The heartbeat increases and the body temperature drops. Fluoroacetate is not a direct enzyme inhibitor, but is a latent inhibitor, requiring conversion to a derivative, fluorocitric acid, which is a potent enzyme inhibitor. Because of the structural similarity of fluorocitric acid to citric acid, it competes with citric acid for the target enzyme, aconitase, thus blocking the enzyme action. In addition, the result of fluoroacetate poisoning in insects and mammals is the accumulation of citric acid. This probably leads to the complexing of calcium by the citrate, thus reducing free calcium levels at critical sites such as muscles involved in respiration. Inorganic fluorine compounds – for example, sodium fluoride (NaF) have rather low toxicity to insects and mammals equally. Fluoride has an unusual and direct effect on the mammalian heart, causing a prolonged increase in the force of contraction. Fluoride has no effect on axonic transmission of nerve impulses. The probable mode of action for inorganic fluoride is the inhibition of a large number of metal-containing enzymes. It forms complexes with those containing iron, calcium, and magnesium, the latter including phosphatases and phosphorylases. Because fluoride may exert its lethal action on any one of a large number of key enzymes, it has not yet been possible to specify the fatal biochemical lesion.

Arsenical Compounds. The arsenicals also show adverse effect as metabolism. Arsenicals kill primarily by inhibiting respiratory enzymes. Three categories of action compose the mode of action for arsenic: (1) uncoupling of oxidative phosphorylation by arsenolysis; (2) combination with various enzymes containing the sulfhydryl group (–SH), especially in pyruvic oxidase; and (3) gross protein precipitation. Arsenic chemically resembles phosphorus and substitutes partially for phosphorus in some reactions. The best example is arsenolysis, which takes place instead of phosphorylysis. Phosphorylysis is essential in forming the high-energy bond of ATP, and oxidative phosphorylation is the major energy- producing step of the cell, but arsenolysis uncouples phosphorylation. In this uncoupling, arsenite is about twice as effective as the arsenate ion.

There are numerous enzymes contain the – SH group, and all are suscepti- ble to arsenicals. Pyruvic oxidase is the most arsenic-sensitive enzyme; other enzymes that are equally important but not as sensitive include cytochrome oxidase, lactic acid dehydrogenase, and glycerol-phosphate dehydrogenase. Inorganic arsenicals can cause gross coagulation of proteins in high concentrations. This effect, however, is probably the same as that seen on – SH groups; but instead of attacking a particular – SH group at an enzyme's active site, it involves the sulfur bonds that maintain the physical configuration or structure of most proteins.

Muscle Poisons

Some pesticides like Ryama and sabadilla cause muscle poisons because of their direct action on muscle tissue. Ryania contains an alkaloid, ryanodine, the active principle of *Ryania speciosa,* grown in South America. Ryanodine is at least 20 times more toxic to mammals than to most insects. Thus, this natural insecticide is far more hazardous to use than many synthetic compounds. The mode of action of ryanodine is that of membrane disruption, and its effect is specific for the excitable membrane of muscle. Poisoned insects show tremendous increases in oxygen consumption, as much as tenfold, followed by flaccid paralysis and death. In the forg, complete rigoir our follows paralysis. In mammals there is progressive rigidity of the muscles with final respiratory failure and death. One theory of ryanodines, effect is that it may interfere with the relaxing system of muscles, which strips calcium ions (Ca^{-}) from contracted muscle and leads to the relaxed state. In addition, however, ryanodine is believed to interfere with metabolism, an effect that may be unrelated to its effects on muscle. Sabadilla comes from the powdered seeds of the lily *Schoeno- caulon officinale* and contains two insecticidal alkaloids, cevadine and veratridine. Especially susceptible are houseflies, household insects, and the Hemiptera and Homoptera, which include the true bugs. Purified cevadine, for example, is approximately 10 times more toxic than DDT to houseflies. In mammals veratridine produces a prolonged rigor in skeletal muscle following the initial twitch, accom- panied by repetitive impulse discharge in muscle fibers. Oxygen consumption increases, but not to the extent noted in ryanodine poisoning. Sabadilla appears to have the same general mode of action in insects as ryanodine, resulting in flaccid paralysis and death.

Alkylating Agents

There are several biologically active substances like alkylating agents found in a broad range of chemical groupings, and replace an active hydrogen in a biologically significant compound by an alkyl group (carbon chain-like structure). Notable among these are the early World War I gases, mustard and nitrogen mustard, and a more recent group of experimental chemosterilants, particularly the aziridines. Several of the halogen-containing compounds, mostly fumigants, are also identified as alkylating agents. Methyl bromide is the most prominent of these . Alkylating agents act in two basic ways. They react directly with cell chromosomes by attacking one or more of the reactive loci on the nucleic acid molecule, and they deactivate essential enzymes, which subsequently cannot fulfill normal functions in the synthesis of nucleic acids.The latter is comparable to the inhibition of –SH enzymes by the arsenicals. There by show the adverse effects.

10

The Physiology of Insecticides Resistance by Insects

It is true that through the centuries man has fought indefatigably to restrain, control, or eradicate his insect enemies, but there is little doubt that, through intensification of this gallant fight with formidable new chemical weapons, man has been instrumental in the past three decades in speeding up the course of certain evolutionary processes which otherwise might have gone unnoticed. For many insect species, the environmental changes of recent years have been most sudden and overwhelming, as one insecticide after another, with complete disregard for selectivity has been poured in their paths – so the insects have responded valiantly with a weapon of their own – a phenomenon well known to all of us as "*resistance.*"

Actually, insect resistance to insecticides is not entirely of recent origin, Melander (1914) is generally credited with the first publication on insect resistance and the now famous quotaion, "Can insects become resistant to spray ?" although observations on resistance had been made as early as 1887. The earlier publications on insect resistance to kerosene, lime-sulfur, hydrocyanic acid, tartar emetic, lead arsenate, cryolite, barium fluorosillicate, phenothiazine, and other compounds, as well as the early work on resistance to chlorohydrocarbon insecticides have been reviewed and well documented by Babers (1949) and Babers and Pratt (1951). Comprehensive monographs on the resistance problem have been published by Brown (1958) and Brown and Pal (1971).

The Origin and Development of Resistance

It is widely accepted that the development of insecticide resistarnce by insects is due to the selection of variants in the population carrying preadaptive genes. In fact, insecticide resistance provides some of the best examples of Darwinian evolution, but at an accelerated pace, since the intensive selection which pre-cedes the development of resistance by insects results, in numerous instances, in the accentuation of many characters which otherwise might have gone unnoticed.

The rate at which resistance develops in population may depend on (a) "the frequency of resistant genes present in the normal population, (b) the nature of these genes (either single or multiple, dominant or recessive), (c) the intensity of selection pressure, and (d) the rate at which the species breeds". In some instances of geographical isolation, confined populations may not contain the prerequisite resistant genes, and therefore do not develop resistance and highly inbred laboratory strains may fail to develop resistance if the gene pool contains no resistance factors. It is quite apparent from toxicological observations that there exists a dynamic variation in tolerance to a poison among individuals of a given population. However, truly sublethal doses of insecticides do not induce resistance in susceptible populations of the house fly; or increase the level of DDT dehydrochlorinase. Rather, the cumulative effect of daily sublethal amounts of DDT, γ BHC, dieldrin, or diazinon renders the house fly more susceptible.

It is also well established that insecticides are not mutagenic. Genetic variability for resistance already exists in natural populations not previously exposed to insecticides. This is also evident from the fact that resistance develops slowly in highly inbred lines. If new mutations were to arise, inbred lines under selection pressure would develop resistance as rapidly as hetero geneous populations. Genetic variability in DDT dehydrochlorinase content is also evident from analysis of individuals in a random population of house flies selected with DDT, showing a clear genetic relationship between DDT dehydrochlorinase and resistance. The acceleration of resistance development in a population may have a nongenetic explanation. It may be the property of the chemistry or the physiology of the insect that when a certain level of resistance is reached a large change in resistance can follow from a small genetic change.

The resistance is biphasic. Phase I is due to selection of variants in the population according to genetic principles, attaining a resistance level commensurate with the gene pool for resistance initially present in the population. Assuming that resistance is due, in most instances, to enzymatic detoxication of the chemical, genetic selection will then result in an increase in enzyme activity. But how does enzyme activity increase beyond the preadap- tive level ? In phase II, acceleration of resistance takes place by induction of preexisting detoxifying enzymes towards enhanced activity, resulting in faster breakdown of the chemical. The inducer is the insecticide itself which is being used for the control of that particular insect. Evidence for phase II comes from extensive studies on the induction of microsomal mixed-function oxidases in mammals and in some insects. This part of the total resistance depends on the presence of the insecticide in the insect's environment. Removal of the insecticide from that ecosystem results in quick reversion of resistance to the preexisting genetic level. Prolonged breeding in complete absence of the chemical will cause further reversion toward susceptibility—the rate of return being dependent on the resistance genotypes, including the incorporation of modifier genes during the selection process. The induction phenomenon corresponds to the expression of a regulatory mechanism affecting nucleic acid and protein synthesis. For instance, DDT has been shown to increase not only the synthesis of messenger RNA but also to produce new RNA species with higher template activity. RNA may also be involved since translation processes dictate the characteristics of protein synthesis.

Neurophysiological Aspects of Insecticide Resistance

The unstabilizing effect of DDT on peripheral nerves of insects is well estabilished. Even if this effect were not the primary cause of DDT poisoning, differences in response among insect strains might shed some light on the resistance mechanism. Thus, Pratt and Babers (1953a) demonstrated that the thoracic ganglia of DDT- resistant (R) house flies are less affected by, and recover more rapidly from the direct action of DDT than ganglia of the susceptible (S) strain (Table1). This difference cannot be due to DDT breakdown during the initial 5 minutes after treatment and must, therefore, involve an intrinsic factor characteristic of the ganglion itself. A similar conclusion was reached by Wiesmann (1955a) and Weiant (1955) who demonstrated stronger tremours in isolated legs or sensory cells of S flies than in those of DDT-R flies after exposure to DDT, benzene, or toluene vapours. Unlike DDT, however, neither aldrin nor dieldrin produced symptoms in isolated legs of the American cockroach *Periplaneta americana*.

DDT lowers the tarsal sucrose threshold of the R strain. The same result occurs either by topical application, injection, or tarsal contact. Hence, cuticular penetation of DDT is not the cause of this difference. Resistance, therefore, must be due to an intrinsic mechanism by which the receptors are protected from the unstabilizing effect of DDT. Similarly, Barton-Browne and Kerr (1967) concluded that resistance to DDT in their strains of flies could be fully expressed at the level of chemoreceptor hairs on the labella and was due to the R fly's ability to recover from the poisoning effect. Nerve sensitivity to lindane and dieldrin is lower in R flies than in S flies, as determined by the longer latent period required for initiation of bursts of impulses. This difference is not as marked with DDT, and this relationship does not hold true for diazinon resistance.

TABLE 10.1

Duration of Leg Tremours Induced by DDT Applied to Thoracic Ganglia

Fly strain	*DDT applied (μg)*	*Files tremoring (%)* 5 minutes	60 minutes	120 minutes
Susceptible	0.57	88	76	–
Resistant	0.57	64	16	–
Susceptible	1.14	100	–	93
Resistant	1.14	60	–	27

An explanation of this latent period would be plausible if the site of the resistance mechanism were at the site of action of the toxicant. An interesting experiment by Earle (1963) indicates that such might be the case. Earle found that a 1-minute exposure to aldrin vapours was sufficient to kill 90% of an S strain of house flies, whereas exposure of 120 minutes resulted in little or no mortality of the R strain. It has been fairly well established that house flies do not metabolize dieldrin; hence, a protective effect such as storage away from the site of action is unlikely to act with such rapidity. It is probable then, as Earle (1963) concluded, that the resistance mechanism resides at the site of action, but that other possibilities might also exist. The possibility of a dieldrin-impermeable neural lamella in R cockroaches *Blattella germanica* was considered by Ray (1963). His results showed no difference in permeability of nerve tissue to dieldrin between S and R cockroaches. The use of crossing experiments, with visible mutants as genetic markers, enabled Tsukamoto *et al.* (1965) to ascertain the lower nerve sensitivity to DDT of F strains of house flies as an incompletely recessive genetic characteristic carried on the second chromosome. This resistance mechan is not the same as the one caused by detoxi- cation since the latter is carried on the fifth chromosome. The genetic factor for low nerve sensitivity to BHC is associated with neither the second nor the fifth linage group.

Electron microscopic and autoradiographic studies on the distribution of dieldrin in the intact nerve tissue of dieldrin-R and diedrin-S *B. Germanica* revealed that the interstrain difference in susceptibility is due to the accumulation of less dieldrin in the R strain. Otherwise, the nervous system of the R strain shows no visually recognizable morphological characteristics distinct from those of the S strain. On the other hand, Eaton and Sternberg (1967) could find no relationship between the DDT content of the central nervous system of susceptible *P. americana* and the pattern of nerve activity observed; also, no [^{3}H] dieldrin deposition was observed in the nerve cell body of dieldrin-R house flies, and no radioactivity was detected in mitochondria or tracheae enmeshed in gangloinic tissue.

Another aspect of a physicochemical nature which might throw some light on nerve sensitivity to toxicants is the binding of insecticides to subcellular components of nerve tissue. Studies by Matsumura and O' Brien (1966) and Matsumura and Hayashi (1966a, b) suggest that DDT and dieldrin from complexes with components of insect nerve. This "charge-transfer" complex with a component of the axon resulting in greatly enhanced efflux of potassium ions from the nerve cord might be related to the mode of action of these compounds. As extension of this work by Hatanaka *et al.* (1967) throws some doubt on the validity of this assumption, especially in the use of Sephadex for demonstrating the formation of stable complexes. Holan (1969) presents evidence that neither DDT nor the active cyclopropane substituents of DDT give UV spectra showing complex formation with pure lipid nerve components. Rather, the author suggests complex formation with protein and not with lipid components. On the other hand, Wilson *et al.* (1971) confirmed O'Brien and Matsumura's (1964) charge-transfer complex by analogy with the UV spectral phenomena of an alcohol-water suspension of DDT and the interaction of the chlorophenyl groups of DDT with tetracyanoethylene.

Many of the above examples would seem to indicate the presence, in resistant strains of insects, of a lipid barrier which prevents the rapid penetration of the toxicant into the delicate nerve structures. Another barrier which may hinder the action of a toxicant is the ionic nerve barrier. Insect nerves are surrounded by a homogenous noncelluar elastic sheath, the neural lamella, which is structurally different from the protective coating of vertebrate nerves. This sheath has been shown by Hoyle (1953) to be impermeable to ionized substances, especially potassium ions, and Winton *et al*. (1958) have provided histochemical evidence for this lipid sheath barrier in *P. americana*. It remained for Twarog and Roeder (1956, 1957) to demonstrate the active role of the Hoyle sheath surrounding nerve fibers and ganglia in protecting the nerve against penetration of various ions and drugs. By desheathing the nerve cord of *P. americana,* acetylcholine, which is inactive when injected into the insect, produces a rapid asynchronous burst of action potentials follwed by synaptic depression and block. It is postulated that this sheath, too, may be responsible for altered nerve responses to foreign chemicals as a mechanism of resistance to insecticides. Howefver, this assumption is not without criticism. Treherne (1962) has indicated that stripping off of the nerve sheath involves much more than the mere removal of a superficial membrane. Removal of this sheath will result in dramatic changes in the chemical composition of the fluid surrounding nerve fibers and other nerve cells due to the disruption of the equilibrium between this fluid and the hemolymph. In particular, the decrease in sodium and calcium ions will affect nerve transmission.

Indeed, a most attractive hypothesis advanced in explanation of the unstabilizing action of DDT is that entry of the toxicant into the lipoprotein structure of the nerve membrane destroys the capacity of the membrane to maintain the normal ionic balance, especially that of calcium, so essential for normal function between nerve and environment. This hypothesis, first stated and supported experimentally by Welsh and Gordon (1947) and Gordon and Welsh (1948) has been elaborated by Mullins (1954,55) in a theoretical model membrane composed of a lattice of cylindrical lipoprotein molecules. These molecules are oriented in such a manner as to provide free interspaces for the passage of ions and small molecules from one side of the membrane to the other. Foreign molecules many penetrate freely through the interspaces and interfere with membrane permeability, thus causing narocosis but no membrane damage, or they may cause distortion of the surrounding lipoprotein molecules, thus increasing the size of adjacent interspaces and leading to sodium ion leaks and excitation. A concise critique of this theory has been given by Kearns (1956). Recently, Mullins' theory has been put into practice in the rational synthesis of hydrocarbon insecticides specifically designed against DDT-resistant strains of insects. The prediction of activity was made on the premise that the substitution of the more chemically stable dichlorocyclopropane for the easily degradable trichloroethane group of DDT would prevent the enzymatic detoxication of the insecticide. This proved to be correct. Based on biological and structural findings, Holan proposes a modification of Mullins' theory as follows: some of the insecticide dis- tributes itself at the lipid-protein nerve membrane interface. There, because of its three-dimensional conformation, the base con- taining the phenyl ring forms a molecular complex with the overlaying protein layer. The cyclopropane ring fits into the channel of a pore in the lipid portion of the membrane, but keeps the pore open to sodium ions with the consequent delay in the falling phase of the sodium ion potential. In this hypothesis, the complexing of DDT analogues with proteins is controlled chiefly by substituents in the rings and is largely independent of insecticidal activity. This explains the tenacity of DDT and its inactive metabolites in the environment.

Effects of Neurohormones

The hypothesis of a lipid barrier which prevents the entry of a toxicant into the nerve structures of resistant insects implies, of course, that entry of the poison into nervous tissue is a prerequisite for insecticidal action. The blood of DDT-prostrate cockroaches *P. americana* contains a pharmacologically active substance which is toxic to both S and R strains of house flies. When inected into house flies it elicits symptoms resembling those of DDT. Prostration ordinarily is not permanent in either stranin, indicating that the toxin is not DDT or a DDT derivative, as is also shown by chemical analysis. The same or a similar neuroactive substance can be produce by electical stimulation and by tetraethylpyrophosphate (TEPP). Similarly, blood from TEPP= or DDT-prostrate *P. americana* which contains high titers of corpora allata and corpora cardiaca hormones stimulates the nerve cord activity of normal cockroaches. The neurohormone is derived from the corpora cardiaca but not from the corpora allata.

DDT appears to induce a hormonal-type effect on the chordotonal receptor cells of *P. americana,* but this effect is not produced by lindane. Also, mechanical or electrical stimulation produces paralysis in cockroach nymphs. The paralyzing agent is carried in the blood. It can be shown that this neuroactive agent is depleted in the corpora cardiaca after subjecting the insect to stress. These neurohormones, whether reponding to stress or inducing spontaneous excitation or depression are not necessarily the same in all situations, a fact which makes the stress theory much more complex.

Extensive studies have not led to the elucidation of structure of the DDT-induced stress factor. It is dialyzable, unstable, but has been collected in sizable quantities form DDT-poisoned crayfish. Spot colour tests suggest that the factor is an aromatic amine containing an ester group. Similarly, the nerve cord and the head of *P. americana* contain the highest concentrations of factor S, a neuroactive substance which excites motor neurone activity in the cockroach at low concentrations. The chemical and biological properties of this substance are indicative of a biogenic amine which appears to be distinct form other known neuropharmacologically active agents.

Stress induced by DDT or immobilization of *P. americana* differs in its effects on oxygen consumption, heartbeat frequency, and loss of weight, but both conditions stimulate the production of a characteristic fluorescent compound which is not produced by organophosphorus insecticides. Neurohormonal effects may have no bearing on insecticide resistance, although Heslop and Ray (1959) have demonstrated that the ability of *P. americana* to withstand stress by prolonged mechanical agitation can be directly correlated with their degree of DDT-resistance. The evidence and role of pharmacologically active substances released during various forms of stress has been reviewed by Sternburg (1963). The *"neurotoxin"* theory could, in principle, be compatible with nerve membrane interference but it does not offer and explanation for the primary lesion of DDT action, which, presumably, is unstabilizing efect on sensory neurones and nerve cells.

Physiological Protective Mechanisms

A. Detoxication in Insects

Insects, like most other organisms, must chemically alter and dispose of a large variety of compounds to maintain their normal body functions. In addition, they must neutralize, inactivate, or eliminate many foreign organic substances, including poisons, in avoiding toxic damage. According to Williams (1959), all foreign organic compounds are susceptible to metabolic attack *in vivo.* These "biotransform- ations" may involve oxidation, reduction, hydrolytic processes, and synthetic or conjugation reactions.

Comprehensive reviews on conjugation mechanisms in insects have been published by Smith (1962). These reactions include the following: 1. Conjugation of benzoic acid with glycine to from hippuric acid as shown in mosquito larvae, silkworms, silkworms and locusts.

2. Formation of ethereal sulfates such as the conversion of aminophenols and naphthols by certain beetles and by house flies.

3. *B-Glucoside formation, e.g;* conjugation of phenols to form-B-glucosides in a similar manner the b-glucuronides are formed in verte- brate detoxication processes. Insects do not synthesize glucuronic acid, although Terriere *et at.* (1961) have demonstrated glucuronide-like metabolites resulting from naphthalene metabolism in house flies and blow flies, and Smith (1955) found an active B-glucuronidase enzyme in the crop fluid of locusts. Glucuronic acid conjugation has also been reported to occur in *Prodenia* larvae and in silkworms.

In most other insects so far examined, including representatives of the orders Thysanura, Orthoptera, Coleoptera, Lepidoptera, and Diptera, phenols, as well as other compounds, combine with glucose to from the corresponding glucoside. The source of glucose for this type of conjugation is uridine diphosphate glucose. Similar conjugation reactions have now been shown to occur with the newer type insecticides, especially the carbamates.

4. *Phosphate conjugation:* Smith and Turbert (1964) detected an acidic conjugate in the excreta of flies treated with 1-naphthol that had properties similar to 1-naphthyl dihydrogen phosphate. This rather unusual type of conjugation has been reported to occur freely in the house fly, blow fly, and the New Zealand grass grub *Costelytra zealandica* treated with various Phenols, the reactions yielding substantial amounts of monoaryl phophates. These authors questions whether previous reports of glucuronide conjugatuion in insects might have been based on the nonrecognition of phosphates.

5. Cysteine and sulfhydryl conjugation: Detoxication mechanisms involving cysteine detected in locusts by Smith (1962). Chlorobenzene is metabolized by locusts and excreted as *o-,m,* and r-chlorophenylcysteine.

6. *Acetylation:* Insects, like mammals, acetylate certain amino compounds to their corresponding acetamido derivatives. For example, 4-6-dinitro-*o*-cresol is first reduced by locusts to 6-amino-4-nitroo-cresol followed by acetylation to 6- acetamido-4-nitro*o*-cresol. Primary amines such as chlorophenyl cysteine conjugates are excreted as the corresponding acetyl derivatives by locusts, and a *Drosophila* mutant excretes hydroxytryptamine as *N*- acetylhydroxytryptamine glucoside.

7. *Methylation:* Certain heterocyclic nitrogen compunds are methylated by coleopterous insects.

8. *Detoxication* involving sulfur compounds: A number of heavy metal poisons such as those containing Pb, As, Ni, Co, etc., are detoxified by clothes moths larvae *Tineola biselliella* through conversion to insoluble sulfides. The sulfur used in these conjugations is derived from the keratin in the wool eaten by the larvae. In other insects and in the blue tick *Boophilus decoloratus,*

B. Metabolic Fate of the Chlorohydrocarbon Insecticides

1. Conversion of the Insecticide to Nontoxic Metabolites

The first study of DDT metabolism in an insect was that of Ferguson and Kearns (1949) who injected DDT into the large milkweed bug *Oncopeltus fasciatus* and found that 80-100% of the dose was metabolized within 90 minutes. The metabolite did not respond to the colorimetric method of schechter *et al.* (1945) and was, therefore, neither 2,2-bis(r- chlorophenyl) -1,1- dictolorocthylene (DDE) hor bir (P- chlorophenyl) acetic acid (DDA). The latter compounds were also metabolized to unknown derivatives. On the other hand, immersion of the insects in boiling water prior to administration of DDT resulted in complete recovery of the unchanged DDT, indicating the involvement of an enzyme mechanism. Perhaps the most challenging and fruitful investigations in elucidating the mechanism of DDT resistance have been those dealing with DDT metabolism in the house fly, *M. domestica*. The house fly deatlt the first blow to our hopes that certain insect control problems should have never gotten out of hand if proper suppressive measures were taken. But the development of resistance also gave impetus to a wide variety of fundamental research in insect morphology, physiology, biochemistry, gentics and behaviour. It also stimulated the chemists to synthesize new compounds and correlate their chemical structure with biological activity.

The early assumptions that control failure with DDT was due to such factors as improper application of the chemical, adulterated insecticide, favourable environmental conditions for fly breeding, etc., were soon discarded with the simultaneous discovery by Sternburg *et al.* (1950) and Perry and Hoskins (1950) that resistance resulted from the ability of the house fly to metabolize DDT to the nontoxic derivative DDE. This finding was soon corroborated by many other investigators who variously showed that only resistant flies metabolized DDT to DDE or that both susceptible and resistant strain could accomplish this detoxication but at different rates. The same type of conversion occurs also with the bromine analogue of DDT.

In one instance large quantities of water soluble conjugates were found in the excreta of DDT- treated R and S flies, and continuing formation of unidentifiable DDT-derivatives were reported by Tahori and Hoskins (1953). Several insect species exhibit natural tolerance to DDT. The Mexican bean beetle, *Epilachnte varivestis,* the red-banded leaf roller, *Argyrotaenia velutinana,* the red-legged grasshopper, *Melano- plus femurrubrum,* and the differential grasshopper. *M. differentialis,* are able to convert DDT to DDE and other metabolites before selection pressure has been applied. The high degree of tolerance in the grasshopper species appears to be due, in great part, to the slow absorption of DDT through the cuticle and the intestinal wall.

Similarly, slower penetration of DDT plays a major role in resistance of the boll weevil, *Anthonomus grandis* and the tobacco budworm, *Heliothis virescens*, but more extensive dehydrochlorination of DDT to DDE is the predominant mechanism in the pink bollworm, *Pectinophora gossypiella*. In other stains of *E. virescens,* rate of absorption and metabolism of DDT, as well as excretion of DDT, DDE, and DDA, determine the extent of resistance. Both susceptible and resistant strains of the spotted root maggot *Euxesta notata* metabolize DDT to DDE at approximately equal rates, but absorption of DDT is much slower in the resistant strain (Hooper, 1965). Here, too, absorption of DDT plays a major role in resistance.

Most of the availbale information on insecticide metabolism has been obtained by investigating interstrain differences within species, but biochemical differences occur also among closely related species. For example, DDT-resistant cabbage worm *Pieris rapae* larvae metabolize DDT to DDE quite freely, whereas a closely related species, *Barathra brassicae,* is unable to accomplish this conversion. In this instance, penetration is practically the same in both species. In other insects, the resistance mechanism is less understood. Thus, fifth-instar larvae of the European corn borer, *Pyrausta nubilalis*, convert significant amounts of DDT to DDE but show only slight tolerance to the insecticide and two strains of the codling moth, *Carpocapsa pomonella,* metabolize DDT to DDE but the rate of detoxication is considered insufficient to account for the resistance. Aside from the resistance potential present in many species, DDT has also had the effect of reducing predator and parasite populations of insects. Interestingly, however, the lady beetle, *Coleomegilla maculata,* an important predator of several insect pests, is highly resistant to DDT. It resistance is due to conversion of DDT to DDE and to excretion of both DDT and DDE in the feces and eggs. Enhanced DDT tolerance has also been noted in *Macrocentrus ancylivorus,* a parasite of the oriental fruit moth, and in *Bracon mellitor,* and ectoparasite of the boll weevil.

Investigations on the metabolism of DDT in some 30 insect species revealed that, in general, DDT metabolism falls into three classes: (1) absorbed DDT remains largely unchanged, (2) much of the absorbed insecticide is metabolized to DDE, and (3) the chief metabolites do not respond to the colorimetric method of Schechter *et al.*(1954). With the advent of newer analytical techniques such as thin-layer and gas-liquid chromatography, it is now possible to add (4) metabolism to 2,2-bis(p-chlorophenyl)-1,1- dichloroethane (DDD or TDE), 2,2-bis (p-chlorophenyl)-1,1,1- trichloroethanol (dicofol), 4,4-dichlorobenzophenone (DBP), bis (p-chlorophenyl) acetic acid (DDA), bis (p-chlorophenyl) methanol (DBH), 4,4-dichlorodiphenylmethane (DCPM), and p-chlorobenzoic acid.

DDT metabolism in susceptible and DDT-resistant body lice, *Pediculus humanus humanus,* pose an even greater dilemma. The resistant (Korean strain tolerates oral doses of 200 ppm in the blood whereas the susceptible strain succumbs to 10-20 ppm. Nevertheless, homogenates of both strains metabolize DDT *in vitro* at approximately equal rates and yield the same metabolites. What then is the physiological basis of this resistance ? Unfortunately, the answer is not known, but one can speculate that an effective nerve barrier in the resistant strain prevents the penetration of DDT into sensitive neurons since rate of enzymatic attack is no different between the two strains.

Another type of DDT metabolism is exemplified by the vinegar fly, *Droso-phila melanogaster* DDT-resistant strains of this species metabolize DDT *in vivo* to the corresponding ethanol derivative 2,2-bis (r-chlorophenyl)-1,1,1- trivhloroethanol (dicofol). This resistance mechanism is controlled by a domianant gene *Rst* (2) DDT, located on the second chromosome. Similar results have been obtained with [^{14}C]DDT in two other strains of *D. melanogaster,* but in addition to dicofol, another metabolite 2,2-bis (p-chlorophenyl)-1,1-dichloroethanol (FW-152) was detected (Menzel (et al., 1961).

Dicofol also is one of the metabolites resulting from the in vivo degradation of DDT by T. infestans (Dinamarca et al. 1962), in addition to DDE and three other polar metabolite. Slower penetration and enhanced detoxication of DDT appear to account for the resistance of fifth- instar nymphs of this species.

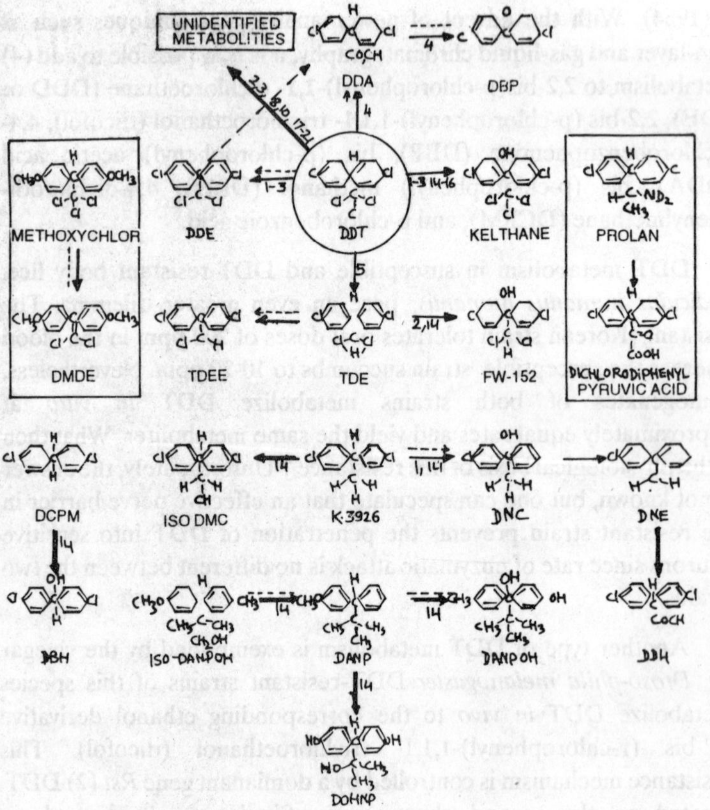

Fig. 10.1. Metabolism of DDT and related compounds by various insect species, adapted and modified from Tsukamoto (1961). Key: 1, *Musca domestica;* 2, *Blattella germanica;* 3. *Periplaneta americana;* 4, *Pediculus humanus;* 5, Anopheline and Culicine spp.; 6, *Malacosoma americana;* 7, *Pyrausta nubilalis;* 8, *Plodia interpunctella;* 9, *Pieris, rapae;* 10, *Epilachna varivestis;* 11, *Argyrotaenia velutinana;* 12, *Melanoplus differentialis;* 13, *Melanoplus femur- rubrum;* 14, *Drosophila melanogaster;* 15, *Drosophila virilis;* 16, *Triatoma infestans;* 17, *Nymphalis antiopa;* 18, *Leucophea maderae;* 19, *Oncopeltus fasciatus;* 20,*Ornithodorus coriaceous;* 21, *Anthonomus grandis.*

Although the weight of evidence supports a causal relationship between dehydrochlorination of DDT and resistance in the house fly, there are several areas of disagreement. Among the latter are that (a) large quantities of unchanged DDT sufficient to kill several S flies remain in the tissues of surviving R flies; (b) DDT metabolism is too slow to account for the resistance of several strains; (c) certain R strains show cross ressistance to compounds such as 1,1-bis (p-chlorophenyl)-2- nitropropane (Prolan) and 1,1-dianisyl neopentane which cannot be dehydrochlorinated. The presence of unchanged DDT need not necessarily imply that the toxicant is in a biochemically active state, for it may be bound or stored in nonsensitive tissues away from the site of action, or it may be distributed in various organs throughout the body (Table 10.2), so that the concentration at any one locus may be below the threshold of poisoning. Rate of DDT-metabolism varies from strain to strain, but even when exposed to sublethal doses it can be shown that R flies metabolize DDT at a faster rate than S flies (Menn *et al.*, 1957).

Cross-resistance is a unique phenomenon in insect toxicology both among closely related compounds and smong those relatively distant in structure. For example, Prolan-R flies metabolize Prolan to neutral and acidic derivatives, but they also dehydrochlorinate DDT via a different mechanism. *Drosophila melanogaster* adults metabolize 1,1-dianisyl neopentane to the trichloroethanol derivative cf. Fig.1) in addition to detoxifying DDT (Tsukamoto, 1961).

i. Metabolism of DDT by Cockroaches. The metabolites of DDT responding to the colorimetric method of Schechter *et al.* (1945) form only a small portion of the total metabolic pool in *P. americana* after topical application of DDT. Approximately 55% of the applied dose is in the form of unidentified metabolites. Injection of [^{14}C] DT in *P. americana* also yields a water-soluble metabolite which appears to be a DDT-derivative conjugated with a carbohydrate fragment. Conversion of DDT to DDE occurs mainly in the fatbody, alimentary canal, and cuticular components of the American roach, but fat-body appears to contribute much to the differential susceptibility to DDT exhibited by the sexes and the developmental stages of this insect.

Application of radiochromatographic techniques to the study of DDT metabolism in P. americana revealed the presence in the excreta of *six metabolites of differing* R f values. One of these metabolites,

Table 2. Distribution of DDT, Including Metabolites , in Various Tissues Following Topical Application of DDT

	DDT and metabolites recovered					
					Applied dose (%)	
	Absorbed dose (%)				*Treated leg*[d]	
Tissue	*a*	*b*	*c*		*S*[e]	*R*[e]
Head	5.8	38.0		Head	5.5	0.5
Wings and legs	11.5		58.6	Thorax	18.0	1.5
Thoracic cuticle	18.5	26.0		Abdomen	11.0	1.5
Abdominal cuticle	33.5					
					Treated labella	
					S	R
Thoracic muscle	16.2	5.0				
Intestinal tract	14.4	5.0		Head	11.0	3.0
Hemolymph	—	13.0	26.5	Thorax	33.0	3.0
Malpighian tubules	0.0	—		Abdomen	30.0	2.5
Sex organs	0.0	10.0				
Never tissue	0.0	3.0	14.8			
	DDE = 100%	Radioactive DDT			DDE = 73%	Radioactive DDT
	DDT = 0		DDT = 23%			

a From Sternburn and Kearns (1950).
b From Lindquist *et al.* (1951b).
c From Tahori and Hoskins (1953).
d From Le Roux and Morrison (1954)
e S, susceptible; R, resistant.

produced by both susceptible and DDT-resistant American roaches, corresponds exactly to the *R*f value of 4,4- dichlorobenzophenone. Small amounts of DDE but larger quantities of DDT and polar metabolites predominate in the excreta of DDT-treated *Leucophaea maderae*. Metabolism of DDT in the German roach *B. germanica* is some- what simelar, with both susceptible and resistant strains converting small amounts of DDT to DDE and to predominantly unidentified metabolites. In addition to DDE. dicofol has been recorded as a DDT meta- bolite in the German roach and both S and R strains metabolize TDE to the ethylene derivative, TDEE.

The paucity of data regarding metabolism studies with cock- roaches stems, perhaps, From the fact that DDT has never been extensively used for roach control. Although resistant populations of *B. germanica* have been obtained in the laboratory by selection pressure (Grayson, 1951, 1953) and have been recorded worldwide, little information is available regarding the biochemical changes brought about by this selection. Nevertheless, it is evident that conversion of DDT to DDE or to dicofol plays only a minor role in metabolism. Detoxication of DDT to polar derivatives which are excreted in the feces seems to be more important, but in neither of the cases cited above can the mechanism of resistance be directly correlated with absorption or detoxication of DDT.

ii. DDT Metabolism by Mosquitoes. Among the numerous species which have developed resistance to insecticides, mosquitoes assume a prominent position. Eighty or more anopheline and culicine species are known to be resistant to one or more of the commonly used insecticides, and there is reason to believe that this list will keep growing as more areas around the world come under present-day mosquito control operations. The first appearance of DDT resistance in culicine mosquitoes was recorded in 1946 and in anopheline species in 1951. Although the resistance problem became acute in certain parts of the world, especially in connection with malaria eradication programmes, little progress has been made in ascertaining the nature of this resistance. DDT resistance has been more extensively investigated in the yellow fever mosquito *Aedes aegypti.* The case of mass rearing, the availability of many resistant strains from different genetic background and geographic distribution, and the introduction of marker genes in larvae and adults make this insect an ideal species for resistance studies.

DDT-resistant larvae of *A. aegypti* and *A. taeniorhynchus* detoxify more DDT to DDE than their respective susceptible counterparts (brown and Perry, 1956). On the other hand, resistant *A. nigromaculis* larve absorb more DDT than susceptible larvae but convert about the same amount of DDT to nontoxic metabolites. Several DDT- resistant *A. aegypti* strains produce much larger amounts of DDE than their susceptible counterparts, while others, notably those of Asiatic origin, show little increase in DDE production In one strain, at least, relaxation of DDT pressure caused an increase in susceptibility to the toxicant but no corresponding decrease in detoxication, as would be expected if the resistance mechanism were primarily detoxication of DDT. DDT resistant *A. aegypti* larvae originating from a strain selected with malathion respond to DDT by excreting long streamers of peritrophic membrane containing large quantities of **DDT** and **DDE**. This hypersecretion might constitute a defense mechanism by removing unabsorbed DDT from the alimentary tract, thus reducing the internal concentration of the toxicant. This, however, is an isolated case involving two strains of Caribbean origin and it does not extend to resistant strains of different geographical distribution. Among culicine species, DDT-resistant *Culex pipiens fatigans* larvae are very efficient in hconverting DDT to DDE, and similar R species of *Clulex pippiens quinquefasciatus* from california contained practically all DDE and only minute amounts of DDT within larval tissues. Two polar metabolites similar to those found in German roach were also detected in the water. In contrast with the above results, a highly DDT-resistant strain of *Culex pipiens molestus* produces only little more DDT than its susceptible counterpart.

The results of Plapp *et al.* (1965) and Plapp and Hennessy (1966) suggest that the mechanism of resistance to DDT in *Culex tarsalis* is unknown and does not involve dehydrochlorination of the insecticide. Kalra *et al.* (1967) also concluded that mechanisms other than dehydrochlorination contribute to DDT resistance in *Culex fatigans*. Alternately, Dimura *et al.* (1965) attribute the DDT resistance of their *Culex tarsalis* strain to enhanced DDE production, whereas enhanced DDT metabolism was characteristic of both R and S larvae of *Culex fatigans.*

In addition to DDE, Hooper (1967, 1968) detected the metabolite DDD (TDE) as a by-product of DDT metabolism by *Culex fatigans,* a finding confirmed by Kalra *et al.* (1967) Earlier DDD was demonstrated in the excreta of the stable *Fly Stomoxys calcitrans* but was presumed to be due entirely to the anaerobic action of bacteria. The degradation of DDT to DDD is an interesting phenomenon since it as been reported in a number of organisms, including mammals, birds, fish, microorganisms (bacteria and fungi), and by the action of ferrous deuteroporphyrins, reduced porphyrins and aqueous media. In addition, DDD has been reported as a product of DDT metabolism by Hela cells by stored wheat grains and by ionizing radiation. DDT resistance in anopheline mosquitoes is less understood. Field-collected adults of *Anopheles atroparvus, A. maculipenis typicus, A. labranchice, A superpictus,* and *A. claviger* from Italy exhibited normal susceptibility to DDT and little conversion of DDT to DDE.

Laboratory colonies of DDT- resistant *A. atroparvus* and *A. stephensi* Produced little DDE in proportion to their resistance and showed no significant difference in DDT metabolism from their susceptible counterparts. Hence, resistance cannot be correlated with DDT dehydrochlori- nation in these species. Only in certain strains of *A. sacharovi* and in *A. sundaicus* have substantial quantities of DDE been recovered so that the resistance mechanism might possible correlate with DDT detoxication.

So far, the available date indicate little correlation between DDT resistance and DDT metabolism in mosquitoes. However, some exceptions, i.e., in some strains of *A.aegypti,* do exist. Lack of agreement in results from various laboratories likewise precludes forming a generalized hypothesis regarding the resistance mechanism in *Culex* species, and the dearth of information on anophelines makes the problem even less amenable to interpretation in that group.

iii. Metabolism of DDT in Vitro. The conversion of DDT to DDE by resistant house flies had, for many years, been regarded by some investigators as the consequence of survival of the insect rather than the cause of its resistance. Adequate proof to the contrary could not be provided until the isolation by Sternburg *et al.* (1953) of the enzyme DDT dehydrochlorinase (DDTase) which, in the presence of glutathione, catalyzes the dehydrochlorination of DDT *in vitro* (Fig.10.2). Under identical experimental conditions only DDT-R strains were shown to contain this enzyme. House fly DDT dehydrochlorinase has been purified by various procedures and its kinetics have been studied spectrophotometrically in some detail with purified preparations in which DDT was dossolved in egg yolk lipoprotein. With this technique it was possible to demonstrate DDT dehydrochlorinase in adult susceptible flies also, particularly in the larval stage.

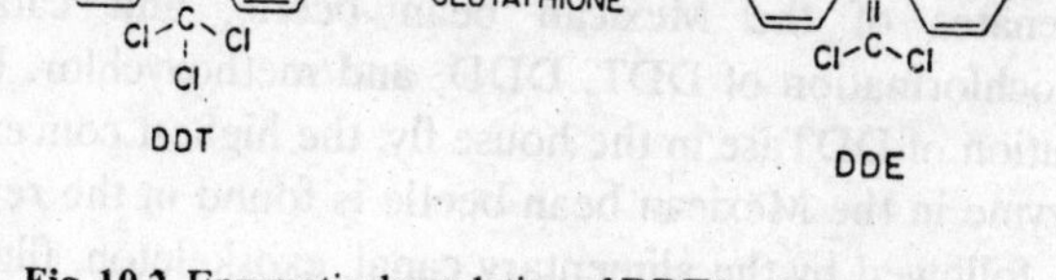

Fig. 10.2. Enzymatic degradation of DDT by the house fly.

Lipke and Kearns (1959b) found a molecular weight of 36,000 for house ;fly DDT ase. Reivestigation of the problem by Dinamarca *et al.* (1969) established the fact that DDT dehydrochlorinase has a tetrameric structure, with each monomer having a molecular weight of 30,000. The tetramer is formed in the presence of DDT, possible because the enzyme is a lipoprotein. Glutathione, which is not used during the reaction, is necessary to maintain the tetrameric structure in the presence of DDT. Recently, DDT isozymes have been described by Goodchild and Smith (1970) but it is not clear if these isozymes are related to the tetrameric form of DDTase.

With present-day gas chromatographic methods the detection of enzyme activity from a single susceptible fly is quite feasible and differences between strains can be made more manifest.

The amount of DDT dehydrochlorinase in house fly eggs is negligible. During metamorphosis of the resistant strain there is a progressive increase in enzyme concentration throughout larval life, and an abrupt drop of as much as 50% activity in the pupal stage. This lower level of activity is maintained throughout the pupal and adult life (Fig. 10.3). In the adult fly, high DDTase titers are found in fat-body and brain tissue, intermediate amounts in cuticle, muscle, and hemolymph, and little or none in the intestinal tract and ovaries.

The occurrence of high levels of DDTase in fat-body is highly significant since, in general, insect fat-body is rich in detoxifying enzymes and rather than being a storage depot, fat-body is an active tissue similar in function to mammalia liver. The high concentration of DDTase in nerve tissue is also important, especially from the resistance stand- point, since it might provide a site for local detoxication. consistent with this hypothesis is the demonstration by Wigglesworth (1956) that the underlying cytoplasmic sheath of insect nerve is exceptionally rich in oxidases, dehydrogenases, and various esterases and, presumably, could protect the sensitive axons by local detoxication. DDT dehydrochlorinase is also present in tissue homogenates of the Mexican bean beetle, and catalyzes the dehydrochlorination of DDT. DDD, and methoxychlor. Unlike the distribution of DDTase in the house fly, the highest concentration of the enzyme in the Mexican bean beetle is found in the reproductive organs, followed by the alimentary canal, exoskeleton, flight muscle, central nervous system, and fat-body.

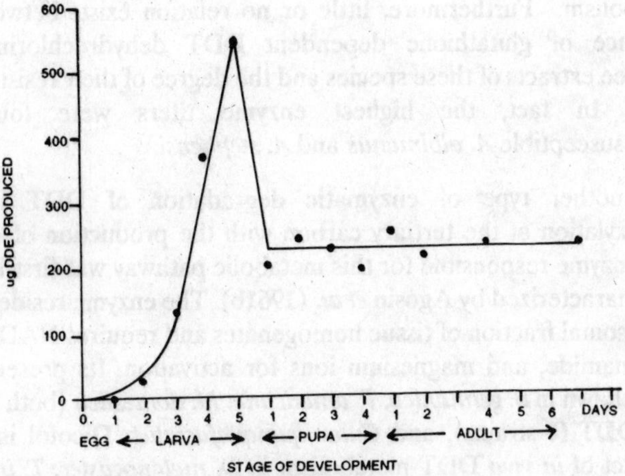

Fig. 10. 3. Distribution of DDT dehydrochlorinase in different stage of development of the DDT-resistant house fly. (From Moorefield and Kearns, 1957.)

Enzymic degradation of DDT in the body louse yields three metabolites: DDE, DDA, and 4, 4-dichlorobenzophenone (DBP). Accordingly, three or more enzymes are involved in these biotransformations, two of which have been isolated and partially purified. The conversion of DDT to DDE is analogous to that of the house fly, but DDE is not an intermediate in the breakdown of DDT to DDA, and the degradation of DDA to DBP most likely involves a microsomal mixed function oxidase. The role of DDT dehydrochlorination in mosquito resistance became more apparent with the isolation of the enzyme DDT dehydrochlorinase from *A. aegypti* larvae by Kimura and Brown (1964). This was made possible by protecting the enzyme with glutathione, nitrogen, and low temperature during the process of preparation. The mosquito enzyme has an optimum pH of 7.4 and, in addition to attacking DDT, dehydrochlorinates DDD and methoxy- chlor. In a large number of *A. aegypti* strains of American origin, the DDT dehydrochlorinase activity was found to be proportional to the resistance level.

Such correlation does not hold true for anopheline species. Relatively large amounts of DDE are formed *in vivo* by both S and R strains of *A. stephensi, A. quadrimaculatus, A. gambiae,* and *A.*

albimanus, but no correlation is evident between resistance and metabolism. Furthermore, little or no relation exists between the presence of glutathione dependent DDT dehydrochlorinase in cell-free extracts of these species and the degree of their resistance to DDT. In fact, the highest enzyme titers were found in DDT-susceptible *A. albimanus* and *A. stephensi.*

Another type of enzymatic degradation of DDT involves hydroxylation at the tertiary carbon with the production of dicofol. The enzyme responsible for this metabolic pathway was first isolated and characterized by Agosin *et al.* (1961b). The enzyme resides in the microsomal fraction of tissue homogenates and requires NADPH, O_2 nicotinamide, and magnesium ions for activation. Its presence has been shown in *B. germanica, P. americana, M. domestica* (both DDT-S and DDT-R strains), and *Culex quinquefaciatus*. Dicofol is also a product of *in vivo* DDT metabolism in *D. melanogaster. T. infestans* and the grain weevil *Sitophilus granarius*. The properties of the microsomal enzyme catalyzing the hydroxy- lation of DDT by *T. infestans* were described by Agosin *et al*. (1969). Evidence from experiments with phenobarbital indicates that more than one enzyme in involved in the metabolism of DDT by microsomal mixed-function oxidases. The *in vivo* availability of NADPH might be a limiting factor in the production of dicofol, since NADPH is also necessary for the synthesis of reduced glutathione, and essential cofactor for DDT dehydrochorinase. Hence, the microsomal hydroxylating system might be in competition with DDT dehydrochlorinase for NADPH.

The role of DDT hydroxylation in resistance was first demonstrated by Morello (1964) who showed that hydroxylation of DDT is blocked by SKF 552-A (B-diethylaminoethyl diphenylpropyl acetate) and iproniazid (2-isopropy-1- isonicotinoyl hydrazine) with a concomitant increase in mortality of resistant fifth-instar *T. infestans* nymphs, while 3- methylcholanthrene increases the DDT-hydroxylating capacity of the insect resulting in somewhat greater tolerance for the insecticide. Earlier, Arias and Terriere (1962) had shown a higher rate of naphthalene hydroxylation in microsomes of DDT-R house flies than in those of DDT-S counterparts, and observation substantiated by Schonbrod *et al*. (1965). Microsomal hydroxylation of DDT was soon corroborated in a number of other DDT-resistant strains of house flies and it now appears to be a major pathway of DDT detoxication in certain insects.

Recently, Kappor *et al.* (1970) compared the metabolism of DDT, methoxychlor, and methiochlor in various organisms in a model ecosystem for evaluating pesticide biodegradability. R and S house flies and the salt marsh caterpillar *Estigmene acrea* were included in this study. Piperonyl butoxide (a methylenedioxyphenyl synergist) was applied to the flies' abdomens 1 hour prior to topical application of the insecticides to indicate the extent of metobolism by the mixed-function oxidase system. The data in Table 10.3 show that methoxychlor is substantially more effective than DDT against the R strain and this can perhaps be related to its lower rete of dehydro-chlorination by DDTase. The effective inhibition of methoxychlor and methiochlor degradation by piperonly butoxide clearly indicates that detoxication of these compounds proceeds via oxidative processes rather than by dehydrochlorination. Methiochlor which is first metabolized by oxidation to 2-(p-methylsulfinyl-phenyl)-2- (p-methylthiophenyl)-1,1,1-trichloroethane and by further oxidation to the bissulfoxide and bissulfone derivatives, becomes than more susceptible to attack by DDTase. Methoxychlor is metabolized by *O*-demethylation to form 2-(p-hydroxyphenl)-2-(p-methoxy-phenyl)-1,1,1-trichloroethane and 2,2-bis (p- hydroxyphenyl)-1,1,1-trichloroethane which are largely eliminated in conjugated form. Hence, both pathways play a role in the detoxication of these compounds.

b. Metabolism of Benzene Hexachloride (BHC). The metabolic fate of BHC (actually, the compound is hexachlorocyclohexane in insects also has been a subject of considerable investigation, and in

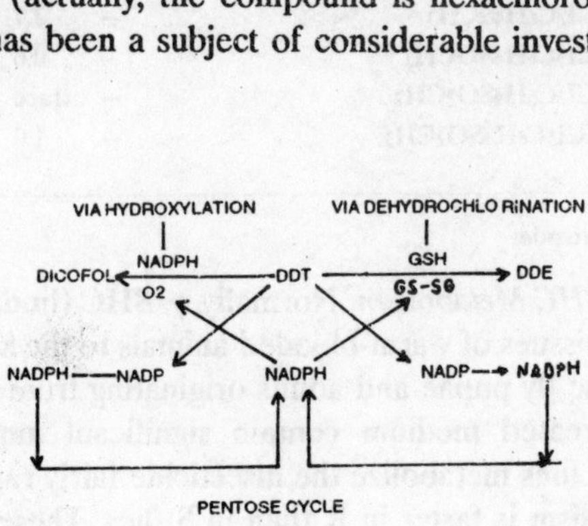

conformity with DDT, the house fly has emerged as the organism of choice, for no other insect to date has shown such an amazing potential for chmical biotransformation of xenobiotics.

TABLE 3.

Toxicity and Metabolism of DDT, Methoxychlor, and Methiochlor in Resistant House Flies[a]

	LD_{50} n/female fly		*% Radioactivity recovered DDT treatment*			
Compondrecovered	*Alone*	*+PB[b]*	*External*	*Internal*	*Excreta*	*Total*
DDT	3.4	0.8	9.1	12.4	1.0	22.5
DDE	–	–	–	12.6	8.7	21.3
Dicofol	–	–	–	11.9	7.6	19.5
Conjugates	–	–	–	23.1	7.5	30.6
			Methoxychlor treatment			
Methoxychlor	0.96	0.09	9.7	8.5	12.0	30.2
O-Demethylmethoxychlor			–	3.5	–	3.5
Conjugages			–	5.9	30.9	3.8
			Methiochlor treatment			
Methiochlor	56	10	50.3	3.7	4.9	58.9
$CH_3SC_6HC_4C{=}{=}CCl_2C_6H_4SCH_3$			–	1.4	0.2	1.6
$CH_3SOC_6H_4HCCCl_3C_6H_4SCH_3$			–	2.7	0.8	3.5
$CH_3SOC_6H_4HCCCl_3C_6H_4SOCH_3$			–	0.6	0.2	0.8
$CH_3SOC_6H_4HCCCl_3C_6H_4SO_2CH_2$			–	trace	0.3	0.3
$CH_3SO_2C_6H_4HCCCl_3C_6H_4SO_2CH_3$			–	1.0	2.4	3.4

b PB, piperonyl butoxide.

i. Rate of BHC Metabolism. Normally, γ-BHC (lindane) does not accumulate in tissues of warm-blooded animals to the same extent as DDT, but house fly pupae and adults originating from larvae reared on a lindanetreated medium contain significant amounts of the toxicant. Adult flies metabolize the insecticide fairly rapidly, and the rate of metabolism is faster in R than in S flies. These observations suggest an increased detoxication capacity in the adult stage, in contrast with the pattern observed with DDT metabolism. In addition to γ-BHC, the α-, β-, and δ-isomers which are less toxic are also metabolized at a faster rate by R than by S strains. This fact clearly demonstrates a greater metabolic capacity in the R strain and a possible causal relationship between metabolism and resistance. The results of Bradbury and Standen (1960) with ^{14}C- labeled γ-BHC administered to four strains of flies add support to this hypothesis by showing a fitness in the relationship

$$\text{In}\, Y = \text{In}\, Y_O - rt$$

Where Y is the amount of γ-BHC present at time t, Y_o the amount picked up in the original 15-minute exposure, and r represents rate of metabolism. In the four strains studied, r increased with an increase in the level of resistance. Oppenoorth (1956) summarized his extensive work on the metabolism of BHC isomers in eight strains of flies as follows: (a) injected γBHC is metabolized by both S and R strains, but at a faster rate by the latter; (b) rate of absorption of the different isomers decreases in the order gd a; (c) rate of metabolism decreases in the order g ~ a d; and (d) good correlation exists between degree of resistance, absorption rate, breakdown capacity, and amount of unchanged chemical found in the flies.

Bradbury (1957) extended his work on the comparative metabolism of γ-BHC in several other species and showed (Table 10.4) that house flies, both S and R strains, are in a class by themselves in their capacity to metabolize the toxicant. Even γ-BHC-resistant *A. gambiae* adults converted no more of the absorbed dose (10% in 24 hours) than their susceptible counterparts, and a similar situation occurred with *Cimex lectularius*. According to Busvine and Townsend (1963), the significance of BHC degradation in resistant house flies can be explained if rate of detoxication is considered in relation to the resistance spectrum which Busvine (1954) showed to be linked with cyclodiene insecticides. Resistance, then, has two components, one associated with cross-resistance to cyclodienes which are not metabolized, and another, due to enhanced degradation of γBHC which is characteristic of the house fly but not of other insects.

ii. Nature of BHC Metabolites. In the presence of alcoholic alkali, BHC isomers readily undergo dehydrochlorination liberating 3 moles of HCI per mole of BHC to yield principally 1.2,4-trichlorobenzene and smaller amounts of 1,2,3,- and 1,3,5- trichlorobenzene.

TABLE 4.

Comparative Metablism of ^{14}C-Labeled g-BHC in Various Insects

		BHC Recoverd (ug/gm)			
Species	*Dosage (ug/gm)*	*External*	*Internal*	*Water-soluble metabolites*	*% metabolized*
Cockroach	42	10	19	9	22.0
Bean weevil	111	7	94	8	7.7
Grain weevil	34	14	13	1	4.2
Locust	100	16	83	4	4.7
Mosquito	213	17	141	10	5.1
Khapra beetle	166	81	45	11	12.9
House fly (S)[b]	193	22	54	99	57.9
House fly (R)[b]	230	13	21	174	80.2

b S, susceptible, R, resistant.

The early work with house flies indicated that trichlorobenzene was, at best, only a minor meta-, bolite of [^{14}C]lindane and that the expired air contained no radio- activity, but the internal tissues contained equal amounts of polar and nonpolar metabolites. Both S and R strains metabolized the α-,β-, and δ-isomers with equal facility, but the γ-isomer was metabolized twice as fast by the R strain.

γ-Pentachloroyclohexene (γ-PCCH), the monodehydrochlorinated product of gBHC, was reported to be a major product of lindane metabolism in house flies. Both S and R strains produce this metabolite, which is further converted to other unidentified polar products. However, Bradbury and Standen (1958) consider γ-PCCH to be a minor metabolite, and isotope dilution techniques indicate that this metabolite is formed in relatively small amounts. A second isomer of PCCH, also produced in small amounts, was recently detected by Reed and Forgash (1968). While the production of γPCCH is perhaps toxicologically insignificant, there appears to be a correlation between the formation of the second PCCH-isomer and lindane resistance.

House flies treated with either lindane, γ-PCCH, or γ-PCCH-isomer yield a common metabolite which was identified by mass spectroscopy and gas-liquid chromatography as 1,2,4,5-tetrachlorobenzene. In addition, 1,2,3,4- tetrachlorobenzene, pentachlorobenzene, 1,2,4- and 1,2,3- trichlorobenzene were also detected. While these metabolites appeared only in small quantities following lindane application, they appeared in much larger quantities (50%) following treatment with the two PCCH-isomers. The relationship of lindane metabolism to organic soluble products and resistance is, at best, only of minor significance since, by far, water-soluble metabolites predominate in all fly strains studied. Water-soluble compounds also account for most of the lindane metabolic pool in mammals

iii. Enzymic Metabolism of BHC in Vitro. As discussed earlier, the rate of detoxication of BHC-isomers by house flies decreases in the orderγ α δ. Alkaline dechlorination to trichlorobenzene decreases in the order α δ γ β (Kauer *et al.*, 1947). Hence, *in vivo* metabolism of BHC does not follow the pattern of alkaline dechlorination, a situation analogous to that found with some DDT derivatives.

Faster metabolism of theα - and δ- isomers by R flies (note exception for α-BHC, Bridges and Cox, 1959) implies a causal relationship to resistance, but the same interpretation for the γ-isomer might be open to objection on grounds that the greater metabolizing capacity of the more toxic isomer is due to the longer survival of the resistant fly. As in other instances, isolation of the detoxifying enzyme *in vitro* is a prerequisite in characterizing the resistance mechanism. An important step in the direction was the finding that alkaline hydrolysis of BHC metabolites produces dichlorothiophenols. From this it is inferred that metabolism of BHC to water-soluble metabolites involves the formation of a C–S bond. The nature of the initial metabolic product suggests a conjugation with aryl mercapturic acid similar to the linking of sulfur with nuclear carbon in mammalian detoxication mechanisms.

Homogenates, acetone powders, and crude soluble enzyme preparations of BHC-resistant flies yield thiophenols as end-products in *in vitro* BHC metabolism. Reduced glutathione is essential for activation of the enzyme. The reaction might proceed by substitution of an aromatic chlorine followed by hydrolysis of the conjugate to yield dichlorothiophenols (Fig. 10.5) ; γ- and δ- pentachlorocyclohexene (PCCH), the monodehydrochlorination products of BHC-isomers, are also converted to water-soluble metabolites by the same enzyme system and at a much faster rate than γ-BHC. This is, perhaps, the reason why some investigators find only traces of PCCH in their studies of BHC metabolism. The water-soluble metabolites arising from PCCH metabolism in grass grubs and blow flies are chromatographically indistinguishable form S-2,4-dichlorophenyl glutathione (et al., 1969). These observations support the contention of Bradbury and Standen (1959) that a pentachlorocyclohexylglutathione is the initial metabolite of γ-BHC degradation, and suggests that the detoxication enzyme belongs to the group of glutathione *S*-aryltransferases and might also include DDT dehydrochlorinase. The scheme outlined in Fig. 10.5 will also permit the production of PCCH by metabolism of the thioether conjugate. PCCH is then further metabolized to water-soluble compounds which are excreted. This process releases the HSR moiety which again can conjugate with additional BHC to repeat the cycle.

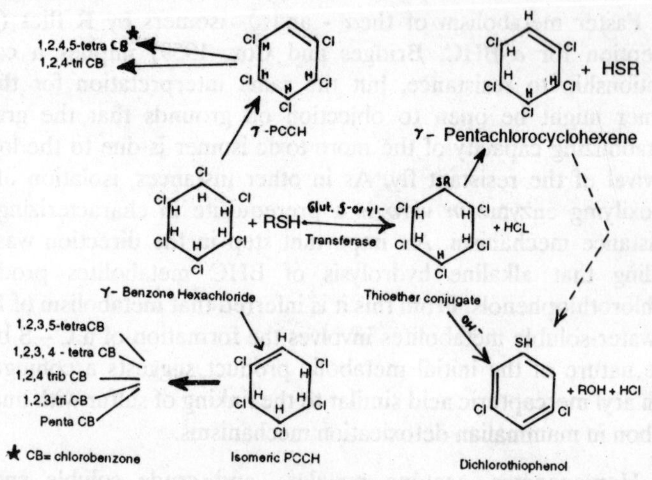

Fig. 5. Metabolism of benzene hexachloride by the house fly. Major metabolic routes are designated by heavy arrows. (From Bradbury and Standen, 1959, 1960; Reed and Frogash, 1968, 1969).

If detoxication by these enzymes is causally related to resistance, then their inhibition becomes of considerable importance. Insect glutathione *S*-aryltransferase and DDT dehydrochlorinase are inhibited by phthaleins and sulfonphthaleins, and these dyes also inhibit the metabolism or γ-and δ-BHC. Furthermore, bromophenol blue inhibits house fly enzymes that detoxify γ-BHC and DDT. Inhibitors such as these might be used to distinguish between isozymes among different species. Indeed, Clark *et al.* (1959), using the inhibitor principle, found distinguishable characteristics between house fly and blow fly enzymes that detoxify γ-BHC and DDT. The properties and distribution of the BHC-metabolizing enzyme show that it is found in the soluble fraction of house fly homogenates, has a molecular weight of 54,000 and is specific in its requirement for reduced glutathione. Enzyme activity is particularly high in house

fly homoegenates as compared with similar preparations from other insect species and from mammals, and the level of activity is higher in the adult fly than in the egg, larval, and pupal stages. This is in contrast with the distribution of DDT dehydrochlorinase showing highest activity in the larval stage. A significant feature of these studies is that no correlation is evident between enzymic metabolism of γ-BHC to water-soluble metabolites and resistance. Neither is the metabolism of γ-BHC to organic-soluble products a significant factor in house fly resistance. As more species are studied the detoxication hypothesis is weakened, except for the fact that rhe house fly stands alone with its tremendous potential for insecticide degradation. Alternatively, a supplementary defense mechanism might be essential for the insect's survival, and together with detoxication, the combination might provide the organism with a formidable weapon to withstand the toxicant.

c. Metabolism of Chlordane and Toxaphene. Very little is known about the metabolic fate of chlordane and toxaphene in ensects. The paucity of data might be due, perhaps, to their limited use in pest control operations and, until recently, to the unavailability of suitable analyitical methods for the detection of metabolites.

Technical chlordane consists of at least five compounds, three of which are relatively toxic to insects. For an understanding of the biochemistry of resistance to this insecticide it is essential to study the metabolism of each component separately. This has not yet been accomplished.

TABLE10.5.

Comparative Rates of BHC- and PCCH-Metabolizing Enzymes in Various Species

Enzyme source	*Sex*	*α-BHC*	*γ- BHC*	*γ-BCCH*	*δ-PCC*
Incects					
Musca domestica (S)[a]	M	92	22	290	3,650
	F	158	38	560	6,390
Musca domestica (R)[b]	*M*	*332*	*59*	*920*	*10,600*
	F	228	24	590	7,240
Drosophila melanogaster	M and F	0	0	23	990
Stomoxys calcitrans	M and F	13	0	63	0
Periplaneta	M	40	2.8	50	–
	F	14	0	44	–
Blattella germanica	M and F	0	0	23	–
Leucophaea moderae	F	0	0	2.4	–
Ostrinia nubilalis (larvae)		0	0	23	–
Ostrinia nubilalis (adults)	M and F	0	0	18	–
Apis mellifera	F	0	0	0	–
Diabrotica virgifera		0	0	0	–
Lachesilla pedicularia (adults)	*M and F*	*0*	*0*	*0*	—
Mammals					
Rattus norvegicus (rat liver)	M	0	0	19	–
Lepus cuniculus (rabbit liver)	M	2.4	0	16	274
	F	3.2	0	13	227
Lepus cuniculus	M	1.5	0	3.7	105
	F	1.1	0	2.7	79

b S, susceptible; R, resistant.

Bioassay procedures have established that chlordane-resistant house flies are capable of metabolizing a large percentage of topically applied chlordane to innocuous derivatives. Chiordane-resistant German roaches, *B. germanica,* metabolize 70% of the absorbed does to unidentified products. However, the remaining unchanged insecticide within the tissues is sufficient to kill more than 20 susceptible roaches. Here, too, it appears that two or more defense mechanisms

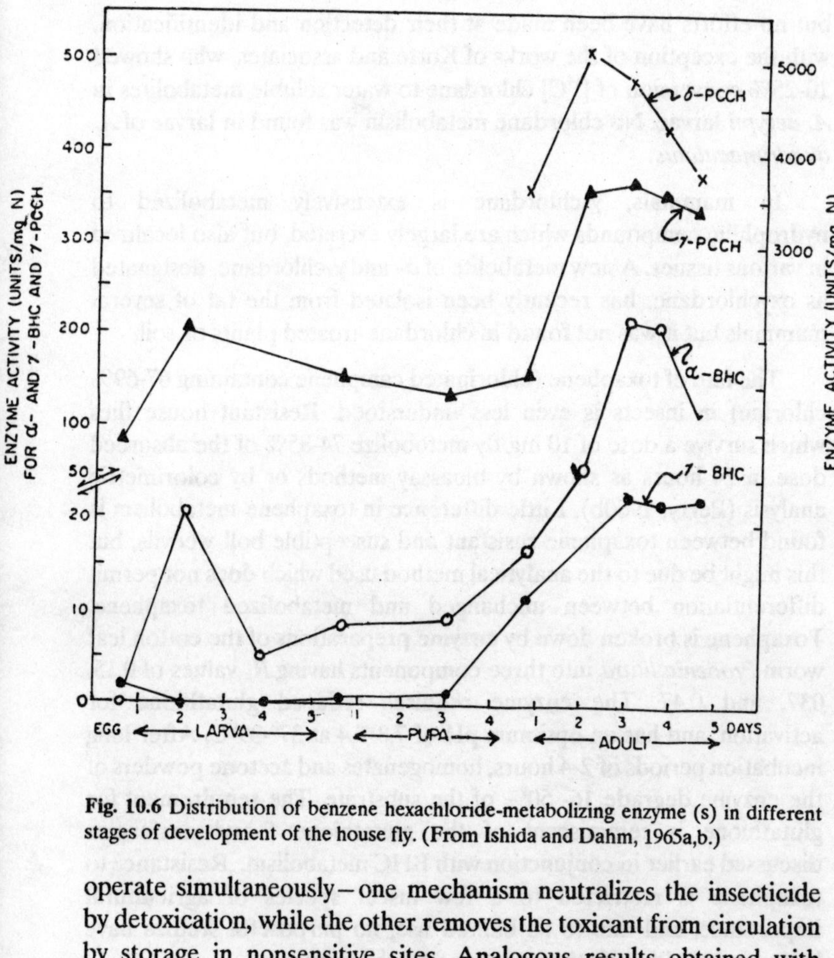

Fig. 10.6 **Distribution of benzene hexachloride-metabolizing enzyme (s) in different stages of development of the house fly. (From Ishida and Dahm, 1965a,b.)**

operate simultaneously – one mechanism neutralizes the insecticide by detoxication, while the other removes the toxicant from circulation by storage in nonsensitive sites. Analogous results obtained with chlordane-resistant house flies are amenable to the same interpretation, but Earle (1963) contends that chlordane detoxication does not contribute materially to the resistance mechanism in house flies.

γ-Chlordane penetrates the fly's cuticle quite readily and accumulates unchanged in the tissues to fairly high within 24 hours after topical application. Thereafter, a progressive decrease in internal concentration follows but the excreta yield no detectable products. Very likely, hydrophilic metabolites are formed by insects, but no efforts have been made at their detection and identification, with the exception of the works of Korte and associates, who showed 10-25% conversion of [^{14}C] chlordane to water soluble metabolites in *A. aegypti* larvae. No chlordane metabolism was found in larvae of *A. quadrimaculatus*.

In mammals, γ-chlordane is extensively metabolized to hydrophilic compounds which are largely excreted, but also localized in various tissues. A new metabolite of α- and γ-chlordane, designated as oxychlordane, has recently been isolated from the fat of several mammals but it was not found in chlordane-treated plants or soil.

The fate of toxaphene (chlorinated camphene containing 67-69% chlorine) in insects is even less understood. Resistant house flies which survive a dose of 10 mg/fly metobolize 74-85% of the absorbed dose in 24 hours as shown by bioassay methods or by colorimetric analysis (Perry, 1960b). Little difference in toxaphene metabolism is found between toxaphene-resistant and susceptible boll weevils, but this might be due to the analytical method used which does not permit differentiation between unchanged and metabolized toxaphene. Toxaphene is broken down by enzyme preparations of the cotton leaf worm *Prodenia litura* into three components having R_b values of 0.15, 037, and 0.47. The enzyme requires reduced glutathione for activation, and has an optimum pH of 7.3-7.4 at 37°-39°C. After long incubation periods of 2-4 hours, homogenates and acetone powders of the enzyme degrade 16- 50% of the substrate. The requirement for glutathione is reminiscent of the glutathione *S*-aryltransferases discussed earlier in conjunction with BHC metabolism. Resistance to toxaphene is restricted to a few insect species of agricultural importance, and due to its limited use, no purposeful studies have been made to identify the resistance mechanism.

d. Metabolism of Endosulfan. Technical endosulfan (Thiodan) consists of 90% of a mixture of two stereoisomers: a high-melting isomer (melting point, 208°-210°C), and low-melting isomer (melting point 108^0-110°C). The remaing 10% contains endosulfan alcohol and endosulfan ether, which are relatively nonoxic. The physical and biological properties of endosulfan were investigate by Lindquist and Dahm (1957) who found the two isomers to be slightly more toxic than DDT to susceptible house flies. Endosulfan is rapidly oxidized to endosulfan sulfate by resistant and susceptible house flies. The oxidized product, which is as toxic as the high-Melting isomer, accumulates much more rapidly in the susceptible strain, suggesting either that the S strain cannot metabolize the sulfate further of that the rate of endosulfan oxidation in the R strain is slower. Two conjugated metabolism. The rate of endosulfan sulfate metabolism appears to be the important factor in house fly resistance to this insecticide.

2. Resistance Due to excretion of the Insecticide of Toxic Metabolites

Metabolism of Prolan. Prolan is one of two constituents of the insecticide Dilan, the other constituent being Bulan.

Selection pressure with Dilan against house flies results in a strain having a higher level of resistance to Prolan than to Bulan or Dilan. The reason for this is not clear, but indications are that Prolan is metabolized at a faster rate than either Bulan or Dilan by the resistant strain.

Penetration of Prolan is fairly rapid, almost half of the appiled dose (8 μ γ/Fly) being absorbed in 24 hours. The amount of unchanged toxicant within the tissues does no exceed 0.5 mg/fly which is only slightly more than the LD_{50} for the susceptible strain. The remainder, or 87% of the absorbed dose, is excreted as (1) a neutral material soluble in common organic solvent, similar to Prolan in infrared, UV, and colorimetric absorption spectra, and as toxic to mosquito larvae as the parent compound; and (2) an acidic derivative extractable with dilute alkali, different from Prolan in photometric spectra and chemical properties (such as loss of NO_2 group from the propane moiety) and much less toxic to mosquito larvae. Rate of excretion of the toxic compond exceeds that of the acidic metabolite during the initial 24 hours after application of the toxicant. This lapse of time is well beyond the period necessary for toxicological evaluation since Prolan acts fairly rapidly. As the time interval increases beyound 24 hours, rate of excretion of the acidic metabolite surpasses that of the neutral compound. Concerning the protective mechanism involved in Prolan resis- tance, it appears that excretion of the acidic metabolite is not of primary importance to the fly's survival, since it is produced in substantial quantities only after the period of greatest dangeı to the fly has elapsed. Excretion of the unchanged insecticide is of far greater importance in the overall protective mechanism.

In vertebrate detoxication of foreign compounds, the end-produts tend to become more water-soluble than their precursors so they can be excreted. Excretion of nonpolar compounds is indeed rare. In one instance, following injection of chlorobenzene in locusts, relatively large amount of the nonpolar chlorobenzene were found in the excreta, along with polar conjugates. In the case of Prolan, both the unchanged fat-soluble compound and the acidic, polar metabolite are excreted simultaneously. The acidic metabolite has tentatively been identified as 3,3-bis (p-chlorophenyl) pyruvic acid. Prolan resistance has several characteristics which distinguish it from other types of resistance. For example, it exhibits cross-resis- tance to the DDT-type compounds and to 1,1-dianisyl neopentane which is devoid of chlorine atoms in the pentane moiety. It also shows croos-resistance to the cyclodiene compounds such as aldrin, dieldrin, heptachlor, etc., and is also highly resistant to 2-(4-chlorophenyl)- 2-(2, 4 dichlorophenyl)- 1,1,1-trichloroethane (*o*-chloro-DDT. The latter compound is refractory to dehydrochlorination due to steric hindrance caused by the *ortho* chlorine atom. Most DDT-resistant strains of house flies succumb to low dosages of this compound because of their inability to detoxify it. However, certain strains of DDT-resistant mosquitoes, A. aegypti and Culex fatigans, resist *o*-chloro-DDT by dehydrochlorinating it and *Culex tarsalis* tolerates this compound by other undnown mechanisms. The intriguing potentialities of Prolan resistance in house flies may be summarized as follows: (1) ability to dehydrochlorinate DDT, both *in vivo* and *in vitro;* (2) detoxication of compounds such as 1,1-dianisyl neopentane, possibly via microsomal hydroxylation reactions; (3) excretion of unchanged insecticide and metabolism to acidic metabolites, such as occur with Prolan; (4) resistance to cyclo- diene compounds—a resistance mechanism largely undetermined; and (5) ability to retain within the tissues without harmful effects large quantities of nondehydrochlorinatable *o*-chloro-DDT which, other- wise, are fatal to many other strains, both resistant and susceptible. It is not known whether this dynamic resistance pattern is characteristic of Prolan resistance in gerneral or is peculiar to this particular strain of house flies.

3. Resistance Due to Storage of Unchanged Insecticide or of Toxic Metabolites.

Metabolic Fate of the Cyclodiene Insecticides. i. Metabolism in Vivo. Early investigations on the fate of cyclodiene compounds in insects demonstrated the biotransformation of heptachlor, aldrin, and isodrin to their epoxides, i.e., heptachlor epoxide, dieldrin, and endrin, respectively. Heptachlor epoxide, isolated and characterized from heptachlor-treated resistant house flies was found to be identical with the authentic compound, melting point of 159°-160.5°C, in colorimetric and infrared spectra, as well as other chemical and biological properties. Heptachlor epoxide is not metabolized further and can be recovered quantitatively when applied to either resistant or susceptible flies.

Susceptible house flies also convert heptachlor to the epoxide. The onset of symptoms of poisoning, after a latent period of 1-2 hours, coincides with the appearance of heptachlor epoxide in the tissues. This observation led to the inference that trans- formation of the parent compound is an activation process not involving a detoxication mechanism. In spite of this "autointoxication" phenomenon, the resistant fly emerges unharmed. Similarly, a latent period is observed with regard to a rise in O_2 consumption in hepta- chlor-injected

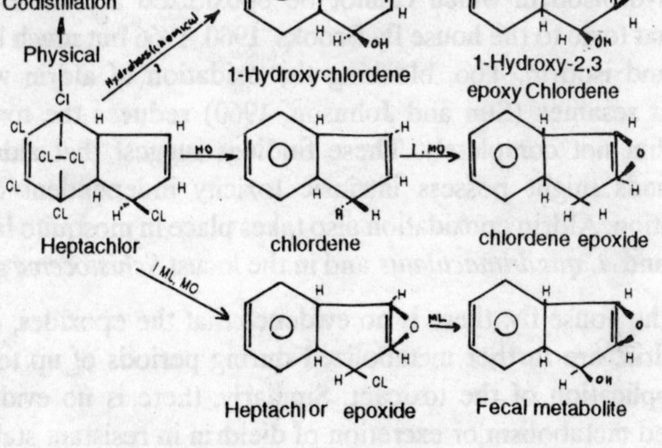

Fig. 10.7. Metabolism of heptachlor by insects (1), mammals (ML)., and microorganisms (MO). (Adapted from Brooks, 1966; Broods and harrison, 1964b, 1965; Miles *et al.*, 1969; Perry *et al.*, 1958.)

cockroaches presumably coinciding with the appearance of the toxic epoxide. The above hypo- thesis is not totally shared by other workers who showed that dihydrohepta- chlor which cannot be epoxidized has, nevertheless, some intrinsic toxicity of its own. Epoxidation of heptachlor also occurs in mammals

(Davidow and Radomski, heptachlor-treated plants and soils and microorganisms. In addition to heptachlor epoxide, soil micro-organism and house flies produce 1-hydroxychlordene, chlordene, chlordene epoxide, 1-hydroxy 2,3- epoxychlordene. Aldrin and dieldrin have been widely used as broad-spectrum insecticides for many years, yet little is known about their mode of action in insects. More disheartening is the fact that many agriculture pest and insect vectors of disease have developed high levels of physiological resistance to these insecticides but the mechanism of this resistance is still elusive. Metabolism of aldrin and isodrin follows much the same pattern as that of heptachlor. Injection of aldrin and isodrin into the American cockroach results in their partial conversion to dieldrin and endrin, respectively occuring chiefly in the digestive tract with the final diposition of the epoxides in various tissues, especially fat-body.

[^{14}C] Aldrin is rapidly metabolized to [^{14}C]dieldrin by S and R house flies and the onset of symptoms of poisoning, after a latent period of 2-3 hours, appears to coincide with the accumulation of dieldrin at critical sites. However, such compounds as dihydroaldrin and dihydroisodrin which cannot be epoxidized are, nevertheless, somewhat toxic to the house fly Brooks, 1960, 1966 but much less that aldrin and isodrin. Too, blocking the oxidation of aldrin with the synergist sesamex (Sun and Johnson, 1960) reduces the toxicity of aldrin, but not completely. These findings suggest that aldrin-type compounds might possess intrinsic toxicity independent of their epoxidation. Aldrin epoxidation also takes place in mosquito larvae *A. aegypti* and *A. quadrimaculatus* and in the locust *Schistocerca gregaria*.

In the house fly, there is no evidence that the epoxides, dieldrin and endrin, are further metabolized during periods of up to 4 days after application of the toxicant. Similarly, there is no evidence of enhanced metabolism or excretion of dieldrin in resistant stable flies *S. Calcitrans*. In the locust, [^{36}CI] aldrin is converted to [^{36}Cl]dieldrin at a very slow rate but no further metabolism occurs. However, [$_{36}$CI]dieldrin is slowly excreted unchanged along with small amounts of water-soluble metabolites. These findings illustrate the extreme biological stability of toxic cyclodiene epoxides in insect tissue. There is also evidence of the production of polar metabolites of cyclodiene insecticides in other insects. Korte *et al.* (1962) and korte and Stiasni (1964)demonstrated presence of several hydrophilic metabolites of [^{14}C]aldrin, [^{14}C]dieldrin, and [^{14}C]telodrin in larvae of *A. aegypti.* The extensive work of Korte and associates, covering several years of investigation on the metabolism of aldrin, dieldrin, dieldrin, endrin, heptachlor, dihydroheptachlor, and chlordane in insects, mammals, and microorganisms was summarized by Korte (1967). In general, most organisms examined, including mosquito larvae, meta- bolized to some extent all the above compounds to hydrophilic metabolites, but there is no evidence that these biotransformations are in any way related to resistance. Production of hydrophilic metabolites is also characteristic of resistant *C.p. quinquefasciatus.* In this instance, treatment of adult mosquitoes with [^{14}C]dieldrin yielded an excretory product resembling 6,7-dihydroxydihdroaldrin (apparently *cis*- aldrin glycol in its chromatographic behaviour. Unfortunately, no comparison was made with a corresponding susceptible strain. Authentication of the dieldrin metabolite as *trans*-6,7- dihydroxy- dihydroaldrin was made by Ludwig and Korte (1965) and Tomlin (1968).

As mentioned earlier, symptoms of aldrin poisoning in the susceptible fly are not manifest until approximately 2 hours after application of the toxicant. The next few hours are characterized by rapid knockdown, at which time the insects exhibit an exhaustive burst of respiratory activity and a decrease in tissue α-glycerophosphate. Therefore, the resistant insect must possess an efficient defense mechanism, not governed by detoxication or storage, which protects it during the initial critical stages of poisoning. Evidence for this comes from investigations on the vapor toxicity of aldrin designed to by pass cuticular penetration and other barriers. The extreme tolerance of the resistant strain to such vapors might indicate that the protective mechanism resides at the site of action of the toxicant. The site of action of the cyclodiene compounds may involve a decreased sensitivity to dieldrin of the thoracic ganglia of the resistant insect, or nerve components of the resistant strain might have less binding capacity for dieldrin than their susceptible counterparts. Alternatively, perhaps one of the metabolites produced from cyclodiens degradation, such as 6,7-dihydroxydihy droaldrin, might be the neurotoxic agent, since the latter can poison the nerve cord *in situ* much more rapidly than does dieldrin.

A number of investigators sought differences between R and S strains of insects relevant to the resistance mechanism. Gerolt (1965) found no difference in translocation of sublethal amounts of dieldrin in R and S house fly body parts. Adult house flies and *A. aegypti* larvae converted 10% and 5% respectively, of a sublethal dose of dieldrin to unidentified metabolites, but there was no difference between R and S insects in this respect. Similarly, the penetration, distribution, and metabolism of [^{14}C] aldrin and the metabolism of isodrin was found to be the same in R and S house flies. Direct measurement of dieldrin penetration through the nerve sheath showed no significant differences between R and S cock-roaches and in general, there were no large differences in rates of uptake of [^{14}C]dieldrin by the central nervous systems of R and S house flies following injection or infusion (Schaefer and Sun, 1967). Earlier, a noninsecticidal ^{82}Br-labeled analogue of aldrin was found to be excreted unchanged in equal proportion by R and S house flies, and a ^{35}Slabeled analogue of dieldrin was partially metabolized to unknown derivatives by both strains.

The distribution, metabolism, and localization of [^{14}C]dieldrin is practically the same in R and S house flies, except for a somewhat slower penetration of the toxicant in R strain initially (Sellers, 1971). Although metabolism and excretion are slow in both strains, degradation is more rapid and greater quantities of water-soluble metabolites are produced in R flies. These factors alone cannot account for the dynamic resistance; hence, an unidentified mechanism must be operative in the R strain.

It was found that certain compounds which are mildly toxic to S flies are markedly synergized by sesamex but the synergist has little effect on the toxicity to R flies even though sesamex inhibits the metabolism of these compounds to the same extent in both strains. The synergistic effect of sesamex implies the inhibition of oxidative metabolism of these compounds as was earlier postulated by Sun and Johnson (1960). Obviously, oxidative detoxication of these compounds is not related to the resistance mechanism resulting from selection of house flies with dieldrin since sesamex does not enhance their toxicity to the resistant strain.

ii. Metobolism in Vitro. Metabolism of the cyclodiene insecticides *in vitro* is mediated by the microsomal mixed- function oxidase enzyme system requiring NADPH (reduced nicotinamide adenine dinucleotide phosphate) and atmospheric oxygen. Microsomal epoxidation of cyclodiene compounds has been investigated in several insects. Ray (1967) showed the epoxidation of aldrin by house fly microsomes and its inhibition by carbon monoxide. Lewis *et al.* (1967) studied the relationship between microsomal epoxidation of aldrin and lipid peroxidation and showed that the house fly contains an endogenous inhibitor of lipid peroxidation. Khan and Terriere (1968) and Khan (1969) examined the *in vitro* epoxidation in various genetically distinct strains of house flies and found that the same enzyme system also hydroxylates naphthalene, as was determined earlier by Schonbrod *et al.* (1968). Both reactions are carried by genes located on the second chromosome which confer resistance to a variety of insecticides.

Aldrin epoxidase, in the southern armyworm *Prodenia eridania* in the corn earworm *Heliothis zea,* and the polyphemus moth *Antherea polyphemus,* occurs chiefly in the alimentary canal and fat- body, whereas in the house fly it occurs mainly in the abdomen. In sawfly larvae *Macremphytus varianus,* aldrin epoxidase and dihydroisodrin

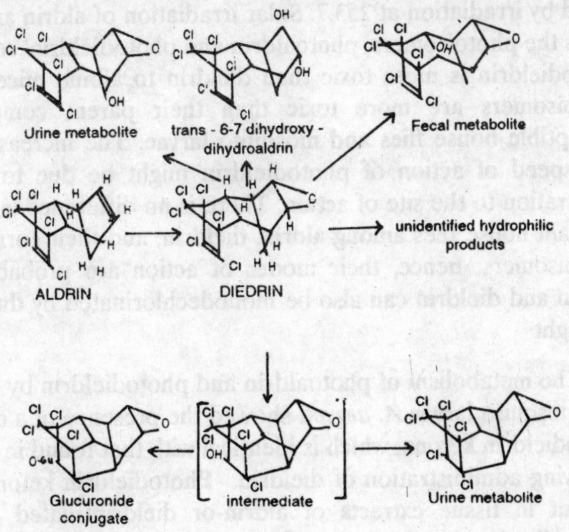

Fig. 10.8. Metabolism of aldrin and dieldrin by insects and mammals.

hydroxylase occur mostly in gut tissue in the process of active feeding, while no activity is found in fully mature harvae.

Among larvae of several strains of five lepidopterous species, i.e., the corn earworm, *H. zea,* tobacco budworm. *H. virescens,* pink bollworm, *Pectinophora gossypiella,* European corn borer, *Ostrinia nubilalis,* and the redbanded leaf roller, *A. velutinana,* aldrin epoxidation was highest in *H.zea* and *H. virescens*. There was no difference in epoxidase activity between diapausing and nondiapausing larvae.

The extensive investigations by Brooks and associates on the oxidative metabolism of aldrin, isodrin, dihydroaldrin, dihydroheptachlor, and a number of related, moderately toxic, cyclodiene compounds have been summarized by Brooks (1966, 1969), and the hydration of dieldrin to *trans*-6,7,- dihydroxydihydroaldrin by the house fly has recently been reported. The epoxide hydrase enzymes are also present in the micro- somes but require no NADPH or oxgen for activation.

iii. Photoalteration of Cylodiene Compounds. In addition to metabolism by living organisms and by enzymic reactions, reactions, several of the cyclodiene compounds undergo photolytic transformations, to birdcage and half-cage structures in nature or are decomposed by irradiation at 253.7. Solar irradiation of aldrin and dieldrin yields the photoisomers, photoaldrin and photodieldrin' respectively. Photodieldrin is more toxic than dieldrin to albino mice and both photoisomers are more toxic than their parent compounds to susceptible house flies and mosquito larvae. The increased toxicity and speed of action of photodieldrin might be due to enhanced penetration to the site of action. There is no difference in toxicity to resistant house flies among aldrin, dieldrin, and their corresponding photoisomers; hence, their modes of action are probably similar. Aldrin and dieldrin can also be monodechlorinated by the action of UV light

The metabolism of photoaldrin and photodieldrin by house flies and mosquito larvae *A. aegypti* showed the presence of a metabolite, photodieldrin ketone, which is identical with that found in male mice following administration of dieldrin. Photodieldrin ketone was not present in tissue extracts of aldrin-or dieldrintreated flies. This metabolite is more toxic to S flies and mosquito larvae than either

Table 10.6. Comparative Rates of Microsomal Aldrin Epoxidase Activity in Various Insect Species

Enzyme source	*Dieldrin formed (pmoles/min/mg protein*
Insects	
Musca domestica (whole fly)	245, 41, 197
Musca domestica (female abdomen) (S)[i]	83
Muscadomestica (female abdomen) (R)[i]	713
Phormia regina (Whole fly)	3
Prodenia eridania	2145
Panthea furcilla	513
Trichplasiani	42
Pseudaletta unipuncta	26
Danaus plexippus	8
Antheraea pernyi	147
Hyalophera cecropia	82
Acheta dimesticus (male)[k]	4
Acheta domesticus (female)[k]	12
Heliothis virescens (field strain)	1090
Heliothis virescens (lab.strains)	250,180
Ostrinia rubilalis	50
Argyrotainia velutinana	40
Pectinophera gossypiella	20
Other species	
Rattus norvegicus (male rat liver)	84, 274
(female rat liver)	21
Lepus curiculus (male rabbit liver)	344
Coturnix coturnix (male quail)	108
(female quail)	9
Salmo gairdneri (rainbow trout)	6

photoaldrin or dieldrin. These results suggest that the greater toxicity of the photo- isomers to insects might be due to the rapid formation of photodi- eldrin ketone which also is more rapid in its toxic action. Mosquito larvae, however, do not form photodieldrin from photoaldrin. The possibility of a ketone formation (photoaldrin detone, Fig. 10 has been suggested but not found.

The cyclodiene insecticide endrin is also converted to photo-isomers by irradiation of the deposited solid at 253.7 nm. Two of these metabolites are a ketone and an aldehyde (Fig. 10.9). Photolysis of endrin in hydrocarbon solvents yields a metabolite identified as the half-cage ketone, 1,8-exo-9,11,11- pentachloropentacyclo- dodecan-5-

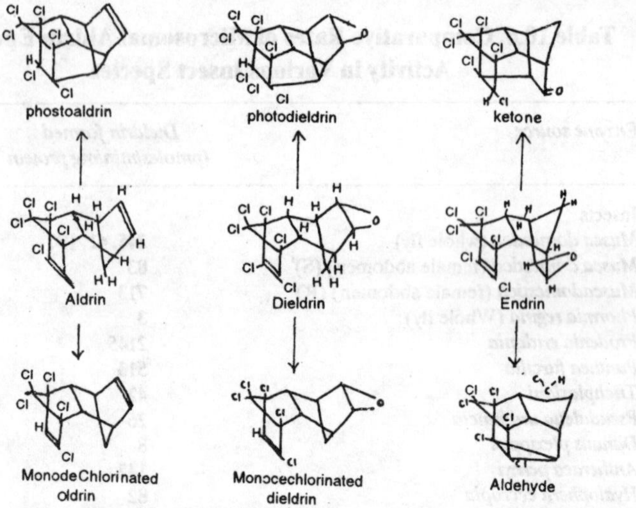

Fig. 10.9. Photodecomposition products of aldrin, dieldrin, and endrin in insects and mammals.

one. This photolytic product of endrin is resistant to chemical degradation and its recent detection in the field might play a role in the buildup of residues in the environment. The toxicity of this compound has not been determined. It is clear from the foregoing discussion that insect resistance to the cyclodiene compounds is independent of detoxication mechanisms whether oxidative, epoxidative, or hydrolytic.

The fact that dieldrin can bind with various components of insect nerve, and that particulate components of resistant cockroaches have less binding capacity with dieldrin than their susceptible counterparts supports the view that this binding property of the nervous system might be related to the poisoning effect and to the resistance mechanism. Deeper probing into nervour tissue of various insects and expanding our knowledge of insect neuropharmacology might yield fruitful results in understanding the mode of action and the mechanism of resistance of the cyclodiene insecticides.

C. Biochemistry of Insect Resistance to Organophosphorus Insecticides

The introduction of organophosphorus (OP) insecticides for control of insect pests had entertained the hope of many entomo-

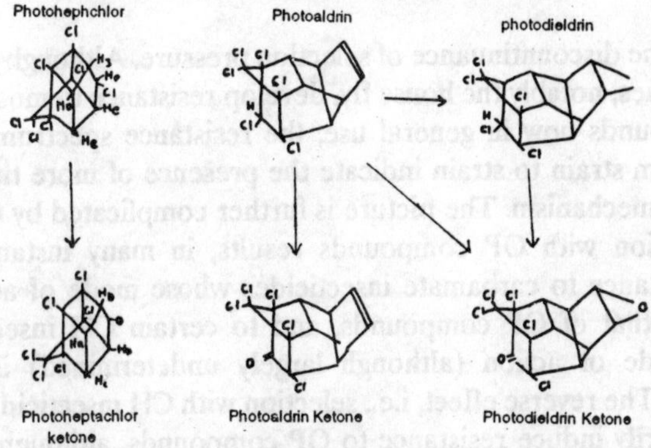

Fig. 10.10. Metabolism of photoaldrin, photodieldrin, and photoheptachlor by insects, mammals and UV ligh. (From Khan *et al.*, 1968.)

logists that the resistance problem generated by the CH insecticides would be overcome. But history has a habit of repeating itself and, despite all our hopes to the contrary, resistance to many of the OP compounds developed rapidly in several insect species. The great achievements in this field of organic chemistry and the dramatic consequences to the science of entomology and pest control stem largely from our more fundamental understaning of the mode of action of these compounds *vis-a-vis* the chlorohydrocarbon insecticides.

To be effective as an insecticide, OP compound must first possess sufficient stability and suitable physicochemical properties to be absorbed and transported to the site of action. A second prerequiste is that the compound must have nucleophilic properties to act as a reactive phosphorylating agent. When these conditions are met, the OP compound will, at the site of action, bind to and inhibit a vital enzyme, cholinesterase.although other esterases have been implicated in the poisoning process 1960 and other modes of action have been suggested.

In contrast to resistance to chlorohydrocarbon (CH) insecticides, OP resistance is biochemically limited to levels peculiar to each particular insecticide, beyond which the resistance is not ordinarily augmented by further selection with the same compound. It is more specific and cross-tolerance to other OP compounds varies considerably from species to species and within strains of a single species. Furthermore, OP resistance is relatively unstable, declining rapidly following the discountinuance of selection pressure. Although various insect species, notably the house fly, develop resistance to most of the OP compounds now in general use, the resistance spectrum varies greatly from strain to strain indicate the presence of more than one resistance mechanism. The picture is further complicated by the fact that selection with OP compounds results, in many instances, in cross-resistance to carbamate insecticides whose mode of action is similar to that of OP compounds, and to certain CH insecticides whose mode of action (although largely undetermined) is quite dissimilar. The reverse effect, i.e., selection with CH insecticides does not ordinarily induce resistance to OP compounds, although recent investigations indicate that where genes for oxidative metabolism of DDT and pyrethrins are present, increased tolerance to OP and carbamate insecticides is quite common. Physiological resistance to OP compounds is characterized by the degree of inhibition of cholinesterase and/or aliesterase enzymes, and by differences in rates of activation and detoxication of the OP compound. In some instances, OP reesistance has been correlated with behavioristic avoidance of the insecticide, such as avoidance of malathion baits but not trichlorfon or dichlorvos baits with lipid content of tarsi and thoracic ganglia or total body lipid, and with rate of cuticular penetration of the OP compound.

In the living insect, the insecticide is subject to the interaction of many or all of the above mechanisme, i.e., penetration, tissue distribution, storage, activation, detoxication, and excretion. In general, death or survival of the organism will depend upon the extent of these interrelationships and the contribution of each factor to the total defense mechanism. However, in some instances, survival may be determined by a balance between rates of activation and detoxication, insensitivity of the OP receptor at the site action, or the production of altered specific detoxifying enzymes.

Metabolic Fate of OP Insecticides. The metabolism of most OP compounds follows two general types of reactivity: (a) metabolic activation of slightly or moderately toxic compounds to more active anticholinesterases, and (b) metabolic degradation leading, in most cases, to relatively innocuous derivatives. Generally, activation reactions are oxidative in nature and are catalyzed by microsomal mixed- function oxidases which reside in the endoplasmic reticulum of the cell and require NADPH and O_2 for their activity. Detoxication mechanisms may involve both oxidative reaction catalyzed by microsomal mixed-function oxidases, and hydrolytic pathways mediated by esterases of which the commonest types are the phosphatases and carboxylesterases (aliesterases).

Metabolic activation is accomplished by desulfuration of esters containing P=S groupings to form the corresponding P=O derivatives, such as occur in the conversion of malath on, parathion, methyl parathion, diazinon, coumaphos, dimethoate, Imidan, azinphomethyl, fenthion, etc., to their corresponding phosphates (oxon derivatives ; oxidation of aliphatic or aromatic thioether

$-S-$ to sulfoxide $-\overset{\overset{O}{\|}}{S}-$ and sulfone $-\overset{\overset{O}{\|}}{\underset{\underset{O}{\|}}{S}}-$ as in demeton, disulfoton, carbophenothion, phorate, etc.

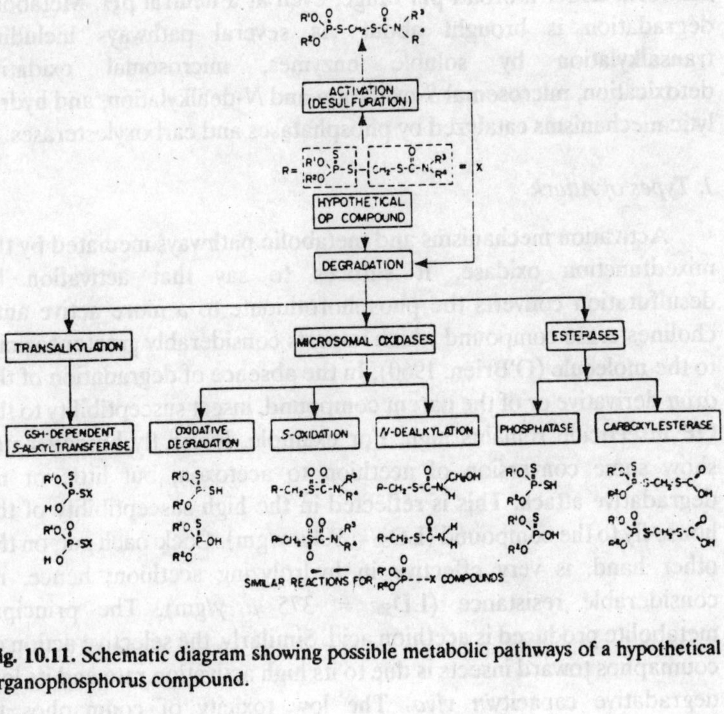

Fig. 10.11. Schematic diagram showing possible metabolic pathways of a hypothetical organophosphorus compound.

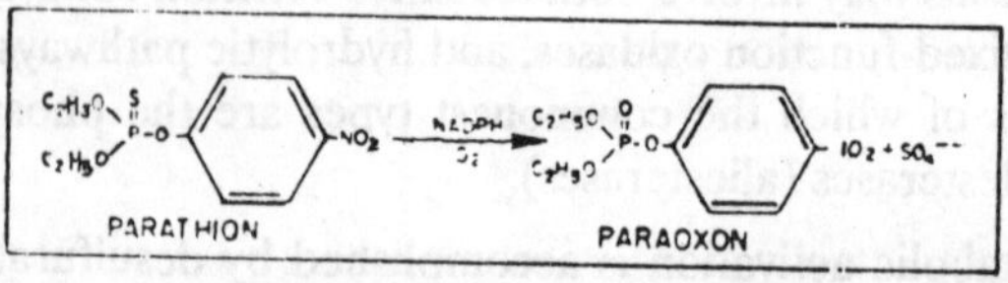

Fig. 10.12. Activation. Desulfuration of parathion. (From Nakatsugawa *et al.*, 1969.)

(although in some instances thioether oxidation may result in products of lower toxicity), and *N*-oxidation of phosphoramides as in the weakly cholinergic compounds schradan and dimefox.

Other types of activation also occur. For example, trichlorfon is converted to the more active anticholinesterase dichlorvos by dehydrochlorination and rearrangement, a process which does not involve the enzyme DDT dehydrochlorinase. However, trichlorfon is enzymatically converted to diclorvos by the digestive juice of the silkworm under a broad pH range, even at a neutral pH. Metabolic degradation is brought about via several pathways including transalkylation by soluble enzymes, microsomal oxidative detoxication, microsomal *S*-oxidation and *N*-dealkylation, and hydrolytic mechanisms catalyzed by phosphatases and carboxylesterases.

1. Types of Attack

Activation mechanisms and metabolic pathways mediated by the mixedfunction oxidase. It suffices to say that activation by desulfuration converts the phosphorothioate to a more active anticholinesterase compound which imparts considerably greater toxicity to the molecule (O'Brien, 1960). In the absence of degradation of the *oxon* derivative or of the parent compound, insect susceptibility to the OP insecticide remains high. For example, house fly homogenates show some conversion of acethion to acetoxon, but little or no degradative attack. This is reflected in the high susceptibility of the house fly to the compound ($LD_{50} = 9.4 \mu \gamma/gm$). Cockroach gut, on the other hand, is very effective in hydrolyzing acethion; hence, its considerable resistance ($LD_{50} = 375\ \mu\ \gamma/gm$). The principal metabolite produced is acethion acid. Similarly, the selective action of coumaphos toward insects is due to its high activation rate and its low degradative capacity*in vivo*. The low toxicity of coumaphos to mammals is due to hydrolytic attack at the P – O coumarynil and the rapid excretion of polar metabolites.

The in vivo activation of parathion to paraoxon has been demonstrated by Metcalf and March (1953). Investigations *in vitro* with microsomal preparations from the rich stem borer cockroach fat-body , and house fly abdomens showed that desulfuration of parathion is accomplished by a mixed-function oxidase. The sulfur atom which is removed in the *in vitro* reaction is bound to the microsomes; *in vivo* it is found as inorganic sulfate. In common with other microsomal oxidases, the activating enzyme exhibits a high degree of nonspecificity and can catalyze the desulfuration of a great number of phosphorothioate insecticides.

2. Degradation

a. Transalkylation. Conjugation reactions utilizing glutathione are believed to be an important mechanism of detoxication of organo-phosphorus compounds since they yield primary metabolites. The conversion of methyl parathion, methyl paraoxon, and fenitrothion the their respective demethyl derivatives was shown to reside in the supernatant fraction and to require reduced glutathione for activation. While deethylation of ethyl parathion proceeds much slower in both insect and mammal. However, Nolan and O'Brien (1970) made an unexpected observation that deethylation of parathion was a major metabolic pathway in the susceptible rather

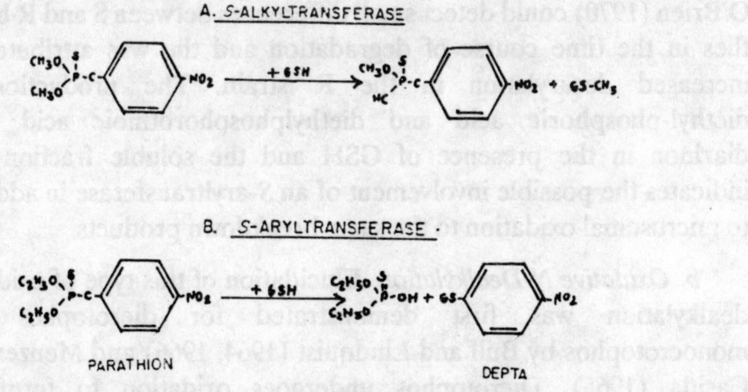

Fig. 10.13. *O*-Dealkylation. (A) Dealkylation of methyl parathion. (From Fukami and Shishido, 1963b, 1966.) (B) Dearylation of parathion. DEPTA, diethylphosphorothioic acid. (From Dahm, 1970.)

than in the resistant house fly. Enhanced dealkylation has been implicated in the mechanism of resistance to parathion and paraoxon in the rich stem borer and Fukunaga *et al.* (1969) described the glutathione transferase activity in the horn beetle and silkworm larvae. Other OP compounds also undergo dealkylation reactions. Thus, *O*-dealkylation occurs with dicrotophos mevinphos fenitrothion and methyl paraoxon dimethoate bromophos and diazinon. In the house fly, demethylation of methyl parathion and fenitrothion is an important detoxication mechanism and proceeds faster in the R strain, but its relative importance increases only slightly with an increase in dosage, apparently due to saturation of other detoxifying reactions. Conceivably, many more OP compounds can undergo dealkylation reactions of this type, but a purposeful search for this degradative process has not been applied in many instances.

The examples cited above indicate that transalkylation reactions require reduced glutathione, that the enzyme resides in the soluble fractions and, most likely, belongs to the group of glutathione *S*-alkyl transferases (GSAT). The substrate specificity for this enzyme favours dimethyl esters of both phosphorothioates and phosphates and yields *S*-alkyl glutathione and mono-*O*-dealkylated derivatives.

Glutathione transferase reactions are also responsible for the transfer of aryl groups. Dahm (1970) reports that the P–O–aryl bond of parathions is cleaved by a soluble enzyme requiring glutathione. At low doses of injected [^{3}H] paraoxon, Nolan and O'Brien (1970) could detect small differences between S and R house flies in the time course of degradation and the was attributed to increased desarylation in the R strain. The production of diethyl-phosphoric acid and diethylphosphorothioic acid from diazinon in the presence of GSH and the soluble fraction also indicates the possible involvement of an *S*-aryltransferase in addition to microsomal oxidation to the same breakdown products.

b. Oxidative N-Dealkylation. Elucidation of this type of oxidative dealkylation was first demonstrated for dicrotophos and monocrotophos by Bull and Lindquist (1964, 1966) and Menzer and Casida (1965). Dicrotophos undergoes oxidation to form the *N*-hydroxy- methyl derivative with subsequent loss of the hydroxymethyl group to yield monocrotophos. The latter is oxidized to its *N*-hydroxymethyl derivative and loss of this group yields the

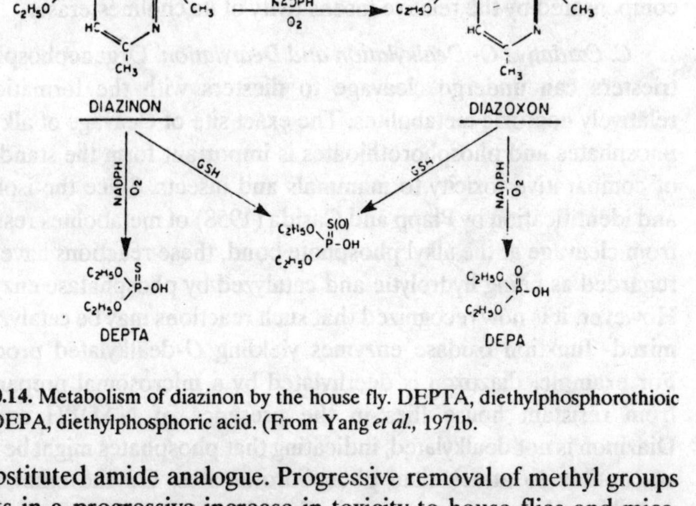

Fig. 10.14. Metabolism of diazinon by the house fly. DEPTA, diethylphosphorothioic acid; DEPA, diethylphosphoric acid. (From Yang *et al.*, 1971b.

unsubstituted amide analogue. Progressive removal of methyl groups results in a progressive increase in toxicity to house flies and mice. Phosphamidon and dimethoate undergo similar dealkylation processes, but no dealkylation products could be detected in dimethoate metabolism by the boll weevil and bollworm. *N*-Demethylation of famphur has been shown in cockroaches and

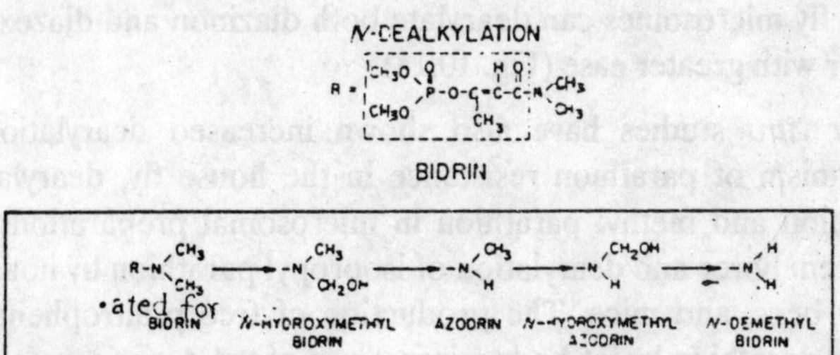

	BIDRIN	N-HYDROXYMETHYL BIDRIN	AZODRIN	N-HYDROXYMETHYL AZODRIN	N-DEMETHYL BIDRIN
TOPICAL LD_{50} mg/kg ♀ HOUSEFLY					
−sesomex	38	14	6.4	30	1 0
+sesomex	1.0	1.2	0.8	3 4	0 9
I.P LD_{50} mg/kg ♀ Mouse	14	18	8	12	3

Fig. 10.15 *N*-Dealkylation. Metabolism and biological activity of Bidrin (dicrotophos)' Azodrin (monocrotophos), and their metabolites. (From Menzer and Casida, 1965.)

milkweed bugs, but the slow rate of degradation in the latter is well compensated by the relative insensitivity of its cholinesterase.

C. Oxidative O- Dealkylation and Dearylation. Organophosphorus triesters can undergo cleavage to diesters with the formation of relatively nontoxic metabolites. The exact site of cleavage of alkylaryl phosphates and phosphorothioates is important form the standpoint of comparative toxicity to mammals and insects. Since the isolation and identification by Plapp and Casida (1958) of metabolites resulting from cleavage at the alkyl phosphate bond, these reactions have been regarded as being hydrolytic and catalyzed by phosphatase enzymes. However, it is now recognized that such reactions may be catalyzed by mixed- funxtion oxidase enzymes yielding *O*-dealkylated products. For example, diazoxon is deethylated by a microsomal preparation from resistant house flies in the presence of NADPH and O_2. Diazinon is not dealkylated, indicating that phosphates might be more susceptible to attack than phosphorothioates. On the other hand, parathion, but not paraoxon is dearylated by microsomal preparations from cockroach fat-body yielding diethyl phosphorothioate and p-nitrophenol. This reaction is inhibited by piperonyl butoxide, giving supporting evidence of tghe involvement of the microsomal enzyme system. Similarly, microsomal preparations from rat liver oxidatively dearylate parathion and diazinon, but not diazoxon wereas house fly microsomes can dearylate both diazinon and diazoxon the former with greater ease (Fig. 10.14).

In vitro studies have also shown increased dearylation as a mechanism of parathion resistance in the house fly, dearylation of parathion and methyl parathion in microsomal preparations of the rice stem borer and dearylation of isopropyl parathion by house flies, honey bees, and mice. The production of free p-nitrophenol from methyl parathion by fat-body microsomes of the American cockroach most likely is due to oxidative dearylation rather than to hydrolysis as shown in the oxidative metabolism of [^{35}S]parathion to diethylphosphorothioic acid (DEPTA) and its inhibition by sesamex. The above examples indicate that cleavage of alkyl or aryl bonds by the mixed-function oxidae system is an important detoxication mechanism in insects and mammals.

d. Sulfur Oxidation. The systemic insecticide demeton consists of two isomers, the thiono and thiol isomers. Both isomers are rapidly metabolized by the mouse and the American cockroach. The biochemical mechanism of primary importance from the toxicological standpoint involves the oxidation of the mercaptosulfur of the ethylthioethl moiety of both isomers, first to the sulfoxide and then to the sulfone. A second important mechanism which involves only the thiono isomer is the oxidation of the thionosulfur to the corresponding phosphate with subsequent oxidation to the sulfoxide and sulfone. A similar pattern occurs with phorate whose chief metabolic products in *P. americana, O. fasciatus,* and *Rohdnius prolixus* are the sulfoxide and/or the sulfone. However, there is no evidence of desulfuration activity as in demeton. In contrast, the southern armyworm *P. eridania* rapidly oxidizes the sulfoxide and sulfone to their corresponding phosphates. The systemic insecticide fensulfothion is metabolized *in vitro* by cockroach gut to the sulfone, *S*-ethyl isomer, and *S*-ethyl isomer sulfone. Disulfoton is metabolized by boll weevils and bollworms to four oxidative products including the sulfoxide and sulfone of disulfoton and isulfoxon. But no free disulfoxon could be detected. In addition, nine hydrolytic products are produced.

Fenthion and fenoxon are oxidized to their respective sulfoxides and to a lesser extent to their sulfones by the German cockroach, the boll weevil, and the house fly. Similar products are found in *C.π.*quinquefasciatus and the presence of dimethyl phosphorothioic acid and dimethyl phosphoric acid are indicative of oxidative degradation rather than thionase and oxonase activity. In *S.calcitrans* and *C. lectularius,* fenthion is oxidized at the thiophosphoryl and

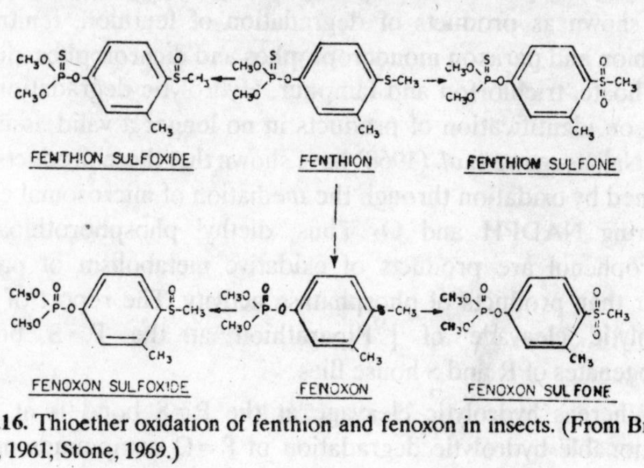

Fig. 10.16. Thioether oxidation of fenthion and fenoxon in insects. (From Braby and Arthur, 1961; Stone, 1969.)

thioether positions, the latter producing the sulfone as the predominant metabolite with subsequent hydrolysis and excretion as cresol conjugates. Oxidation of the thioether moiety is more rapid following activation of fenthion to fenoxon. All the examples discussed above are studies of thioether oxidation *in vivo*. As yet, no data are available on the *in vitro S*-oxidation in insects. However, by analogy with similar *in vitro* metabolic pathways with carbamate insecticides, one can assume that such reactions with OP compounds also are catalyzed by microsomal mixed-function oxidases.

3. Hydrolytic Reactions

a. Phosphatases. Historically, and until recently, degradation of insecticides was though to occur primarily by hydrolytic routes with the commonest type being attack by phosphatases (Health, 1961; O'Brien, 1960, (1967). These enzymes hydrolyze the phosphorus ester or the anhydride bond at P – O – C, P – S – C, P – F linkages, etc., and yield hydrolytic meta- bolites of the leaving group (the side chain). The weight of evidence suggests that phosphates rather than phosphorothioates are preferred substrates for enzymic hydrolysis. Since the isolation and identification by Plapp and Casida (1958) of hydrolytic products of ronnel, dicapthon, chlorthion, methyl parathion, parathion, and diazinon from the American cockroach, numerous other investigators have reported such hydrolytic products with other insects and with diverse OP compounds. In many instances, resistance has been attributed to more rapid hydrolysis and excretion of conjugated products of hydrolysis. Hydrolytic metabolites have been shown as products of degradation of fenthion, fenitrothion, parathion and paraxon monocroptophos and dichrotophos, diazinon, dimethoate, trichlorfon and famphur. Hydrolytic degradation based solely on identification of products in no longer a valid assumption since Nakatsugawa *et al.* (1968) have shown that these products can be obtained by oxidation through the mediation of microsomal enzymes requiring NADPH and O_2 Thus, diethyl phosphorothioate and p-nitrophenol are products of oxidative metabolism of parathion rather than products of phosphatase activity. The report of *in vitro* hydrolytic cleavage of [^{32}P]parathion at the P = S bond by homogenates of R and S house flies.

Whereas hydrolytic cleavage at the P = S bond is at present questionable hydrolytic degradation of P = O compounds might be more of a reality. Stemming from the work of Welling *et al.* (1971), it appears that enhanced hydrolytic action leading to the formation of diethylphosphate and p-nitrophenol may account for the resistance to paraoxon in a strain of house flies, in accord with previous reports, but that an additional resistance factor due to increased oxidative activity may be present. Diazinon resistance has been the subject of numerous investigations and some claims of enhanced hydrolytic activity as a mechanism of resistance have been previously noted. Investigations at the subcellular level indicate the presence of at least three detoxication mechanisms in several, genetically marked, strains of house flies: (1) cleavage of diazinon and diazoxon to diethylphosphorothioic acid and diethylphosphric acid, respectively—a mechanism which resides in the microsomes and requires NADPH and O_2 (curiously, the addition of glutathione also enhances this reaction). Degradation by this route is significantly higher in the resistant strain and the reaction is inhibited by sesamex, an inhibitor of microsomal oxidases, but not by *S,S,S*-tributyl phosphorotrithioate, (TBTP) an inhibitor of aliesterases; (2) deethylation (transalkylation) of diazinon and diazoxon occurs entirely in the soluble fraction of strains with gene *a* (low aliesterase) and requires reduced glutathione for activation, with further decomposition to monoethyl phosphorothionate and monoethyl phosphate. Deethylation is inhibited by TBTP but not by sesamex. This reaction does not occur in rat liver preparations and (3) Microsomal degradation of diazoxon but not of diazinon to two unidentified derivatives. This mechanism is also inhibited by sesamex but not by TBTP (Table 10.7). The overall metabolism of diazinon and diazoxon by the soluble fraction (GSH) and the microsomal fraction (NADPH, O_2) is at least four- to fivefold greater in R than in S strains of house flies. It appears that diazinon resistance is a multifacet phenomenon with varied metabolic patterns and related to the genetic background of the insect. A new product of diazinon metabolism, i.e., hydroxydiazinon, has recently been found in kale and in guinea pigs and sheep this metabolite has not been reported in insects.

b. Carboxylesterases (Aliesterases). It is generally agreed that there are practically no differences between OP- resistant and OP-suceptible insects in their cholinesterase activity or in the sensitivity of this enzyme to inhibition by OP-compounds. Some observations with resistant house flies drew attention to the possible role of aliesterases, such as enzymes hydrolyzing methyl or phenyl butyrate, in OP-resistance. Subsequent work showed that homogenates of six resistant strains of different geographical origin had considerable lower aliesterase activity than their susceptible counterparts.

TABLE 10.7
Pathways of Diazinon Metabolism in Relation to Genotype Composition of Susceptible and Resistant Strains of House Flies

Strain	*Degree of resistance*	*Type of resistance*	*Located chromosome*	*Detoxication mechanism*[b]
ocra SRS	1	None (susceptible)	–	A
ac; ar; bwb, ocra SRS	1	None (susceptible)	–	A
29	10-15	Gene *a* (low aliesterase)	II	A B
393	10-15	Gene *a* (low aliesterase)	II	A B
466.500	10	Sesamex-inhibited	V	A C
SKA	400	Gene *a* (low aliesterase)	II	A B C
		+ Penetration factor	III	
		+ sesamex-inhibited	V	

b A, cleavage of diazinon and diazoxon to diethylphosphorothioic acid and diethylphosphoric acid, occurring in microsomal fraction. B, deethylation of diazinon and diazoxon (GSH-dependent *S*-alkyltransferase), occurring in soluble fraction. C, degradation of diazoxon but not of diazinon to unidentified metabolites, occurring in microsomal fraction.

There are instances, however, where aliesterase activity is not correlated with resistance in house flies. For instance, Oppenoorth and van Asperen (1961) found an OP-resistant strain with relatively high aliesterase activity, and Franco and Oppenoorth (1962) found a susceptible strain with esterase levels equal to those of the resistant strains. In contrast with previous reports on house flies, lower levels of aliesterase activity are not found in OP-resistant strains of either *Culex tarsalis* or *A. aegypti*. Aliesterase activity in *D. melanogaster* bears no relation to insecticide resistance (Ogita, 1961), and in the green rice leafhopper Hayashi and Hayakawa (1962), Kojima et al., (1963), and Kasai and Ogita (1965) found more aliesterase activity in malathion-R strains than in their S-counterparts. Taken collectively, these and other reports indicate that OP-resistance is not always associated with low aliesterase activity.

Oppenoorth and van Asperen (1960) then proposed the "mutant aliesterase" theory, which postulates the existence, in many OP-resistant strains of house flies, of a mutant allele *a* whose function is to produce a modified aliesterase, or OP-hydrolyzing enzyme, replacing the normal aliesterase in the susceptible strain which is under control of the wild-type gene *a*. This theory has been criticized by O' Brien (1966, 1967) on grounds that the rate of detoxication found is insufficient to account for the high degree of resistance, and that enzymztic action was not proven due to lack of indentification of the hydrolytic products. In addition, the finding of Stegwee (1959) that the aliesterase of house flies could be inhibited by as much as 81% without causing ill effects strengthens the conclusion that aliesterases are of little consequence in the poisoning by OP compounds. In spite of the uncertainties concerning the role of these enzymes, carboxylesterases have been shown to be important in the detoxication of OP compounds. In particular, one of the most studied insecticides, malathion, has attrachted considerable attention because of its low mammalian toxicity and its high insecticidal activity. The rates of metabolic attack upon different parts of the molecule vary with different organisms. Thus, the mouse and chicken degrade malathion to several metabolic products, whereas degradation is less extensive in the American cockroach. Activation to malaoxon occurs in both insects and mammals, but the balance of activation and hydrolytic activities accounts satisfactorily for the greater toxicity of malathion to the cockroach.

Degradation of malathion involves at least two major pathways; that involving attack on P–S and S–C bonds, and that concerned with carboxylesterase hydrolysis of the diethyl succinate moiety. Oxidation to malaoxon proceeds at about equal rates in R and S house flies, but the resistant strain degrades malaoxon more rapidly via phosphatase action. In *Culex tarsalis* a more important difference between R and S larvae is the much higher carboxylesterase activity of the R strain a characteristic which is genetically controlled, EPN, a carboxylesterase inhibitor, synergizes malathion against R larvae, thus establishing the importance of carboxylesterase activity. The carboxylesterase of *Culex* was later partially purified and found to be a low molecular weight protein of 16,000. Since the malathion-R strain cannot hydrolyze carboxymethyl malathion (which contains $COOCH_3$ in place of $COOC_2H_5$) it may be inferred that the carboxylesterase is

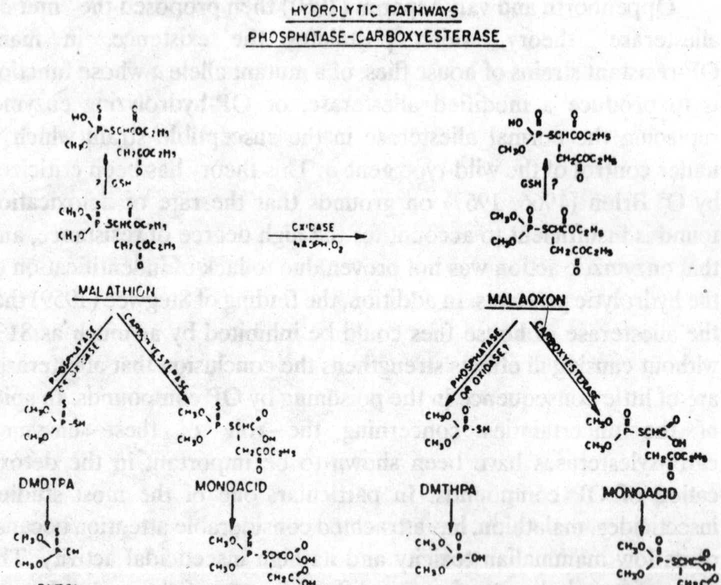

Fig. 10.17. Carboxyesterase and phosphatase. Metabolic pathways of malathion in insects. DMDTPA, dimethyldithiophosphoric acid; DMPTA, dimethylphosphorothioic acid; DMTHPA, dimethyl phosphorothiolate; DMPA, dimethylphosphoric acid.

highly specific for carboethoxy groups, a specificity also seen in malathion-resistant German cockroaches. Like *Culex* and house flies, malathion-R cockroaches show little cross-resistance to other OP insecicides.

Detoxication ractions involving carboxylester hydrolysis yield the monoacid derivative of malathion. The enzyme is very active in mammals. The main interstrain difference between malathion-R and malathion-S house flies or blow flies *Chrysomya putoria* is the superior ability of the R strains to degrade malathion to the monocarboxylic acid.

Of particular interest is the difference in malathion tolerance in two species of *Dermestes* beetles. *Dermestes maculatus* is much more resistant to malathion than *D. lardarius,* but this resistance does not extend to a number of other OP compound investigated. It appears that enhanced degradation via carboxy- lesterase action is the likely mechanism of resistance involved in *D. maculatus* and in *Tribolium castaneum*. In both works, triphenyl phosphate synergized malathion by inhibiting carboxylesterase but not phosphatase activity. Electrophoretic evidence implies the existence of at least two malathion-degrading enzymes, one of which is characteristic of the resistant insect only, and an arylesterase band has been detected in A. aegypti which is not inhibited by paraoxon but instead hydrolyzes it. This band is more intense in parathion selected than in malathion-selected strains. Whereas most studies implicate a higher hydrolytic activity as a cause of malathion resistance, some investigators find it difficult to reconcile the small differences in oxidative activity or in hydrolytic breakdown with the high malathion tolerance of the R strains. In *A. aegypti*, for example, resistance to malathion is attributed to decreased penetration and enhanced excretion of the toxicant rather than to increased hydrolysis. Since malathion is used extensively for control of agricultural pests, it is of interest to note that this insecticide is also metabolized by plants. Isolation of degradation products from plants include the monoacid derivative, indicating the presence of a carboxylesterase enzyme.

c. Amicases. Carboxymidases also participate in the degradation of OP compounds containing a carboxyamide group, $CONR_2$ (where R is H or an alkyl group), although this type of hydrolytic cleavage has been studied more extensively in animals and plants than in insects. Among the OP compounds which are susectible to this type of attach in insects, dimethoate takes a leading role.Three factors appear to be involved in the considerable tolerance of fifth, instar bollworm larvae for dimethoate: (a) large amounts of dimethoate are converted to the oxygen analogue; (b) the latter and the parent compound are rapidly hydrolyzed to nontoxic derivatives; and (c) excretion of metabolic products is very efficient. The major site of enxymatic attack appears to be at the amide bond yielding thiocarboxy derivatives (*O,O*-dimethyl-*S*- carboxy- methyl phosphorodithioate). In addition, cleavage of the S – C bond of either dimethoate or the thiocarboxy derivative yields large amounts of dithioate (*O,O*- dimethylphosphorothioic acid). Other metabolites of interest are thioate (O,O-dimethylphosphorothioic acid) and dimethyl phosphate. The dimethylthiocarboxy derivative (O-methyl- *O*-hydrogen- *S*-carboxymethyl phosphorodithioate) which is a major metabolite in mammals is either absent or is detected only in minute quantities in bollworm larvae. Metabolism in boll weevils is similar to but considerably slower than in the bollworm.

In contrast, a comparative study of dimethoate metabolism in several insects indicates that total hydrolytic degradation does not correlate with sensitivity of the insect to the toxicant. Thus, the house fly, which is the most susceptible among the species tested, degrades dimethoate at the slowest rate, but the same slow rate of degradation takes place in the extremely tolerant milk- weed but. Rate of degradation of dimethoxon (the P=O analogue), as well as the small amounts of carboxyamidase products recovered, also show no correlation with toxicity. The greater susceptibility of the house fly can best be explained in terms of its much greater cholinesterase sensitivity (100-fold) to dimethoate. Carboxyamidase activity appears to be totally absent in the pattern of dimethoate metabolism by homogenates of the house fly, American cockroach, rice borer, and green peach aphid, and degradation by phosphatase action is greatest by cockroach fat-body.

In cotton leafworm larvae, a major site of enzymatic attack on dimethoate occurs at the P–O–alkylester bonds followed by hydro-

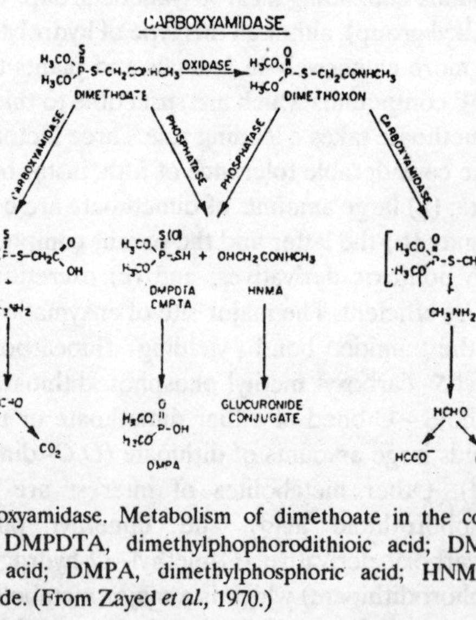

Fig. 10.18. Carboxyamidase. Metabolism of dimethoate in the cotton leaf worm *Prodenia litura*. DMPDTA, dimethylphophorodithioic acid; DMPTA, dimethyl phosphorothioic acid; DMPA, dimethylphosphoric acid; HNMA, α- hydroxy-*N*-methyl acetamide. (From Zayed *et al.*, 1970.)

lysis of the P – S bond to yield thiophosphoric acid. Furthermore, the production of *O,O*- dimethyl-*S*-carboxy- methyl phosphorodithioate as a metabolite indicates the presence of carboxyamidase activity. In a later study with the cotton leafworm using ^{14}C-labeled dimethoate at both – OCH_3 groups or at the – $NHCH_3$ group, zayed *et al.* (1970) showed the presence of several oxidative and hydrolytic pathways, as well as products of carboxyamidase activity. The major polar metabolite resulting from phosphatase attack at the S – C bond is believed to be α-hydroxy-*N*-methyl acetamide ($HO - CH_2 - CONHCH_3$) which is excreted as a glucuronide conjugate.

An interesting amidase cleavage occurs with Imidan. Mammals, insects, and plants metabolize Imidan *in vivo* via hydrolytic pathways principally to phthalamic acid and phthalic acid. Ring hydroxylation is either absent or present in trace amount, Imidoxon (the P = O analogue) is a product of microsomal enzyme metabolism of Imidan in the cockroach and in the house fly. Metabolism of the phosphate-bearing moiety has not been followed, but it is likely to involve microsomal oxidase attack yielding *O,O-dimethylphosphorodithioic acid acid.*

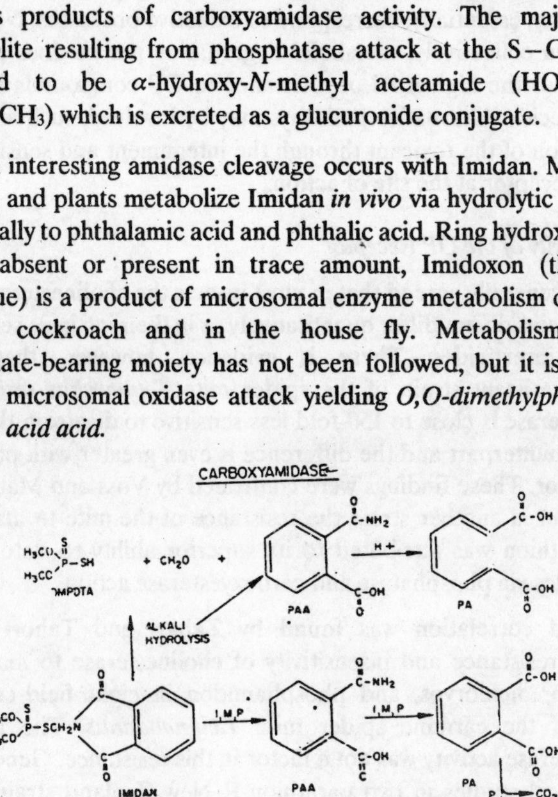

Fig. 10.19. Carboxyamidase. Hydrolytic cleavage of Imidan by amidases of insects (I), mammals (M), and plants (P). PAA, phthalamic acid; PA, phthalic acid; BA, benzoic acid; DMPDTA, dimethylphosphorodithioic acid. (From McBain *et al.*, 1968.)

To recapitulate, it appears that, in general, the mechanisms responsible for insect resistance to OP compounds involve microsomal oxidations, dealkylation through mediation of glutathione, and in some instances degradation by hydrolytic pathways such as phosphatases, carboxylesterases and amidases. However, hydrolytic reactions are subject to reinvestigation and reintepretation inasmuch as they may, essentially, correspond to oxidative processes. Nevertheless, taken collectively, metabolic degradation per se does not jully account for the high level of resistance to OP compounds in many insect species. Alternative pathways may be involved, such as rate of penetration of the toxicant through the integument and sensitivity of the OP receptor at the site of action.

4. Sensitivity of the OP Receptor

It is generally agreed that in most insects the cholinesterases of S and R strains do not differ quantitatively or in their intrinsic sensitivity to OP insecticides. There is evidence, however, that in a demetonresistant strain of the spider mite *Tetranychus urticae* the cholinesterase is close to 150-fold less sensitive to diazoxon than that of its S counterpart and the difference is even greater with paraoxon as inhibitor. These findings were confirmed by Voss and Matsumura (1964), but in another strain the resistance of the mite to malathion and parathion was attributed to its superior ability to detoxify the insecticides via phosphatase and carboxyesterase action.

Good correlation was found by Zahavi and Tahori (1970) between resistance and insensitivity of cholinesterase to malaoxon, malathion, dichlorvos, and phosphamidon in eight field-collected strains of the carmine spider mite *T.cinnabarinus*. The level of cholinesterase activity was not a factor in this resistance. Genetic and biochemical studies in two parathion-R New Zealand strains of *T. urticae* and the Leverkusen strain imply that parathion resistance due to lowered cholinesterase sensitivity is governed by a single gene inheritance completely dominant over its normal allele. The resistance gene appears to be involved in determining a part or the whole structure of the aberrant cholinesterase enzyme which has a decreased sensitivity to OP inhibitors.

The strong OP resistance developed by parathion selection in a susceptible strain of *T. urticae* proved to be due to a completely dominant gene, but the cholinesterase activity of the R strain and its sensitivity to malaoxon inhibition was no different from that of the parent S strain. The selection process resulted in increased detoxication since homogenates of the R strain were considerably more active in converting malathion to water-soluble metabolites. Apparently, the gene for cholinesterase insensivity was absent in that particular strain of mites or else the selecting agent was instrumental in determining the resistance characteristics of the progenies.

Mechanisms of OP resistance in some strains of the cattle tick *Boophilus microplus* also appear to be related to decreased cholinesterase sensitivity. Thus, Lee and Batham (1966) demonstrated that the R strain possessed at least one chlolinesterase which reacted more slowly with OP and carbamate compounds than did the cholinesterase of the S strain. This resistant enzyme accounted for 40% of the total activity and its susceptibility to inhibition varied with the type of inhibitor, the greatest difference occurring with the P = O analogue of carbophenothion. A strain of the blue tick which was resistant to arsenic and to chlorohydrocarbon insecticides had normal levels of cholinesterase and the enzyme was fully susceptible to the OP inhibitors.

Roulston *et al.* (1966) demonstrated that a susceptible strain of the blue tick metabolized coumaphos in the usual pattern of oxidation and hydrolysis, and that cholinesterase was the susceptible target. In a similar study with R and S strains treated with coumaphos, diazinon, and dioxathion, Schuntner *et al.* (1968) found no interstain differences in rates of penetration, oxidation to toxic metabolites, or hydrolytic reactions, even though the R strain was 12.4-fold more resistant to diazinon. However, there was a slower *in vivo* response of the R strain cholinesterase to the OP inhibitors, which is consistent with the resistant enzyme hypothesis and parallels the type of cholinesterase insensitivity found in some spider mites. There also are indications of cholinesterase insensitivity in mammals and insects. For example, in comparing the toxicity and metabolism of paraoxon in the frog, mouse, and cockroach, Potter and O'Brien (1963) concluded that the insensitivity of the frog to paraoxon is due to its insensitive cholinesterase and is not connected with detoxication rates. The milkweed bug which degrades famphur at a very slow rate compared to the cockroach and mouse is compen- sated by a relative insensitivity of its cholinesterase to famoxon. Similarly, in an investigation of the penetration and metabolism of dimethoate in several insects, Unchida *et al.*, (1965) found that the cholinesterase of the milkweed bug was 100-fold less sensitive to the inhibitor than that of the house fly.

D. Biochemistry of Insect Resistance to Carbamate Insecticides

The carbamate insecticides share with other classes of toxicants the now inescapable fact that, given sufficient time and selection pressure, resistance to these compounds becomes the rule rather than the exception. Carbamate insecticides differ from OP compounds in that they are competitive rather than irreversible inhibitors of cholinesterase, i.e., the carbamate complexes with the enzyme without necessarily reacting with it chemically Evidence is also available that, in some instances, a chemical reaction occurs yielding a carbamylated enzyme.

Most of the carbamates synthesized to date are patterned after the closely related, pharmacologically active compounds physostigmine (eserine) and prostigmine (neostigmine). According to Winton *et al.* (1958) and Metacalf *et al.* (1960), the requirement for toxic action of a carbamate are "(a) structural complementarity to acetylcholine, the normal substrate for the enzyme cholinesterase, (b) sufficient stability to hydrolytic attack by cholinesterase to permit the compound to act as a competitive blocking agent for cholinesterase rather than as a substrate, and (c) proper lipoid solubility and absence of a permanent electrical charge to permit penetration into the lipoid sheath surrounding the insect nerve." In general, the stronger the ionization of the carbamate the lower is its insecticidal activity even though the compound might be a potent *in vitro* cholinesterase inhibitor. In addition to a specific acetylcholinesterase, nonspecific aliphatic esterase enzymes might be involved in the mode of action of, or resistance to the carbamates, as has been demon strated for the OP compounds.

Many of the commercially available carbamates are highly effective against a variety of resistant insects, but in spite of their effectiveness and their unique mode of action, the carbamates, too, are susceptible to attack by insect enzymes. In laboratory selection experiments with a variety of carbamates a high or low level of resistance was induced in several generations depending on the insect species and the selecting agent. The apparent discrepancies between *in vitro* inhibition of cholinesterase and the toxicity of certain compounds, e.g., Pyrolan, to the American cockroach *P. americana* and the forest maybeetle *Melolontha melolontha*, suggest that enzymatic destruction of the carbamate *in vivo* is responsible for the resistance. Similarly, Georghiou and Metcalf (1961b) reported that 3-isopropylphenyl *N*-methyl carbamate is more rapidly metabolized by R than by S house flies. Carbaryl (Sevin) has been one of the carbamates most widely studied in insects beginning with the work of Eldefrawi and Hoskins (1961). Carbaryl-R and -S house flies metabolize [^{14}C]carbaryl to a polar metabolite which appears in the tissue and the excreta. The difference between the two strains is quantitative rather than qualitative. The mikweed bug *O. fasciatus,* on the other hand, metabolizes carbaryl very slowly, and this is reflected in the high susceptibility of this insect to carbaryl (LD_{50} = 0.5 μg/bug). The German cockroach *B. germanica* presents a more complicated picture. At least six metabolites of carbaryl are formed, on of which is 1-naphthol. These transformation indicate a common characteristic among the species investigated; hence, the authors conclude that hydrolysis at the esteratic linkage might be the initial step in carbaryl metabolism. In S and R strains of German roaches, Ku and Bishop

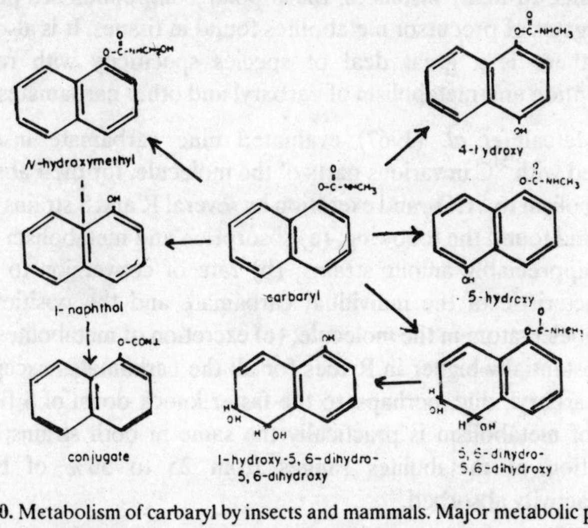

Fig. 10.20. Metabolism of carbaryl by insects and mammals. Major metabolic pathways in insects are designated by heavy arrows. (From Dorough and Casida, 1964; Leeling and Casida, 1966.)

(1967) obtained almost complet metabolism of carbaryl to 1-naphthol and 1-naphthol conjugates.

More recent studies indicate that oxidation may be the primary pathway of carbaryl metabolism in insects and mammals. American cockroaches and house flies metabolize carbaryl via ring and side-chain hydroxylation reactions yielding 4-hydroxy-, 5-hydroxy-, *N*-hydroxymethyl-, and 5,6-dihydro 5,6-dihydroxy-1-naphthyl *N*-methylcarbamate derivatives as 1-Naphthol and 1-hydroxy-5,6-dihydro-5, 6-dihydroxynaphthalene are minor hydrolytic products. Some of these metabolites have been detected after topical application or injection of carbaryl to the house fly, Stable Fly *S. calcitrans,* boll wevil *A. grandis* and the rice weevil *Sitophilus oryzae* to the boll weevil and bollworm *H. zea* larvae and adults and to the cotton leaf worm *P. litura*. Identical metabolites are produced *in vitro* after incubation of carbaryl with house fly abdomen homogenates or microsomes, fat-body and other tissues of the blow fly *Calliphora erythrocephala*, whole tissues and tissue homogenates of the cabbage looper *Trechoplusia ni*, as well as by plants and mammals. Nonpolar carbaryl metabolites are produced in varying degrees by al species examined, by water-soluble metabolites are excreted in substantial amounts. In many instances, these polar compounds are present as conjugates of precursor metabolites found in tissues. It is also evident that there is a great deal of species specificity with regard to absorption and metabolism of carbaryl and other carbamates.

Metcalf *et al.* (1967) evaluated nine carbamate insecticides labeled with ^{14}C in various parts of the molecule, for their absorption, metabolism to $^{14}CO_2$, and excretion by several R and S strains of house flies and found the following; (a) sbsorption and metabolism does not vary appreciably among strains; (b) rate of conversion to $^{14}CO_2$ is characteristic of the individual carbamate and the position of the ^{14}C-labeled atom in the molecule; (c) excretion of metabolites in feces is substantially higher in R flies for all the carbamates except Banol and carbaryl, due, perhaps, to the faster knock down of S flies since rate of metabolism is practically the same in both strains; and (d) excretion of metabolites ranges from 25 to 50% of the total radioactivity absorbed.

In general, total detoxication of carbamate insecticides to CO_2 and H_2O takes place through the following pathways:

a. Hydrolysis. Carbobnyl-labeled carbamates yield phenol or oxime and *N*-methylcarbamic acid. The latter is metabolized to carbonic acid and eventually to CO_2 and H_2O.

$$RO^{14}CNHCH_3 \xrightarrow{HOH} ROH + HO^{14}CNHCH_3$$

$$HO^{14}CNHCH_3 \longrightarrow HO^{14}COH + CH_3NH_2$$

$$HO^{14}COH \longrightarrow {}^{14}CO_2 + H_2O$$

Hydrolytic reactions occur with several carbamates but are now recognized to be a minor importance. Thus carbaryl yields 1-naphthol; carbofurn is hydrolyzed by house flies and the salt marsh caterpillar *E. acrea* (Metcalf *et al.*, 1968) to the 2,3-dihydro-2,2-dimethyl-7-hydroxybenzofuran, 3-keto-Furadan phenol, and 3-hydroxy-Furadan phenol, a large part of the metabolites appearing in conjugated from Landrin hydrolysis is of minor imprtance in the house fly but is more prevalent in mice. Hydrolytic products of aldicarb metabolism in the boll weevil and tobacco budworm include the sulfoxide and sulfone derivatives of 2-methyl- 2(methyl-thio)-propionaldehyde oxime.

b. N-Dealkylation. This pathway yields $^{14}CO_2$ from *N*-methyl-labeled carbamates

$$ROCNH^{14}CH_3 \xrightarrow{[0H]} ROCNH^{14}CH_2OH$$

$$ROCNH^{14}CH_2OHROCNH_2 + H_{214C==O}$$

$$H_{214C==} \xrightarrow{O2} {}^{14}CO_2 + H_2O$$

N-Dealkylation occurs most readily with methiocarb and *m*-isopropylphenyl methylcarbamate, and least with Niagara 10242, aminocarb, and carbaryl. Aminocarb is attacked on the ring

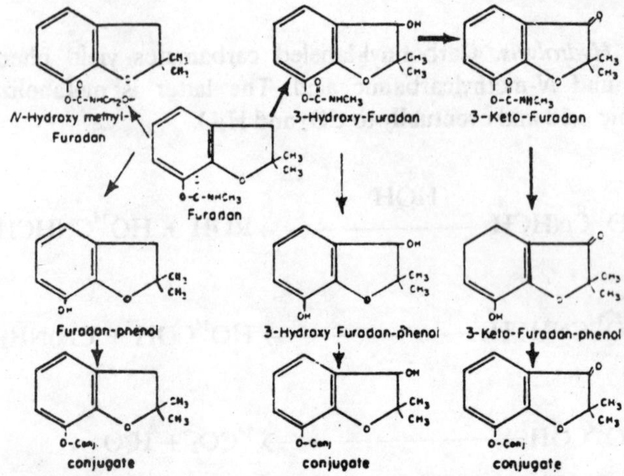

Fig. 10. 21. Metabolism of Furadan (carbofuran) by insects and mammals. Manjor metabolic routes in insects are designated by heavy arrows. (From Dorough, 1968; Metcalf *et al.*, 1968)

methylamino group and forms the 4-methylamino, 4-amino, 4-methylformamido, and 4-formamido derivatives, and Zectran forms an analogous series of metabolites. Production of these metabolites is greater in homogenates of R flies than in those of their S counterparts. All these metabolites are less toxic to house flies than the corresponding parent compounds. On plants, the major photooxidation pathway of aminocarb and Zectran involves extensive oxidation of the dimethylamino moiety with production of the same type metabolites, except that the methylamino and amino analogues are of relatively high toxicity. Propoxur (Baygon) is demethylated by house flies and mosquitoes via *N*-hydroxymethyl propoxur but this is a minor metabolic pathway. There is evidence of some *N*-demethylation of aldicarb in bollweevils and tobacco budworms.

c. O-Dealkylation. This reaction is particularly prominent with propoxur which is dealkylated at the isopropoxyphenyl moiety producting *O*-depropyl propoxur and $^{14}CO_2$. Most of the *O*-dealkylated products come from the intact carbamate rather than from hydrolyzed precursors.

d. S-Dealkylation. This ordinarily takes place after sulfoxidation and sulfonation of the carbamate, such as occurs in aldicarb.

$H_{214CSC(CH3)}{}_2CH = = NOCNHCH_3$ [OH] aldicarb sulfone

$H_{314CSO2}C(CH^3)_2CH = = NOCNHCH_3$ [OH]
S-hydroxymethyi derivative

$HO^{14}CH_2SO_2C(CH_3)_2CH = = NOCNHCH_3$ *S*-dimethyl derivative

$HOSO_2C(CH_3)_2CHNOCNHCH_3 + H_{214C==O}$
$H_{214C==O}$ ${}_{14}{}^{CO}{}_2 + H_2O$

e. Ring-C Hydroxylation. This is a major pathway in carbamate detoxication but there is no CO_2 formation because neither insects nor mammals can convert the ^{14}C in an aromatic ring to $^{14}CO_2$. Ring hydroxylation of propoxur to 5-hydroxy propoxur is the predominant detoxication pathway in the house fly and major metabolic product in propoxur-resistant *C. π.* fatigans *f. Sulfoxidation.* In most instances, sulfoxidation of a carbamate results in the formation of more toxic

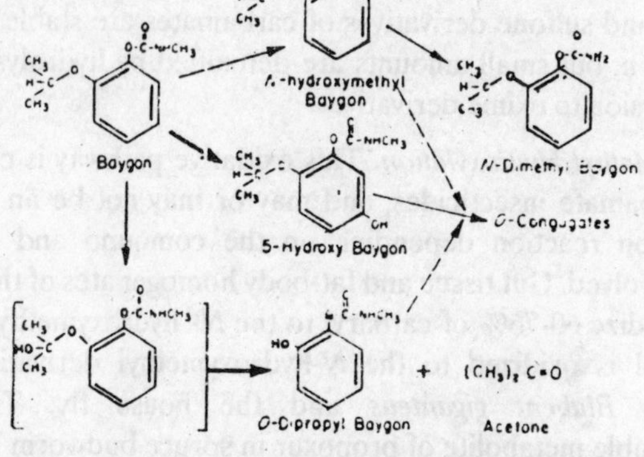

Fig. 10.22. Aromatic ring hydroxylation. Metabolism of Baygon (propoxur) by insects. Major degradative pathways in the house fly are designated by heavy arrows. (From Shrivastava *et al.*, 1969.)

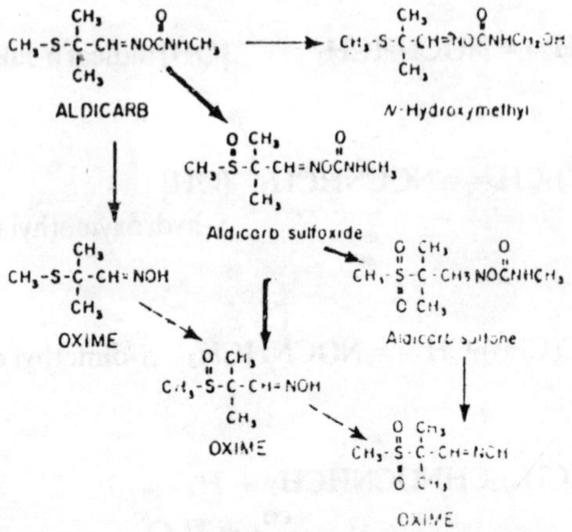

Fig. 23. Oxidative metabolism of aldicarb (Temik) in insects. Major oxidative pathways are designated by heavy arrows. (From Bull *et al.*, 1967; Metcalf *et al.*, 1967.

metabolites (activation). For example, the mojor route of *in vivo* metabolism of aldicarb by house flies and by boll weevils and tobacco budworms involves oxidation of the sulfur atom to the sulfoxide and sulfone derivatives which are more toxic than the parent compound. However, the sulfoxide and sulfone metabolites of methiocarb are less toxic to the house fly than the parent compound. House fly abdomen homogenates also convert methiocarb to the sulfoxide analogue. The sulfoxide and sulfone derivatives of carbamates are stable to further degradation, but small amounts are detoxified by hydrolysis such as the conversion to oxime derivatives.

g. N-Methyl Hydroxylation. This oxidative pathway is common to many carbamate insecticides, and may or may not be an important detoxication reaction depending on the comound and the insect species involved. Gut tissue and fat-body homogenates of the cabbage looper oxidize 60-76% of carbaryl to the *N*- hydroxymethyl derivativ and Banol is oxidized to the *N*-hydroxymethyl derivative by the cockroach *Blaberu giganteus* and the house fly. The major organosoluble metabolite of propoxur in spruce budworm larvae and in *C.p. fatigans* is *N*- hydroxymethyl propoxur while it is only a minor detoxication product in the house fly. *N*-Methyl hydroxylation at the 2- and 5-*N,N*-dimethyl moieties appears to be the major factor in the

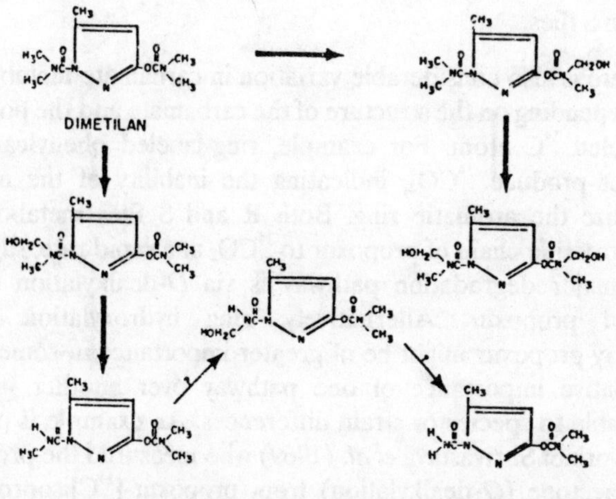

Fig. 10.24. *N*-Methyl hydroxylation. Metabolism of dimetilan. Major metabolic pathways in insects are designated by heavy arrows. (From Zubaire and Casida, 1965.)

detoxication of dimetilan in the German cockroach, the American cockroach, and the house fly.

In the house fly N-methyl hydroxylation of propoxur, Zectran, and carboruran constitutes only a minor detoxication pathway.

Ring methyl hydroxylation is the major oxidative route in the metabolism of Landrin by living house flies and by the mixed-function oxidase system of house fly abodomens.

Relation of Resistance to Metabolism

Most of the studies related above show no evidence of qualitative differences in carbamate metabolism beween resistant and susceptible insects. However, several factors must play important roles in the differential toxicity of carbamates to various species and among strains of the same species, including; rates of absorption, metabolism, conjugation, and excretion. Among nine carbamates investigated little difference in absorption was found between R and S strains of house flies. As far as metabolism to $^{14}CO_2$ is concerned, there is very little difference in total $^{14}CO_2$ Production between R and

S house flies. However, R flies metabolize most carbamates at a faster rate than S flies.

There is also considerable variation in carbamate metabolism to $^{14}CO_2$ depending on the structure of the carbamate and the position of the labeled ^{14}C atom. For example, ring-labeled phenylcarbamate does not produce $^{14}CO_2$, indicating the inability of the insect to metabolize the aromatie ring. Both R and S flies metabolize the isopropoxy side chain of propoxur to $^{14}CO_2$ at a rapid rate, suggesting that a major degradation pathway is via *O*-dealkylation to from depropyl propoxur. Alternatively, ring hydroxylation yielding 5-hydroxy propoxur might be of greater importance in some strains. The relative importance of one pathway over another might be attributable to species or strain differences. An example is provided by the work of Shrivastava *et al.* (1969) who measured the production of [^{14}C]acetone (*O*-dealkylation) from propoxur-[^{14}C]isopropyl and $^{14}CO_2$ (hydrolysis) from propoxur-[^{14}C]carbonyl in 13 insect species. The ratio of carbamate estercleavage to *O*- dealkylation was highest in the American cockroach and spruce budworm, indicating a pre-dominantly hydrolytic attack, intermediate in the German cockroach and the yellow mealworm, low in the honey bee, codling moth, blow fly, and milkweed bug, and lowest in the yellow fever mosquito and the house fly, the latter pointing toward *O*- dealklation as a more significant detoxication route.

Differences in rate of carbamate metabolism are well documented, especially in R and S house flies. More rapid metabolism occurs in resistant strains with carbaryl, 3-isopropylphenyl N-methylcarbamate, aldicarb, Isolan and ten carbamates, i.e., propoxur, carbaryl, 3-isopropylphenyl methylcarbamate, 3,5- diisopropylphenyl methylcarbamate, Banol, methiocarb, aminocarb, zectran, Isolan, and dimetilan. Also, propoxur-resistant *.p. quinquefasciatus* and *C.p. fatigans* (Shrivastava *et al.,* 1970) metabolize propoxur at a faster rate than their susceptible counterparts.

Propoxur-resistant *A. albimanus* from EI Salvador exhibit a tremendous rise in the LC_{50} upon further laboratory selection, from 0.53 ppm to 500 ppm. Cross-resistance to fenitrothion, other OP and carbamate insecticides, and DDT, is evidence of enhanced enzyme activity. The precise mechanisms of this high level of resistance are

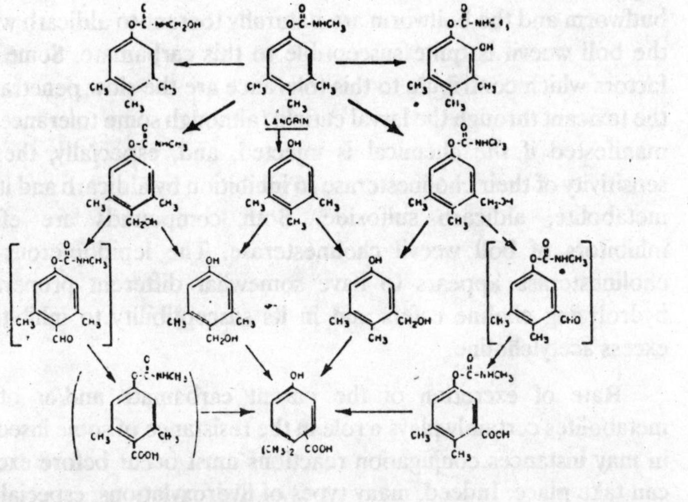

Fig. 25. Ring methyl hydroxylation. Metabolism of Landrin by insects, mammals, and plants. Heavy arrows designate manor metabolic routes in the house fly; brackets: intermediate compounds not indentified but consistent with reaction series; parentheses: unidentified metabolite based on its anticipated chromatographic position. (From Slade and Casida, 1970.)

not well understood, especially since piperobly butoxide and other synergists do not enhance the toxicity of propoxur except at very high insecticide concentrations. Reduced penetration might be one of the factors. Microsomal mixed-function oxidases (MFO) of *A. albimanus, A. quadrimaculatus, A. aegypti, A. triseriatus, C.p. faigans, C. peus,* and *C. tarsalis* catalyze the hydroxylation of benzene, naphthalene, and furan rings in propoxur, carbaryl, and carbofuran, respectively; *O*-and *N*-dealkylation of propoxur; and formation of the sulfoxide and sulfone of aldicarb. Only small amounts of hydrolysis products are produced. Enzyme activity is highest in the propoxur-selected *C.p. fatigans.*

The mosquito MFO system differs from that of other insects in buffer concentration, pH optimum, and incubation temperature, but has the same characteristics with respect to endogenous inhibitors. On a weight basis, adult mosquitoes show higher enzyme activity than larvae. Resistance to carbamates is not always due to more rapid degradation of the toxicant by the resistant insect. The tobacco budworm and the bollworm are naturally toerant to aldicarb whereas the boll weevil is quite susceptible to this carbamate. Some of the factors which contribute to this tolerance are the slow penetration of the toxicant through the larval cuticle (although some tolerance is also manifested if the chemical is injected, and, especially, the lower sensitivity of their cholinesterase to inhibition by aldicarb and its toxic metabolite, aldicarb sulfoxide. Both compounds are effective inhibitors of boll weevil cholinesterase. The lepidopterous lerval cholinesterase appears to have somewhat different properties in hydrolyzing choline esters and in its susceptibility to inhibition by excess acetylcholine.

Rate of excretion of the parent carbamate and/or of toxic metabolites certainly plays a role in the resistance of some insects, but in may instances conjugation reactions must occur before excretion can take place. Indeed, many types of hydroxylations, especially ring hydroxylations which are very important in dettoxication, undergo conjugation reactions before the compounds are excreated. Alternatively, excretion may be impeded in the suceptible insect because of a biochemical lesion causing its knockdown. Most of the evidence supports the contention that carbamates are oxidatively metabolized promarily by microsomal enzymes. However, a partially purified soluble tyrosinase prepared from house flies catalyzes the hydroxylation of certain methyl carbamates. Tyrosinase is also found in house fly microsomal preparations. According to Kuhr (1969) and Wilkinson (1968) microsomal tyrosinase does not seem to be important in carbamate metabolism in general , and the soluble fraction, which contains six times as much tyrosinase as the microsomes, fails to metabolize carbaryl or phenyl methylcarbamate. There appears to be some substrate and strain specificity in this regard. In conclusion, carbamate insecticides are readily attacked by insect enzymes and resistance to these compounds follows much the same pattern as that of the organophosphorus insecticides. Hydrolysis at the esteratic linkage seems to be of minor importance in the overall metabolic pathway. Oxidative degradation by the mixed-function oxidase system includes *O*-,*N*-, *S*-, and *C*-dealkylation; *N*-methyl hydroxylation; ring hydroxylation and sulfoxidation. Rates of absorption, metabolism, conjugation, and excretion play a vital role in carbamate resistance by insects. Since the same metabolites are produced by S and R insects, the rate at which these metobolites are produced and excreted determines the ability of the insect to survive the poisoning effect.

E. Metabolic Fate of Pyrethrins, Rotenone and Nicotine

1. Pyrethrins and Related Compounds

It is a well-established fact that some insects are rapidly paralzed when exposed to pyrethrins, but that a large proportion of the individuals recover completely within 24 hours. In fly spray tests, it is usually expected that a pyrethrin formulation giving complete knock-down within 10 minutes will ordinarily show only 60-70% mortality at the end of 24 hours. This reversal of knockdown and paralysis indicates that insects possess an efficient detoxifying mechanism capable of attacking the pyrethrin molecule at certain reactive sites.

Pyrethrins, like DDT, have a negative temperature coefficient, i.e., their toxicity diminishes with an increase in temperature. At higher temperatures (about 35°C), hydrolysis was presumed to be so great that the insects did not accumulate a lethal dose. Following Swingle's (1934) observation that pyrethrins were ineffective as a stomach poison, Acree *et al.* (1936) suggested that hydrolytic enzymes such as esterases might be involved in the breakdown of pyrethrins.Detoxication of pyrethrins was first demonstrated in the southern armyworm *P. eridania* by bioassaying tissue extracts of the treated insect against mosquito larvae. *In vitro* degradation was highest in fat-body follwed by skin and muscle, digestive tract, and hemolymph.

Most of the earlier work on pyrethrin metabolism centered around the hypothesis that hydrolysis of the ester linkage was the mechanism of detoxication. This was based chiefly on the demonstration of the production of a small amount of an unidentified acid resulting from incubation of cockroach and house fly acetone powder preparations with pyrethrins, and on the incomplete identification of chrysanthemumic acid and keto-alcohols resulting from the *in vivo* metabolism of [^{14}C]pyrethrins by the American cockroach. The latter's findings that 3-12% of the [^{14}C] pyrethrins does was converted to $^{14}C_2$ could not be confirmed later with house flies, but significant amounts of [^{14}C]pyrethrins and [^{14}C]allethrins were shown to be metabolized to nontoxic, nonpyrethroid derivatives, and this metabolism was blocked by the synergist piperonyl cyclonene.

Bridges (1957), extending this work, found more extensive metabolism of [^{14}C]allethrin *in vivo* than by lipase extracts and abdomen homogenates *in vitro,* and suggested that metabolism was not due to hydrolysis. Similarly, Hopkins and Robbins (1957) found a rapid rate of [^{14}C]allethrin metabolism in DDT-resistant house flies with approximately 44% of the absorbed dose excreted as polar metabolites and resembling allethrolone on paper chromatography. Only traces of chrysanthemumic acid and unchanged allethrin were detected. These authors ponted out that modification of the intact ester rather than hydrolysis of the ester linkage might provide a detoxication pathway. Recently, Hayashi *et al.* (1968) showed the production *in vivo* of large quantities of allethrolone or of *N*-hydroxymethyl tetra- hydrophthalimide in house flies exposed to [^{3}H]allethrin and [^{3}H] phthalthrin, respectively. According to Yamamoto (1970) none of these metabolic pathways have been conclusively proven.

The first breadthrough in the present concept of pyrethrin degrardation was the finding by Chang and Kearns (1964) that hydrolysis plays only a minor role in pyrethrin detoxication by intact house flies. More than 96% of the absorbed dose of [^{14}C]pyrethrin I and [^{14}C]cinerin I was detoxified in 4 hours after topical application. Five unknown metabolites were detected plus a small amount of free chrysanthemumic acid which never exceeded 2.6% of the applied dose. Three of the metabolites showed an intact chrysanthemumic acid-ester linkage. The authors concluded that the detoxication process apparently is initiated on the keto-alcohol moiety leaving the acid-ester linkage intact.

The present concept of pyrethrin metabolism is that oxidative breakdown rather than hydrolytic cleavage is the main metabolic pathway. Evidence for this comes from the works of Yamamoto and Casida (1966) and Yamamoto *et al.* (1969) who first succeeded in defining the metabolic path way in insects. The synthesis of ^{14}C-labeled pyrethroids, stereochemically pure and of high specific activity (Yamamoto and Casida, 1968), obviated the difficulties encountered in previous studies. Additional evidence of oxidative metabolism was obtained *in vitro* by using homogenates of resistant house fly abodomens fortified with NADPH, similar to the system used by Tsukamo to and Casida (1967). In this sytem [^{14}C]allethrin, labeled in either the acid or the alcohol moieties, yields more than ten, autoradiographically identical metabolites, indicating that each metabolite contains both the intact acid and alcohol moieties. A major metabolite, representing half of the total radioactivity, proved to be *trans*-allethrenoic acid or demethylallethrin II. Other metabolites are characterized as *trans*-allethrinol and *trans*-allethrinal. Low enzyme levels produce allethrinol metabolites whereas high enzyme titers oxidize the intermediates to allethrinoic acid. There is little, if any, indication of a modification of the allethrolone moiety, in contrast to earlier findings by other workers. Similar metabolic reactions occur with [^{14}C]pyrethrin I, dimethrin, and phthalthrin, and oxidative degradation of these compounds is more extensive in resistant than in susceptible house flies.

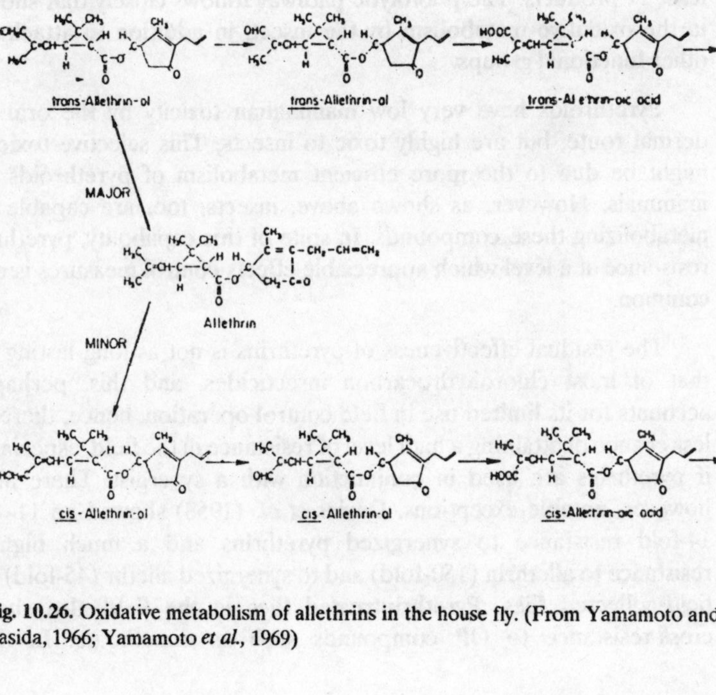

Fig. 10.26. Oxidative metabolism of allethrins in the house fly. (From Yamamoto and Casida, 1966; Yamamoto *et al.*, 1969)

In summary, the major metabolic pathway for the enzymic detoxication of the pyrethroids involves oxidation of the *trans*-methyl group of the isobutenyl moiety to the hydroxymethyl derivative which, acted upon by other enzymes, either forms conjugates, or is further oxidized to the corresponding aldehyde and acid. A minor pathway involves similar reactions at the *cis*-methyl group, and no hydrolysis or modification of the alcohol moiety is evident.

All the metabolic products obtained are less toxic to house flies than their parent compounds. Synergists containing the methylenedioxyphenyl group, particularly pipperonyl butoxide, block the initial hydroxylation at the isobutenyl-methyl group, thus stabilizing the molecule and preventing its detoxication. On the other hand, the isobutenylmethyl moiety is not essential for toxicity but compounds lacking this grouping are also synergized, indicating, perhaps, that other sites than the isobutenylmethyl group might be susceptible to oxidative attack.

Pyrethrin I, Phthalthrin, allethrin, and dimethrin deposited as thin films on glass readily undergo photodecomposisition yielding at least 11 products. The photolytic pathway follows closely that shown in the oxidative metabolism by the insect, in addition to attack on other functional groups.

Pyrethroids have very low mammalian toxicity by the oral or dermal route, but are highly toxic to insects, This selective toxicity might be due to the more efficient metabolism of pyrethroids by mammals. However, as shown above, insects, too, are capable of metabolizing these compounds. In spite of this capabolity, pyrethrin resistance at a level which appreciable effects control measures is not common.

The residual effectiveness of pyrethrins is not as long lasting as that of most chlorohydrocarbon insecticides, and this, perhaps, accounts for its limited use in field control operation; hence, there is less chance of attaining a high level of resistance in the field, especially if pyrethrins are used in conjunction with a synergist. There are, however, notable exceptions. Davies *et al.* (1958) showed an 11- to 14-fold resistance to synergized pyrethrins and a much higher resistance to allethrin (150-fold) and to synergized allethr (45-fold) in field-collected flies. Pyrethrintreated flies in the field also show cross-resistance to OP compounds such as dimethoate. In the laboratory, selection of house flies with pyrethrins of DDT results in enhanced resistance to pyrethrins, synergized pyrethrins, allethrin, barthrin cyclethrin , furethrin, in addition to DDT and other insecticides (Fine, 1961).

Pyrethrin resistance has also been reported in cockroaches *B. germanica* bed bugs *Cimex hemipterus* blue ticks *B. decoloratus*, body lice *P. humanus humanus*, granary weevils *S. granarius* and other insects. It is evident that under favourable conditions of selection pressure, resistance to pyrethroids may develop as readily as with other insecti- cides. Whether or not detoxication of these compounds in various insects is causally related to resistance has not yet been established.

2. Rotenone

It has been common knowledge for many years that rotenone depresses the oxygen uptake of many organisms. Specifically, rotenone inhibits mitochondrial respiration by blocking the enzyme system involved in the coupled oxidation of reduced NAD ($NADH_2$) and reduction of cytochrome b. In insects, this type of inhibition results whether the enzyme is derived from susceptible species or from those naturally resistant to rotenone. Thus, the selective toxicity of rotenone (being moderately toxic to mammals and highly toxic to fish and some insects) does not involve the primary site of action but appears to depend on the distribution pattern and rate of detoxication by various organisms.

Rotenone is extensively metabolized by the microsomal mixed-function oxidase system of mammals and insects. In the presence of NADPH and O_2 microsomes prepared from cockroach fat-body and midgut and from house fly abdomens metabolize rotenone principally to rotenolone 1, 6, 7, dihydro 6,7,dihydroxyrotenone and to a minor extent to 8-hydroxyrotenone, 8'-hydroxyrotenolone 1, 6',7-dihydro-', 7'-dihydroxyrotenolone I and II, rotenolone II derivatives, and midgt soluble fractions block these reactions, but there is little or no interference from the soluble fraction of house fly homogenates. Methylenedioxyphenyl synergists inhibit rotenone metobolism in accord with the mode of action of these compounds as inhibitors of microsomal oxidations. Since rotenone metabolites are less toxic than the parent compound, hydroxylation of rotenone and conjugation of polar derivatives might provide a pathway for resistance development to this compound. So far, this metabolic pathway has not been associated with rotenone resistance in insects.

3. Nicotine

Nicotine is one of the most toxic alkalods to insects, yet several species such as the tobacco hornworm, tobcco, budworm, tobacco, wireworm, green peach aphid, and others feed liberally on tobacco plants and show no symptoms of intoxication. The ability of the green peach aphid *Myzus persicae* to feed on a plant capale of nicotine biosynthesis without suffering harmful effect is due to its ability to selectively feed in the phloem of the plant and avoid the nicotine-containing xylem. On the other hand, the toabcco hornworm *Protoparce sexta* excretes and egests intact nicotine and other ingested alkalods before a toxic dose can accumulate.

In addition to these protective mechanisms, detoxication pathways also account for nicotine tolerance in some insects that do not feed on tobacco. Thus, *B. germanica* and *P. americana* meabolize nicotine principally to cotinine which it essentially a nontoxic metabolite, but nine other unidentified metabolites prevail in the southern armyworm *P. eridania*. Similar comparative experments showed no metabolism of nicotine by the tobacco budworm and the cabbage looper. Only unaltered alkaloids found in the tobacco plant were present in the excreta of these insects. The tobacco wieworm, cigarette beetle, differential grasshoppe, and the house fly metabolize nicotine to 1,2,4 and 3 other alkaloids, respectively. In all cases, at least 70% of the total metabolic products corresponds to continine. In contrast with tobacco- feeding insects, nicotine and the metabolites are not excreted by the house fly.

The possiblity that nicotinophilic microoorganisms associated with the tobcoo hornworm of with the host plant may be capable of detoxifying nicotine was investigated by Guthrie and Apple (1961). The evidence obtained supports the contention that detoxication of nicorine by microorganisms is not a major factor in nicotine tolerance by this insect. It is of interest to note that nicotine metabolism by mammals yields principally the same product, continine, and this oxidative conversion takes place in the microsomal fraction of the liver.

F. Metabolic Fate of Chemosterilants

The first report of resistance to a chemosterilant was made by Hazard *et al.,* (1964) who selected *A. aegypti* larvae with apholate and obtained a 4- to 5-fold tolerance after 11 generations of selection. In the same way, Klassen and Matsumura (1966) and Patterson *et al.* (1967) were able to induce higher levels of resistance to metepa and apholate, respectively, in the same species. Furthermore, homogenates of metepa-resistant larvae metabolized metepa more rapidly than their susceptible counterparts. George and Brown (1967) obtained a moderate tolerance to hempa in *A. aegypti* after five generations of larval selection. However, this tolerance disappeared in the sixth generation of selection due to the accumulation of inheritable rcessive genetic defects resulting from inbreeding. A similar loss of tolerance to metepa occurred in house flies, perhaps for the same reason. After 10 generations of selaction with metepa and 60 generations with apholate, Morgan *et al.,* (1967) were unable to induce resistance to these chemosterilants in the house fly. On the other hand, Abasa and Hansen (1969) showed evidence of resistance to apholate after 15 generations of selection of a diazinon- resistant strain of house flies.

Among the reasons advanced for enhanced insect tolerance of chemosterilants, penetration and degradation are prominent factors. Working with S larvae and adult *Culex tarsalis* and with S and OP-R adult house flies, Plapp *et al.* (1962) obtained complete degradation of [^{13}P]aphoxide (metepa) by *culex* in 48 hours, and 50% degradation by the house fly in 2 hours. Degradation rates were similar in the S and R strains and were more rapid after injection than after topical application. The principal metabolite was thought to be inorganic phosphate. The finding that R flies did not metabolize the chemical at a faster rate indicates that different enzymes might be involved in the metabolism of metepa and OP compounds. Similarly, absorption and excretion of metepa proceed rapidly in mosquitoes and house flies and the difference in the doses needed to sterilize the stable fly and the screw-worm fly is related to the absorption, excretion, and metabolism of metepa by the two species. The principal metabolite was found to be phosphoric acid and in greater quantity in screw-warm fly.

Excretion of the chemosterilant appears to be an important factor in insect chemosterilization. Boll weevils excrete 48% of the injected dose of [^{14}C]tepa in 3 days, and 3.4% is trapped as $^{14}CO_2$, indicating that some metabolism is taking place. Excretion of [^{14}C]tepa also plays an active role in the house fly. Excretory products are of a nonaziridinyl metabolites.

Thiotepa is rapidly oxidized to tepa by the German cockroach, house fly, and stable fly. The boll weevil oxidizes 63% of the absorbed material in 4 hours, but after that period the rate of conversion diminishes markedly. Because tepa is one of the most potent chemosterilants, the oxidation of thiotepa to tepa might be considered an activation process.

Hempa, a nonalkylating chemosterilant, has been the subject of several investigations. The only major metabolite found in hempa-treated male house flies and their excreta was the mono-demethylated product, pentamethylphosphoric triamide (PMPT). This metabolite has a much lower sterilizing activity than the parent compound. Demethylation reactions are suggestive of microsomal mixed- funtion oxidase involvement. This possibility was put to test by Akov and Borkovec (1968) and Akov *et al.* (1968) who compared the rates of conversion of hempa to PMPT by homogenates and microsomal preparations of susceptible and carbamate-resistant house flies. Their results clearly indicate the superior ability of the resistant strains to metabolize hempa.

TABLE. 10.8
Metabolism of Hempa by Microsomal, Preparations from Resistant and Susceptible House Flies

Fly strain	*PMPT produced (nmoles)*	
	Males	*Females*
Susceptible	0.9	1.3
Isolan-resistant	8.6	9.7
Baygon-resistant	5.2	11.0

In addition to PMPT, R but not S flies contained small amounts of abother metabolite, tetramethylphosphoric triamide (TMPT). These metabolites are inactive products, but the sterilizing activity of hempa is of the same magnitude to both S and R strains. This led to the assumption that oxidative demethylation of hempa to PMPT might proceed via an unstable methylol intermediate which has a sterilizing capacity equal to hempa. The important metabolic step would then be the conversion of the active methylol derivative to PMPT rather than the disappearance of hempa. Demethylation of hempa is inhibited by SKF 525-A and by methylenedioxyphenyl synergists.

It is well-known that in many insect species, especially the house fly, the female has a higher capacity for insecticide metabolism than the male. Diet also plays a role in insecticide metabolism. Akov *et al.*, (1968) produced evidence that female flies maintained on sugar metabolized hempa 2.5 times faster than males. Milk-fed flies of both sexes sustained increased metabolic activity, again higher in the female.

In a similar study with S and R house flies, Chang and Borkovec (1969) determined that hempa is metabolized faster by R flies and that the latter contain a larger proportion of tetramethyl and trimethylphosphoric triamides. Tropital (a methylenedioxyphenyl synergist) suppressed the demethyation of hempa, and the relative distribution of metabolites was substantially changed, especially in R flies, including the appearance of an unidentified new peak.

Other chemosterilants, i.e, *N, N, N, N*,- tetramethyl-π-piperi-dinophosphonic diamide and *N, N*-tetramethyl- enebis (1-aziridinecarboxamide) rapidly disappear from the tissues of treated boll weevils, blow flies, and screw-worm flies, particularly when the chemicals are injected.

V. SUPPLEMENTARY PROTECTIVE MECHANISMS

A. Structural Aspects and Insecticide Penetration

The first report of DDT resistance in an insect was coincidentally associated with distinct morphological characters, for Wiesmann (1947) Attributed DDT resistance in the house fly to darker pigment-ation of the cuticle, stiffer tarsal bristles, and thicker and more pigmented tarsi, pulvilli, and articular membranes of tarsal joints. Other morphological characteristics may segregate resistant from susceptible strains of insects but many resistant strains also are quite indistinguishable from their susceptible counterparts. On a statistical basis, for example, 16 morphologica characters did not correlate with DDT resistance in 9 house fly strains.

Practically every publication on insecticide metabolism *in vivo* contains information on penetration of the toxicant. Many resistant strains of insects have been shown to absorb the insecticide either at a faster rate, at a slower rate, or at an equal rate to susceptible strains. In many instances where topical application is used, absorption of the insecticide is biphasic, i.e., penetration is highest during the first few hours after application and then levels off to a slower and more constant rate, but penetration also follows firstorder kintics in some insects, such as DDT-resistant codling moths. Permeability of the nervous system to dieldrin or to diazoxon is not a factor in resistance. Hence, it might be said that permeability of the cuticle is a variable factor, well-randomized among S and R strains of varied origins.

In some instances, however, reduced rate of absorption constitues a major factor in resistance. For example, Forgash *et al.*, (1962) and Farnham *et al.* (1965) found reduced permebility to be a significant factor in house fly resistance to diazinon. Similarly, reduced absorption is largely responsible for *A. aegypti* tolerance to malathion DDT synergist-combinations in house flies and mosquitoes, pyrethrin-resistance in house flies, DDT-resistance in the spotted root maggot *E. notata* and diazinon- selected house flies. Hollingworth *et al.*, (1967) classify slower penetration as a secondary factor in house fly resistance to OP compounds.

Insect larvae, especially lepidopterous larvae, absorb topically applied insecticides at a slow rate. Evidently this accounts for DDT-tolerance in the corn earworm and the tobacco budworm and bollworm. Observations by Plapp and Hoyer (1968) indicate that reduced absorbtion constitutes the mode of action of a gene *(R-tin)* conferring resistance to organotin compounds and acting as resistance intensifier for other insecticides. Thus, reduced absorption allows detoxication mechanisms which might be present a longer time to act, consequently enhancing the overall resistance by combining the two types of genes. A similar phenomenon occurswith the gene *pen* (for penetration delaying) which alone confers little resistance, but when introduced into a strain with a factor for deethylation of OP compounds, the interaction of the two genes enhances the resistance markedly. Some resistant strains of house flies are capable of surviving repeated knockdown by intermittent exposures to DDT while others are paralyzed immediately but recover when removed from contact with the insecticide. Still others are highly resistant to knockdown.

Resistance to knockdown was first demonstrated by Harrison (1951) to be due to an incompletely recessive gene to which Milani (1954) gave the symbol *kdr* and assigned it to chromosome II. DDT-resistant strains possessing this gene depend on other factors than dehydrochlorination for their resistance. Delayed knockdown due to reduced penetration is also found in the resistant SKA strain. From a biochemical standpoint, recovery form knockdown may result from detoxication of the insecticide and its removal from the site of action or from incomplete and reversible enzyme inhibition such as occurs with certain carbamates.

The above examples illustrate the fact that slow rate of penetration can confer partial protection to the insect, and in those insects which possess high levels of detoxifying enzymes, slower penetration can enhance their resistance many-fold.

As far as the writers are aware, only two instances have been recorded. Where natural immunity to DDT in an insect (*Trogoderma* larvae) is associated with complete lack of metabolism of the toxicant. Gupta *et al.* (1971) observed *Trogoderma* larvae crawling on pure DDT crystals for a week without being affected in the process. This is not due to lack of penetration, for as early as 6 hours and as late as 72 hours after application there remains unchanged in the tissues 33-40% of the amount of DDT applied, with no evidence of DDT metabolites. These amounts are even higher in various tissue of the adult, including the nervous system.

B. Lipid Content and Dietary Factors

1. Lipids of Susceptible and Resistant Insects

It is well-established that many insecticides, especially those of the chlorohydrocarbon type, and some of their metabolites are deposited in the adipose tissue of insects. Consequently, numerous attempts have been made to correlate the lipid content of whole insects or of individual tissues with physiological resistance to insecticides. A positive correlation has been shown in some instances.

Thus, Munson and Gottlieb (1963) and Munson *et al.* (1954) concluded that the higher lipid content of DDT- tolerant cockroaches protects the insect from intoxication by preventing DDT from reaching the site of action. Wiesmann (1957) found more lipid in a DDT-resistant strain of house flies than in its susceptible counterpart.

DDT-R house flies have been characterized as having a higher lipid content of the tarsi, epicuticle, and thoracic ganglia; higher iodine numbers of lipids; higher cholesterol levels in various tissues as well as by histological differences from their susceptible counterparts. Boll weevils surviving the toxic effects of dieldrin and toxaphene usually contain more lipid than those than succumb to the insecticides and the greater tolerance of the alfalfa weevil to heptachlor with increase in age correlates favourably with an increase in lipid content.

Dieldrin-R *A. aegypti* larvae have a higher neutral lipid and phospholipid content than the normal strain with close correlation between total lipid and the LC_{30} among various substrains. Since similar differences occur between individuals of a given strain it is unlikely that lipid content per se is associated with the resistance mechanism. A similar conclusion was reached by Kalra (1970_a) who measured total lipids, fatty acids, and phopholipids in several R and S strains of *C. p. fatigans*. Larvae of the khapra beetle *trogoderma granarium* which are practically immune to DDT contain three times as much total body lipid as the more susceptible adults. It is not clear, however, if this factor is related to resistance since the comparison involves two different stages of development.

On the other hand, a comparison of several chlordane and dieldrin-R strains with S strains of house flies failed to show any consistent interstrain differences in lipid content or in the quality of lipids as revealed by infrared spectroscopy. Neither did dieldrin-R *A. gambiae* nor DDT-R *A. aegypti* contain more lipid than their susceptible counterparts. There are variations in total sterols among resistant and susceptible house flies, but it is the susceptible NAIDM strain which contains the largest amount of cholesterol. Higher lipid content might be envisaged as solubilizing more insecticide and, perhaps, enhancing its deposition in various tissues, tissues, thereby reducing the critical concentration at the site of action. At best, this condition can afford only slight protention to the insect.

2. Dietary Factors in Insecticide Tolerance

The relation of nutritional factors to insecticide resistance follows much the same pattern as lipid content. A well-documented review. (Gordon, 1961) discusses the hereditary and nonhereditary effects of nutrition on insecticide tolerance. Most of the examples given agree with the concept of vigour tolerance which delineates them from the more fundamental physiologica resistance. For example, aphids and redspiders often show two-to threefold variations in susceptibility to insecticides depending on the plant they feed on. The two-spotted mite and the carmine spider mite are more susceptible to malathion if fed on bean plants grown in nutrient solutions containing high levels of phosphorus (and high levels of nitrogen in the former only). In contrast, high lovels of ascorbic acid and increased lipid level in the diet confer resistance to DDT in tobacco budworm larvae. House fly resistance to knockdown by DDT can be increased by substituting milk for sugar in the adult diet or by increasing the fat content of milk. On the other hand, milkfed flies contain no more DDT-dehydrochlorinase than flies maintained on a sugar diet.

Recent advances in mammalian and insect pharmacology have added much to clarify the effect of dietary factors on drug and insecticide metabolism. For example, a high sucrose level in the diet of rats markedly depresses mocrosomal hydroxylation of several drugs by the mixed-function oxidase (MFO) system. Similarly, house flies maintanined only on a sucrose diet metabolize the chemosterilant hempa at a much slower rate than milk-fed flies, indicating an increased enzyme activity in the latte. Along the same line, Perry and Buckner (1970) demonstrated that milk-fed house flies contain more cytochrome P-450 than those fed sugar, and that the LD_{50} for various insecticides is 1.5 to 2.3 times higher for milk-fed than sugar-fed adults. Thus, the deffect of dietary factor on tolerance to certain insecticides and other chemicals might well be associated with an increase, in cytochrome P-450, the terminal oxidase in the microsomal enzyme system.

Other attempts have been made to correlate dietary factors with rates of MFO activity in various species. Gordon (1961) comments on larval feeding stages of polyphagous holometabolous insects as showing high and generalized tolerance to contact insecticides, this probably being the result of selection for endurance to a variety of prolonged biochemical stresses, especially exposure to secondary plant substances. This hypothesis was experimentally tested by Krieger *et al.*, who found good correlation between aldrin epoxidase activity in gut microsomes and the number of plants on which certain species are known to feed. On the other hand, Brattsten and Metcalf (1970), who examined 54 insect species from eight orders and 37 families, found a wide variation in susceptibility to carbaryl and carbaryl plus piperonyl butoxide which, in many instances, was not related to diet. The 10,000-fold range in susceptibility at the LD_{50} level and the 300-fold range in synergistic ratio should discourage any toxicologist from the ever-present temptation to generalization.

C. Miscellaneous Chemical Differences

A logical extension of efforts toward understanding the mechanisms of insecticide resistance is the seeking of chemical differences which might be peculiar to either susceptible or resistant strains of a given species. Several such attempts have been reported. Chromatographic patterns of ninhydrin-positive components of homogenates of S and R strains of *M. domestica, Fannia canicularis,* and *Phaenicia sericata* revealed a number of differences among the strains, none of which could be correlated with insecticide resistance. Instead they seemed to correlate more closely with geographical origin.

Exposure to DDT causes a significant decrease in the amino acid content of the S strain hemolymph, but not in that of the R strain of house flies (Reiff, 1956b). However, among several strains of house flies tested, the amounts of some individual free amino acids differ from on strain to another, but these differeces do not appear to be related to selection by or resistance to diazinon or to DDT. Clearance of injected proline after administration of DDT is considerably faster in DDT-poisoned cockroaches. This depletion may be due to excretion or to accelerated metabolism of proline caused by DDT.

Reversal of symptoms of DDT-poisoning results in the gradual restoration of normal proline levels in the hemolymph. Identical results are observed with TDE-poisoned cockroaches and tobacoo hornworms, but other insecticides fail to produce this reversible proline depletion. Furthermore, in cockroaches injected with [^{14}C] proline, DDT increases the level of radioglutamine in the hemolymph and nerve cord while decreasing the radioproline content. On the other hand, DDT, dieldrin, DEP, Baygon and *n*- valone, but not rotenone, reduce the free proline concentration of the central nerve cord of *P. americana* and, in some instances, it results in the accumulation of free α-alanine and glutamine. Malathion causes a reduction in glycine, glutamate, proline, and glutamine levels of susceptible German roaches but, except for glutamine, has little effect on amino acid depletion in the resistant strain. An increase in free alanine occurs also in DDT-R *A. aegypti* larvae after exposure to DDT while the alanine level of the S strain remains comparatively low.

In the house fly, DDT interferes with the metabolism of formate, glycine and proline. After injection of [^{14}C]formate, more uric acid and allantoin and less proline are found in DDT-treated house flies than in untreated controls, but such is not the case with OP and carbamate-treated insects. Relevant to house fly resistance to DDT is the observed DDT and thiocarbo- hydrazine inhibition of proline degradation and purine synthesis in the R strain while the reverse is true in the S strain. Thus, it might be inferred that, in the S strain, the toxic action of DDT is hampered by DDT inhibition of the production of neuroactive amines or toxins, which, presumably, can be derived from proline. It is of interest, too, that in fourth-instar *T. infestans* larvae (the DDT-resistant stage) DDT causes a significant increase in free proline and in total proteins.

DDT also causes an increase in the rate of utilization of pentose phosphate cycle intermediates in *P. americana* in DDT-R *T. infestans* and *M. domestica*. With the introduction of genetically pure, mutant house fly strains, Plapp (1970b) was able to show a twofold increase in the level of glucose-6-phosphate dehydrogenase in DDT-R or dieldrin-R house flies. This factor is under genetic control and can be transferred from one population to another by appropriate genetic crosses. The differences between S and R strains of insects cited above strengthens even further the conclusion that biochemical variability among species and strains is the rule rather than the exception. If biochemical defense mechanims are rather specific in nature, as contrasted with a "multicomponent type" resistance which constitutes a summation of many independent variables, we should then seek those mechanisms which show consistency rather than variability.

D. Ecological and Behavioural Aspects of Resistance

1. Bionomics

Numerous attempts have been made to correlate insecticide resistance with certain biological factors such as egg production, hatching period, egg viabillity, length of developmental stage of egg, larva, pupa, and adult; pupal and adult weights; sex ratios; and preoviposition period. A detailed discussion of these factors would appear superfluous, especially in view of the fact that most of these associations have proven to vary considerably among species and among strains of a single species. The only characteristic for which significant differences are apparent in some instances is length of the larval stage. Several DDT-R strains of house flies have been found to have a 1-to 4-day longer larval stage than S counterparts but in an equal number of investigations no such differences were discernible. Gagliani (1962), on the other hand, found a shorter larval cycle in a DDT-and chlordance-resistant strain. Similarly, life cycle changes brought about by insecticide pressure on *B. germanica* appear to be the result of population selection not associated with resistance. Furthermore, changes in length of larval period, weight, and size brought about by external or genetic manipulation have no striking influence on DDT susceptibility in R and S *D. melanogaster*, but selection of *D. melano- gaster* larvae for a short or a long larval period results in a lower LC_{50} for DDT in the short larval line but no change from the control in the long larval line.

A more significant finding is that late-emerging house flies are more resistant to DDT than those emerging early. Selection of late-emerging adults for three generations or lengthening the larval stage by lowering the temperature confers some DDT resistance to the offspring. A six to eightfold resistance to DDT and related compounds was obtained by lelecting the late pupaters for 15 generation in the absence of a chemical. After 15 generations of selection for early and late pupation in *M. domestica vicina*, the LD_{50}

for DDT of the early strain diminished by a factor of two, whereas that of the late strain increased sixfold. However, there was no change in the LD_{50} when BHC was used as the test compound. It is puzzling that selection for the late pupation should be confined to DDT resistance and certain DDT analogues only and not extend to other type insecticides such as lindane, cyclodienes and pyrethrins.

Changes in reproductive potential following application of an insecticide have been the subject of extensive investigation. Kilpatrick and Schoof (1956) obtained a sizable increase in fly production 4 weeks after treatment of outdoor privies with aldrin, dieldrin, chlordane, and BHC, but a reduction of fly breeding with application of DDT. Knutson (1959) also observed an increase inbiotic potential after 4 years of field spraying with dieldrin to control house flies and similar results were obtained by Gratz (1966) in laboratory selection and field control of *M. domestica vicina*. In the latter case, the increased biotic potential lasted for only one or two generations after relaxation of selection pressure, suggesting that dieldrin resistance is not under genetic control. *Sitophilus granarius* weevils exposed to low concentrations of DDT produced 20% more offspring than unexposed controls.

Greater reproductive potential in S than in R strains of house flies is of common occurrence. This phenomenon shows positive correlation with ovarian size and ovarian nitrogen content.In some instances, sublethal doses of insecticides were found to suppress house fly fecundity and ovarian development and removal of OP-resistant *A. aegypti* from insecticide contact was found to restore the fecundity and fertility levels by stimulating protein syntheis. The nature of ovarian suppression by DDT involves a variety of pathological conditions as discussed and illustrated by Lineva (1962) and Beard (1965). According to Derbeneva-Ukhova *et al.* (1966) development of resistance by sublethal exposure to DDT follows three consecutive phases: (1) a short period of increased tolerance; (2) a period of susceptibility during which ovarian suppression is at its highest; and (3) a period of resistance enhancement during which disfunction of ovogenesis disappears.

Alternatively, the reproduction potential of *A. aegypti* and *C. pipiens* is increased by sublethal exposure to DDT, and din *A.*

albimanus the fertility rate is higher in the dieldrin-R strain. Other workers find no deleterious effects on oogenesis after feeding adult house flies sublethal doses of *p,p'*- DDT, *o,p*-DDT and DDE for several generations, or after exposure of *A. aegypti* to sublethal doses of DDT. In fact, in the F_1 generation, treated females produced more eggs than did the controls.

2. Avaidance Mechanisms

Behavioural resistance is defined as "the ability of insects, through protective habits or behaviour, to avoid lethal contact with a toxicant". A classical example of this type is said to be the ability of certain strains of the codling moth *C. Pomonella* to penetrate apples sprayed with lead arsenate, presumably due to their acquired habit of rejecting the first bite prior to penetrating the fruit. There have been reports of changes in behavioural patterns of house flies, such as fewer flies resting on DDT-sprayed walls and ceilings or nocturnal migrantion from sprayed houses to nearby unsprayed vegetation. Selective avoidance of malathion baits by certain strains of malathion-R flies have been mentioned in Section IV, C. In contrast, resistant potato flea beetles *E. cucumeris* actually feed on DDT-treated foliage more readily. Thus, the change in behaviour here is in the opposite direction. Behavioural changes in mosquitoes have been a source of considerable trouble in mosquito control, expecially in connection with the worldwide malaria eradication operations of the World Health Organization. This sapect of behaviour has been reviewed in detail by Mattingly (1962) and in a most recent most recent monograph on resistance by Brown and Pal.

Behaviouristic resistance or avoidance is characterized by induced exophily which, in most circumstances, results from the increased irritability of adult moxquitoes contacting a residual type insecticide deposit. The results of Trapido (1952) with *A. albimanus* in Panama are a classical example of the phenomenon. Sing the malaria cradication programme began, there have been numerous reports linking increased irritability of anopheline mosquitoes with control failures, but one suspects that many such behavioural changes are not the result of the selective effect of the insecticide; instead they may depict the normal reation of the insect to the toxicant. In this light, the detailed experiments of Gerold and Laarman (1067) and Gerold (1970a,b) are of particular interest. These authours attempted to show whether one could select mosquitoes for their ability to escape contact with DDT, in other words, selection for irritability. *Anopheles atroparvus* were selected for 32 generations on the basis of their escape behaviour from contact with DDT deposits. The percentage escaping differed widely between males and females. Neither DDT nor the solvent used (Risella oil) was found to be essential for the escape reaction, but both factors enhanced the response. The selected mosquitoes showed greater average flight speed, more direct flight, and greater flight activity. Escape was not the direct result of enhanced activity. Irritation by DDT accounts for the increased flight activity, but ability to escape is though to be a separate component of behaviour. Susceptibility to DDT was similar in both strains. Observations on the resting habits of house flies convinced Keiding (1965) that complex factors including external (environmental) and internal (physiological) may influence the activity and choice of resting places. Do changes in behaviour result in physiological resistance to insecticides ? The blow fly *Phormia terrae novae* obtains a larger dose of DDT than *M. domestica* when both are exposed to a DDT deposit. This is due to the fact that *Phormia* crawls over the surface almost continuously whereas *Musca* flies up frequently and spends much time in the air. The house fly begins to avoid contact with the trated surface whereas the blow fly increases the duration of contact. The slow development of resistance in *Phormia* is due to its behaviour with little chance of survival after contact with DDT. Resistant flies of both species are less active and less irritated by DDT. Similarly. OP-selected strains of the spotted root maggot *E. notata* survive longer periods of contact with OP compounds in plastic tubes because they spend less time on the treated papers than on the ends of the tubes. The end-resters are as susceptible to the toxicant by topical application, indicating no in crease in physiological resistance. However, parallel selection with DDT and dieldrin confers true physiological resistance. The authors conclude that strains with increased irritability have greater physiological susceptiblity due to decreased detoxication rate, whereas the opposite is true in the less irritable individuals.

A critical examination of the literature by Muirhead-Thomson (1960, 1968) failed to reveal any convincing evidence of the existence of behaviouristic resistance in mosquitoes, i.e, instances in which irritability and avoidance have appeared or have been intensified as a result of continuous insecticide pressure. The experimental approach to measuring irritability has been hindered by the fact that behaviour can readily be influenced by many external factors other than the insecticide itself, such as temperature, mosquito age, length of exposure, type of testing apparatus, mosquito species and strain, insecticide-resistance or susceptibility, etc. These factors have recently been reviewed by Kaschef (1970) and Brown and Pal (1971) and included many literature citations.

IV. THE ROLE OF SYNERGISTS IN INSECTICIDE RESISTANCE

Synergism is the phenomenon exhibited by a combination of two or more compounds such that the physiological effect produced is significantly greater than the summation of their individual effects. As related to insect toxicology, the synergist is almost always used at high dosages but below their direct toxic level and, for this reason, most synergists have been labeled as nontoxic compounds.

The practical application of synergism has been most successful with pyrethrin insecticides and related compounds and, undoubtedly, is also useful with other insecticides, but a purposeful application of this principal in pest control operations has not been widespread. It is now well recognized that synergists enhance the activity of insecticides by inhibiting the enzymes which are reponsible for their *in vivo* degradation. It is clear from the preceding discussions on insecticide metabolism that resistance to many of the commonly used insecticides is associated to a large extent with the ability of the resistant strain to detoxify the compound at a faster rate than the suceptible insect, although other factors also play a role. In most instances, these differences are quantitative in nature, so that the action of the synergist is to reduce the amount of insecticide being detoxified. The concept of interference with metabolic detoxication was provided by Lindquist *et al.* (1947) who showed that pretreating house flies with piperonyl cyclonene before application of pyrethrins resulted in enhanced knockdown and subsequent mortality. However, application of pyrethrins at an extended interval before application of the synergist had no such effect.

The biochemical significance of synergistic action has its origin in the suggestion by Wilson (1949) that piperonyl butoxide and piperonyl cyclonene most likely damage the mechanism responsible for the detoxication of pyrethrins. Synergists may be divided into two major groups: the first group is characteristic of microsomal mixed-function oxidase inhibitors and includes diverse groups of compounds such as those containing the methylenedioxyphenyl (MDP) moiety, methylenedioxynaphthalenes, substituted phenyl-2-propynyl ethers, propynylnaphthyl ethers, benzothiadiazoles, and miscellaneous compounds such as SKF 525-A (2-diethylaminoethyl-2-2-diphenylvalerate hydrochloride), Lilly 1847 (2,4-dichloro-6-phenylphenoxyethyl diethylamine), octachloro- dipropyl ether, and other experimental compounds; the second group comprises mainly analogue synergists which are structurally related to the insecticides with which they are combined.

With the widespread occurrence of house fly resistance to many of the chlorohydrocarbon insecticides, intensive efforts were made to find chemical adjuvants which might enhance the activity of these compounds against the resistant strains. In the course of searching for such compounds Sumerford *et al.* (1951a,b) and Speroni (1952) found 1,1-bis (*p*-chlorophenyl ethanol (DMC), a nontoxic structural analogue of DDT, and various related compounds to effectively synergize DDT against DDT-resistant house flies. Subsequently, a large number of compound were screened for synergistic activity with DDT, and the most active ones proved to be structural analogues of DDT with modifications mainly at the tertiary carbon moiety or derivatives of diphenylamine and benzenesulfonanilides. The vast number of compounds evaluated as DDT synergists are summarized by Sumer- ford (1954). *N,N*-Dibutyl *p*-chlorobenzene sulfonamide, designated as "WARF anti-resistant," had been marketed as a DDT synergist but attained only limited success.

The synergistic activity of these compounds with DDT is undoubtedly associated with a decrease in DDE formation *in vivo*. It has been suggested that DMC may act by competing with DDT for a site on the DDT-detoxifying enzyme, pointing out the similarity between DDT dehydrochlori- nation and DMC dehydration. DMC is metabolized by the house fly to 1,1-bis (*p*- chlorophenyl) ethylene and then to bis (*p*-chlorophenyl) acetic acid similar mechanism might be envisaged for a number of other compounds but, certainly, not all DDT synergists favour the same type of simple competitive reaction kinetics. DDT analogues which are active *in vivo* also block the activity of the enzyme DDT dehydrochlorinase *in vitro* but the synergists are not attacked in the process, as they should be if the reaction were strictly of a competitive type.

Considerations of structure-activity relationships of DDT-synergists indicate that many compounds structurally related to DDT in having the bis (*p*-chlorophenyl) moiety intact are very active synergists *in vivo*. Substitution of various radicals at the *p*-Cl position reduces synergistic activity markedly. Whether such structural analogues of DDT can ever be of usefulness in practical control operations against resistant insects is doubtful, since house flies selected with DDT-DMC or with DDT-WARF antiresistant develop resistance to the combinations and cross-resistance to other compounds.

In *A. aegypti,* as in house flies, DMC inhibits the dehydrochlorination of DDT, especially in the resistant strains, thus largely restoring the toxicity of DDT. Similarly, WARF antiresistant decreases the LC_{50} levels of resistant *A. aegypti* larvae by 6-to 60-fold, but has a negligible effect on susceptible larvae. Unfortunately, selection of a susceptible strain with the synergist DDT mixture results in a 20-to 40-fold increase in resistance in six generations, On the other hand, DMC does not synergize DDT against resistant *Culex fatigans* (Kalra, 1970b), but this is not due to lack of DDT dehydro chlorination by this strain. Lack of synerfgism by DMC was also noted in DDT-resistant *Culex tarsalis* infents and *Pediculus humanus*. TDE labeled with ^{14}C is dehydrochlorinated to some extent by *Cluex tarsalis*, but a considerable amount of FW-152 [1,1- bis (*p*-chlorophenyl)-2-,2-dichloroethanol] is extracted from the larvae and from the aqueous medium, indicating the involvement of a microsomal oxidase. Probably this is the reason why DMC fails to synergize DDT in this strain.

DMC also fails to synergize DDT against a pyrethrin-resistant strain of *S. granarius* which is also cross-resistant to DDT and Prolan. However, piperonyl butoxide, sesamex, and other MDP synergists increase the effectiveness of these insecticides against the R strain but not against the S strain. Curiously, WARF anti-R also synergizes DDT against the R strain which metabolize DDT mainly to dicofol, but this is not an isolated case since it occurs also with DDT-R *T. infestans* (Fine *et al.*, 196).

Table 10. 9

Effect of Selection with DDT: WARF-Anti-Resistant Combination on Resistance to Insecticides

	Strain A (Diazinon)	*Parent strain (U) (no selection)*			
	↓	↓	↓	↓	↓
	(A-140)	*(U-32)*	*UDS-83)*	*(F-80)*	*(U-95)*
Insecticide	*Diazinon*	*NO selection*	*DDT + WARF-anti-resistant*	*Diazinon selected*	*NO selection*
Diazinon	70x	10x	52x	72x	6x
Malathion	6	2	4	6	2
Ronnel	15	5	15	15	4
Diethoate	3	0.9	3	4	0.7
Isolan	5	7	13	7	4
Dimetilan	8	1.4	8	8	1.5
Pyrethrins	5	1.5	3.	3.	1.1
DDT	6000	80	p	6000	60
Lindane	15	30	34	32	5
Chlordane	5	–	6	–	–
Dieldrin	4	–	8	–	–
Heptachlor	4	–	7	–	–

b Degree of resistance = (LD_{50} R-strain)/(LD_{50} S-strain).

In view of the importance of the oxidative pathway in DDT metabolism (Agosin et al., 1961b) are reminiscent of Oppenoorth's

(1965) findings that the DDT-resistant house fly strain Fc contains little dehydrochlorinase activity which is unrelated to resistance and is not synergized by DMC, whereas the resistance is abolished by sesamex, indicating the presence of an oxidative metabolic pathway. Alternatively, in *Culex fatigans* at least, another resistance mechanism might be operative since neither DMC nor piperonyl butoxide synergize DDT against the resistant strain.

Autosynergism

As mentioned earlier, synergists act primarily by inhibiting the degradation of insecticides to less toxic derivatives, thus restoring to a measurable degree the original potency of the parent compound. Conceivably, such protection against degradation might be obtained by structural modifications of the toxicant at certain reactive sites with no imparirment of its toxic action. Such a compound would then have a "Built-in" synergist and the effect might be termed "autosynergism".

Structural modifications fo this type were attempted in a series of DDT analogues substituted by an additional halogen in the ortho position on one of the benzene rings. The substituted chlorine atom in the ortho position causes steric hindrance of the hydrogen at the tertiary position, thus making the molecule refractory to dehydrochlorination.

Highest insecticidal activity against DDT-R strains of huse flies was obtained with 2-(2,4- dichlorophenyl)-2- (4-chlorophenyl)-1,1,1-trichloroethane (*ortho*-chloro-DDT) but all the *o*-halogen-substituted derivatives were somewhat less toxic than *p,p'*-DDT to S strains. Exceptions were noted by Perry (1964) and Hoyer and Plapp (1966) who found a Prolan-resistant strain and a DDT-resistant strain of house flies, respectively, to be almost immune to *o*-chloro-DDT. Similarly, DDT-R *A. aegypti* larvae and *C. tarsalis* larvae were found to be highly resistant to *o*-chloro-DDT, but resistance in *C.p. fatigans* was only twofold. In neither of these cases was o-CI-DDT metabolized extensively, the maximum in 24 hours being 25% in house flies and 15% in mosquitoes, and DMC proved to have only a mild or no synergistic effect with *0*-chloro-DDT In is noteworthy that DDT dehydrochlorinase from resistant *A. aegypti larvae actively dehydrochlorinates* *o*-chloro-DDT to *o*-chloro- DDE, and this

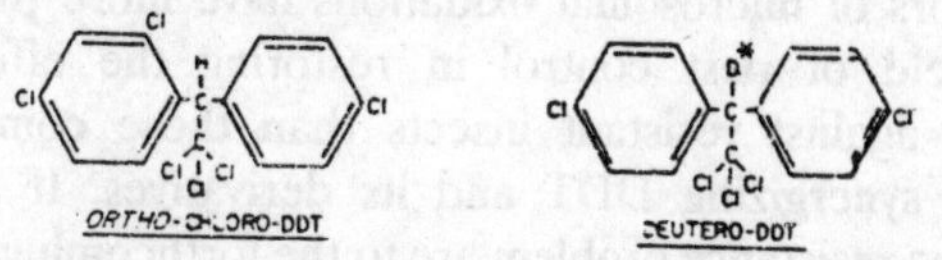

	HOUSEFLY LD50 µg/FLY			MOSQUITO (A. AEGYPTI LARVA) LC50 ppm		
	p,p'-DDT	o-Cl-DDT	DEUTERO-DDT	p,p'-DDT	o-Cl-DDT	DEUTERO-DDT
SUSCEPTIBLE	0.2	0.3	0.15	0.006	0.04	0.009
RESISTANT	0.50	1.0	5.0	4.0	6.0	0.3

Fig. 10.27. Comparative toxicity of *p,p'*-DDT, *o*-Cl- DDT, and deutero-DDT to house flies and mosquito larvae. From Perry, 1966).

conversion might be viewed as conferring resistance to this strain. These contrasting findings emphasize the danger of generalization which is frequently tempting when one tries to formulate a hypothesis correlating resistance with a biochemical factor common to all stecies or strains.

Following the same line of reasoning, substitution of deuterium (^{2}H) for hydrogen at the tertiary carbon in the DDT molecule could produce an isotope effect on dehydrochlorination thereby reducing the rate of DDT breakdown and increasing its toxicity. Indeed [^{2}H]DDT was found to be highly toxic to eleven DDT-R *A. aegypti* strains and to several other resistant mosquito species but not as toxic as *p,p'*-DDT to R *C. fatigans* larvae. On the other hand, [^{2}H]DDT proved to be only slightly more toxic than *p,p'*-DDT to S and R house flies and was shown to be metobolized with equal facility as DDT by the R strains. Comparative toxicities of *p,p'*-DDT. *o*-chloro-DDT and [^{2}H]DDT to mosquito larvae and house flies are shown in Fig. 10.27.

Tritiated DDT ([^{3}H]DDT) is considerably more refractory than *p,p'*-DDT to dehydrochlorination by alkali ($kH/k3^{H = 12.5)}$ and *in vitro* by homogenates of DDT-R house flies ($kH/k3H$ = 10.8. Accordingly, [3H]-DDT should be more toxic than *p,p'*-DDT to R flies. This could not be demonstrated because of the extremely low ^{3}H content less than 1% of the DDT molecule.

A purposeful application of the principle of analogue symergism or of autosysnergism on a rational basis has not been practice in the past, nor has this field been adequately explored. It is safer to predict that inhibitors of microsomal oxidations have more promise in the practical field of pest control in restoring the effectiveness of insecticides against resistant insects than those competitive type compounds synergizing DDT and its derivatives. If any practical answers to be resistance problem are to the forthcoming, the eventual solution may come sooner by a systematic exploration of different classes of synergists and a fundamental understanding of their mode of action.

11

Air Pollutants

Pollution of the atmosphere has been an undesirable spinoff of man's activities presumably since the cavemer first lit fires. These problems increased in magnitude with increasing urbanization. Man dug coal from the ground and used it to heat his clustered dwellings, thus creating an atmosphere of sulfurous smoke and filth above his cities. From the thirteenth century onward periodic efforts were made to forbid the burning of coal in London, but on the whole man resigned himself to acceptance of a polluted atmosphere as a part of urban life. Industrialization and technologic development added a second dimension to pollution of the atmosphere. Power plants burned fossil fuel to generate electricity to light homes and operate machines. Steel mills grew up along river banks and lake shores. Oil refineries rose in port cities or near oil fields. Smelters roasted and refined metals in areas near great mineral deposits. Synthetic chemistry came of age, and factories were built to produce the raw materials needed to manufacture the many things we now take for granted in our daily lives.

In the process of these developments too little thought was given to the effect of waste products on the environment. Some people even went so far as to equate pollution and prosperity, pointing with pride to belching stacks as a symbol of economic development. Cities sprawled and grew, as those who could afford it moved to more desirable areas and commuted daily to the heart of the city. The automobile thus came to add a third dimension to pollution of the atmosphere. "Smog" is really an old word, conined to describe the mixture of smoke and fog that hung over cities like Lodon. In current parlance, however, it has come to refer to the eye-irritating photo-chemical reaction products auto exhaust that blanket cities such as Los Angeles when meteorologic conditions produce a stagnant air mass. Over 20 years ago there was already evidence that the date of appearance of plant damage typical of photochemical smog near great cities of the world could be correlated with the date on which the consumption of gasline of gasoline passed a critical value.

Air pollution is thus by no means a new problem, although it is a problem of vital current interest. The realization has dawned that a polluted atmosphere is not just another nuisance we must sit back and accept. Efforts have been made on a national as well as on an international level to arrive at air quality standards as a rational basis for control measures. In order to do this it is necessary to integrate and interpret the result of research in many disciplines.

Chemistry provides methods for the determination of concentrations of pollutants in the atmosphere and information on their chemical interactions with one another. Meteorology gives information on conditions that can lead to stagnant air masses in which pollutants can accumulate and studies of the dispersion of pollutants from their sources. Engineering supplies the developments in technology needed to control pollution at the source. Toxicology provides information on the physiologic, biochemical, and pathologic effects of known concentrations of pollutants on experimental animals or human subjects. Epidemiology provides information to correlate the health of populations with the known levels of pollution to which they are exposed. Plant pathology yields information on the effects of various pollutants on vegetation, either under field growing conditions or under experimental exposures in greenhouse conditions. Economics attempts on the one hand to assess the cost of pollution in terms of corrosion of materials, the loss of income from damaged crops, and other effects on the economy, and on the other hand to evaluate the cost of pollution control.

Types and Sources of Pollutants

In terms of tons of meterial emitted annually into the air, five major pollutants account for close to 98 per cent of the pollution. These are carbon monoxide (52 per cent),sulfur oxides (18 per cent), hydrocarbons (12 per cent), sulfer oxides (18 per cent), hydrocarbons

(12 per cent), particulate matter (10 per cent), and nitrogen oxides (6 per cent). In individual localities the picture could vary widely from these figures. In the vicinity of a smelter, for example, sulfur oxides would be the major pollutant. Downwind from a steel mill, particulate matter would account for a greater percentage of the total pollution than suggested here. In areas where the automobile is the main source of pollution, carbon monoxide, hydrocarbons, and nitrogen oxides would be higher and sulfur dioxide would be higher and sulfur dioxide would be lower than indicated.

In discussing air pollution, the distinction is often made between two general types of pollution. The first is characterized by sulfur dioxide and smoke resulting from incomplete combustion of coal and by conditions of fog and cool temperatures. It is the sort of pollution typified by Dickens's "London particular." Because of its chemical nature it is termed reducing type of pollution. The second is characterized by hydrocarbons, oxides of nitrogen, and photochemical oxidants. It results from the atmospheric reaction products of automobile exhaust and occurs with particular frequency and intensity in areas such as the Los Angeles basin, where intense sunlight causes photochemical reactions in polluted air masses trapped by a meteorologic inversion layer. Because of its chemical nature it is termed oxidizing type of pollution. It is also called photochemical air pollution.

In terms of tons of material emitted annually, five major sources account for about 90 per cents of the pollution. These are transportation (60 per cent), industry (18 per cent), electric power generation (13 per cent), space heating (16 per cent), and refuse disposal (3 per cent). Again, these figures would be subject to wide variation depending on local sources of pollution. They co-emphasize the importance of the automobile, which has become such an integral part of our American way of life, as a major pollution source.

Acute Health Effects of Air Pollution

From time to time there have been situations in which the level of air pollution has risen to concentrations that are definitely hazardous to man's health and even to his life. In rare instances, such incidents have involved the accidental release of a specific chemical.

One such incident involved the accidental release of chlorine gas into the subway tunnels of Brooklyn. Another in Poza Rico, Mexico, involved the release of lethal concentrations of hydrogen sulfide from a malfunction of a new installation in an oil refinery. One case was of a chronic rather than an acute nature Sufficient beryllium was released from a manufacturing plant to cause beryllium disease in persons residing near the plant. In these incidents the relationship of cause and effect is straightforward. A single chemical was present in a concentration known be toxic.

In general, when we refer to acute episodes of air pollution, we have in mind three chassic episodes. The first occurred in the Meuse Valley in Belgium in 1930, the second in Donora, Pennsylvania, in 1948, and the third in London in 1952. These incidents have much in common. Meteorologic conditions were those of inversion: the normal situation, in which the lower layers of the atmosphere are warmer than the upper layers, has become inverted and hence the normal mixing and dilution that occur as warm air rises and cold air falls cannot take place. The result is an essentially stagnant air mass to which multiple sources continue to add pollutants. The analogy is often made to a"pot with the lid on it." These conditions prevailed for three or four days, during which time the concentration of pollutants rose to concentration well above the normal levels for these heavily polluted areas. The Meuse Valley and Donora are industrial areas and in them, as in London, coal was the main fuel for domestic heating. The pollutions was therefore of the reducing type, characterized by smoke and sulfur dioxide. In the Meuse Valley and in Donora no measurement were made of the actual levels of pollution. During the London for the instruments that routinely recorded daily averages of smoke and sulfur dioxide indicated that on the worst day the concentration of smoke was 4.5 mg/m^3 and of sulfur dioxide was 1.34 ppm. Since these are daily averages, short-term peaks would have reached levels above these values. These concentrations of these single substances are not in the range the toxicologist would consider as lethal. In the Meuse Vallley 65 people died. In Donora 20 people died. In London 4,000 deaths were attributed to the fog incident. It is ironic to note that 16 years earlier the prediction was made that if an incident like that in the Meuse Valley were to occur in London, some 3,200 deaths would occur.

In all three incidents many people became ill as a result of the pollution. The mortality and morbidity occurred mainly among the elderly and among those with preexisting cardiac and/or respiratory disease. These persons were unable to cope adequately with the added stress imposed by breathing the heavily polluted air.

Retrospective investigation suggests the occur rence of earlier acute incidents in these localities. There have been more recent incidents in which meteorologic inversion over polluted urban areas has led to increases in mortality and morbidity. Another incident in London in 1962 caused an excess of 400 deaths over the normal value for the period and several milder incidents have been reported for New York. Studies in England have led to the suggestion that when concentrations of smoke and sulfur dioxide reach values of 0.75 mg/m^3 for smoke and 0.25 ppm for sulfur dioxide, excess mortality is observed.

There is also clear-cut evidence that more ordinary day-by-day fluctuations in pollution levels have adverse effects on sick people. By giving chronic bronchitic patients a very simple diary in which they recorded whether they felt better or worse than usual, and then relating these entries to daily air pollution data in London, a striking correlation was obtained. It was plain the patients felt worse on days of greater pollutin. These studies were continued after the British Clean Air Act had reduced the levels of both pollutants. It was then possible to conclude that when the 24-hour mean concentrations for smoke and sulfur dioxide were respectively below 0.25 mg/m^3 and 0.19 ppm, the patients showed no response.

The acute effects discussed have been associated with the reducing type of pollution. There is less clear-cut evidence to associate the photochemical oxidant type of pollution with acute effects on human health. There is of course no question of the ability of this type of pollution to produce severe eye irritation that can be correlated with the concentration of pollution present. Studies made of daily mortality among persons 65 and over in Los Angeles County indicated that this was strongly influenced by a heat wave but was not altered consistently by variations in oxidant concentrations. There has been no adequately demonstrated relationship between daliy hospital admissions and variations in concentrations of photochemical oxidants. Admissions for conditions considered "highly relevant" (allergic disorders, inflammatory disease of the eye, acute upper respiratory infections, influenza and bronchitis) were, however, found to show significant correlations with oxidant levels, carbon monoxide and ozone. It is possible to show a high degree of correlation between diminished performance of high-school cross- country track runners and increases oxidant levels occurring in the hour before the meet.

Chronic Health Effects of Air Pollution

Ambient air pollution concentrations occurring in many cities of the world contribute to the occurrence and/or aggravation of disease in urban populations. The diseases that fall into this category are the following: acute nonspecific upper respiratory disease (i.e., the "common cold") chronic bronchitis, chronic obstructive ventilatory disease, pumonary emphysema, bronchial asthma, and lung cancer.

Chonic bronchitis is characterized by excessive mucus secretion in the bronchial tree. A chronic or recurrent productive cough is present. In order for the condition to be officially defined as "chronic" these symptoms should have been present on most days for a minimum of three months in the year and for not less than two consecutive years. There appears to be little question that there is a relationship between chronic bronchitis and both cigarette smoking and air pollution. The effect of smoking is by far the greater of the two as contributing factor to respiratory disease. Unless careful data are included on smoking histories it is impossible to assess the contribution made by air pollution. This does not mean, however, that air pollution makes no contribution. The following indices have been shown to be associated with chronic bronchitis mortality: population size of the community, amounts of fuel burned in large cities, levels of annual sulfur dioxide, levels of dust fall as well as airborne dust levels, decreases visibility. The following have been shown to be associated with aggravation of symptoms of chronic bronchitis: decreased visibility, outdoor ambient air as contrasted with less polluted indoor air, temporal changes in concentrations of smoke and sulfur dioxide.

In chronic obstructive ventilatory disease the movement of air in and out of the lungs is impeded by partial closure of the airways. After eliminating the smoking factor, it was found that residents of a plluted area showed lower values for one-second forced expiratory volume (FEV_1) than did residents of an area of low pollution. The difference was small but measurable and was attributed to air pollution. Japanese studies have shown that the airway resistance of schoolchildren living in polluted areas is greater than that of similar children living in nonpolluted areas.

The fact that pulmonary emphysema seems to be increasing, especially in urban areas, points toward air pollution as a possible etiologic factor. If patients with emphysema are placed in a room in which the smog typical of Los Angeles can be removed from the air, after 24 hours they have both subjective relief and objective improvement, which can be measured by pulmonary function tests. Such patients have a decreased oxygen neen when they are breathing cleansed air than when they are breathing untreated community air.

Bronchial asthma is produced by many aeroallergens of natural origin, which are dispersed by natural forces rather than from man's activities. Sometimes these natural allergens can be introduced by man as air pollutants, for example, castor bean dust from factories processing the material, or material from grain handling and milling. Various studies have suggested that asthmatic attacks are associated with higher levels of pollution.

The cause of cancer is, as far as we know, a multiplicity of factors. Among the compounds known to occur as urban air pollutants are some, such as benzopyrene, that have known carcinogenic potency. The adsorption of carcinogenic substances on inert particulate material could prolong residance time at sensitive sites in the respiratory tract. Many air pollutants have an irritant action on the mucous membranes of the respiratory tract. There is experimental evidence that when benzopyrene is inhaled by rats whose respiratory racts have chronic irritation, bronchogenic carcinoma results. Expermental evidence also exists that when ozonized gasoline was inhaled by mice that had been infected with influenze virus, epidermoid carcinomas were produced. There is an urban-rural gradient in incidence of lung cancer, whech is real when corrected for the effects of cigarette smoking. Kotin and Falk say, "Chemical, physical and biologic data unite to form a constellation that strongly implicates the atmosphere as one dominant factor in the pathogenesis of lung cancer."

General Considerations of Reducing Type Pollution

The acute air pollution incidents made plain that under certain meteorologic conditions the reducing type pollution characterized by sulfur dioxide and smoke was capable of producing disastrous effects. This stimulated toxicologic research on experimental animals and human subjects. Special emphasis, perhaps too much for too long, was give to studies of sulfur dioxide alone.

Recognition of the critical importance of atmospheric interactions among components of the sulfurous pollution complex was long overdue when it finally found general acceptance. Sulfur dioxide does not remain unaltered in the atmosphere. Instead, a portion of it is converted to sulfuric acid, ammonium sulfate, and other sulfates. The conversion ot sulfuric acid can be brought about by soot or by trace metals such as vanadium or manganese. Recent evidence indicates that stable sulfite complexes may be formed in the presence of metals such as copper or iron. Gas-particle interactions are compli- cated and on the whole incompletely understood. Toxicologic studies suggest that some, but not all, of the sulfur-containing particulate materials are more irritant than sulfur dioxide. Research to permit definitive identification of the sulfur aerosols in the atmosphere has lagged behind our needs. Until this gap is closed it will be very difficult, if not impossible, to do definitive epidemiology.

In 1969 the National Air Pollution Control Administration issued two documents; *Air Quality Criteria for Sulfur Oxides* (USDHEW, 1969a) and *Air Quality Criteria for Particulate Matter* (USDHEW, 1969b). These reviewed the state of knowledge as of that time and were intended to relate observed effects to concentrations of the pollutants. Criteria Documents were the legal prerequisites for Air Quality Standards and formed the data base for the standards that were promulgated. Currently all of the criteria documents are scheduled for revision and updating by the EPA.

Millions of dollars are involved in control measures designed to meet these air quality standards. Thus, it is indeed valid to inquire as to the adequacy of the data base upon which current standrds rest. The proposed increase in the use of coal further emphasizes this question. In the decade since the criteria for sulfur oxides and for particulate matter were issued, an almost unbelivable amount of time has been spent in reviewing and rereviewing the literature and in identifying and reidentifying gaps in our knowledge. A cynic might wish more effort had been spent in filling in the major gaps. The report of a committee chaired by Dr. David Rall provides an excellent and balanced summary. It has thoughtfulness, brevity, and complete lack of bias to commend it.

Toxicology of Sulfur Dioxide

Studies of Mortality and Lung Parhology

Early studies of the mortality produced by sulfur dioxide utilized mice, rats, guinea pigs and insects. The concentrations required to kill animals are so high that these studies have little relevance to air pollution problems.

Chronic exposure of animals to sulfur dioxide produces a thickening of the mucous layer in the trachea and a hypertrophy of goblet cells and mucous glands that resembles the pathology of chronic bronchitis. The important point is that such changes can be produced by irritant exposure alone without the intermediary of infection. Infection is of unquestioned importance in the etiology of chronic bronchitis, but experimental evidence indicates that it is not an essential factor in the development of the excessive mucous cells characteristic of this disease. An excellent discussion of the respiratory mucous membrane and its response to irritant agents by Jeffrey and Reid (1977) is recommended reading for students interested in pulmonary toxicology. In experiments done by Dalhamm (1956), daily exposures of rats to 10 ppm sulfur dioxide for 18 to 67 days produced a thickening of the mucous layer in the trachea of rats. This layer, normally about 5 μm, increased in the exposed animals to about 25 μm. The rate of transport of the mucous layer was decreased and remained so for a month after the end of the exposure. In acute *in vivo* exposures of the trachea of rats to sulfur doxide, 12 ppm caused cessation of the ciliary beat in four to six minutes. The cilia regained mobility a few minutes after the exposure ceased. In the chronic exposures, however, the frequency of the beat of the cilia was normal. They were unable to move the thickened mucous layer efficiently, thus slowing clearance mechanisms. Normally beating cilia under a mucous blanket too thick for them to move almost suggests the end result of a self-defeating protective mechanism.

Reid (1963) exposed rats five hours a day, five days a week for three months, to 300 to 400 ppm sulfur dioxide. Mucus-secreting cells increased in number in the main bronchi, where they are normally frequent, and appeared in the peripheral airways, form which they are normally absent. Cessation of exposure arrested the increase in cells but the excess persisted for at least three months after the termination of exposure. One group of rats obtained from a supplier who had attempted to breed a strain less susceptible to bronchiectasis showed a lesser response to sulfur dioxide exposure.

Mawdesley-Thomas *et al.* (1971) exposed rats for ten periods of six hours to 50, 100, 200, and 300 ppm and examined them 72 hours postexposure. In the trachea the highest dose produced considerable epithelial damage and almost complete destruction of goblet cells. A doserelated increase in globet cells, however, occurred at lower concentrations in the trachea and at all concentrations in the bronchiolus. Excess mucus that reaches the alveolar spaces must be removed by the alveolar macrophages. Exposure to sulfur dioxide did not increase their number of macrophages but did increase their metabolic activity, as indicated by a dose-related increase in acid phosphatase. This increase could be detected at a concentration of 25 ppm.

Dogs exposed for four to five months to 500 to 600 ppm, for two hours, twice a week showed similar alterations in histochemistry and histology of the respiratory mucous membranes. The magnitude of the pathology suggests that the very high concen- trations were probably not necessary to produce critical alterations, the more so as other workers foun that daily exposures of dogs to 1 ppm for a year produced detectable effects indicating a slowing of tracheal mucous transport. In rats, daily exposures to a total of 70 to 170 hours to 0.1, 1.0, and 20 ppm interferred with the clearance of inert particles. The most marked effects were seen with lower doses administered over a longer period of time.

Following continuous exposure of guinea pigs or monkeys for periods of up to a year or more to concentrations of 0.1 to 5 ppm, no evidence of pulmonary pathology was detected. Unfortunately the more sensitive techniques discussed above for evaluation of alterations in respiratory mucous membranes were not included in the protocol of these studies.

Studies on Absorption and Distribution of Inhaled Sulfur Dioxide

On the basis of its solubility in water, one would predict that sulfur dioxide would be readily removed during passage through the upper respiratory tract. This prediction can be tested experimentally by making measurements of the drop in sulfur dioxide concentration when an airstream passes through the upper respiratory tract of larger animals, such as dogs or rabbits, or in human subjects. Indirect assessment of this factor may be made by comparison of the response to given sulfur dioxide concentrations inhaled through the nose or through a tracheal cannula in animals or by comparison of the response to nose and mouth breathing in human subjects.

By the use of $S^{35}O_2$, Strandberg (1964) was able to examine the absorption by the upper respiratory tract of rabbits over a concentration range of 0.05 to 700 ppm. At higher concentrations removal was 90 per cent or greater; this is in agreement with the findings of other workers on dogs and human subjects. At concentrations below 1 ppm, however, only 5 per cent or less was removed by the upper respiratory tract. These data fit with the observation that guinea pigs breathing through a tracheal cannula to bypass removal by the upper respiratory tract showed an increased response to concentrations of 2, 20, or 100 ppm but no difference at a concentration of 0.4 ppm. This fitting together of data obtained by different methods on different species makes a strong case for the fact that at levels pertinent to air pollution, sulfur dioxide is not efficiently removed by the upper respiratory tract.

The penetration of sulfur dioxide to the lungs is greater during mouth breathing than during nose breathing. An increase in flow rate markedly increases the penetration. In dogs breathing orally, 99 per cent of 1 ppm was removed orally at a flow rate of 3.5 liters/minute. Increasing the flow rate tenfold decreased the removal efficiency to 33 percent. These data are of significance in connection with increased uptake in persons exercising and/or mouth breathing during incidents of heavy pollution.

Studies using $S^{35}O_2$ have shown that inhaled sulfur dioxide is readily distrbuted throughout the body. This also occurs when only an isolated segment of the trachea or the upper airways is exposed. Systemic absorption occurs from these sites, although to lesser extent than when the lungs are exposed. It is possible that the lungs release some of the gas absorbed in this manner, since radioactivity could be detected in exhaled air samples collected at the carina or below when the lungs had had no sulfur dioxide exposure. The gas was presumably carried to the lungs from the pulmonary capillaries.

Inhaled sulfur dioxide is only slowly removed from the respiratory system ofr a week or more follwing exposure. Some of the S^{35} appears to be attached to protein.

Studies of Pulmonary Function

A basic physiologic response to the inhalation of sulfur dioxide is a mild degree of bronchial constriction, which is reflected in measurable increase in flow resistance. This increase in resistance has been demonstrated in guinea pigs, dogs, cats, and human subjects. A method for studying the response of guinea pigs provides a biologic assay for examining such factors as dose- response relationships, comparison of irritant potency of pollutants, and the effect of aersols on the response to irritant gases. The method is sufficiently simple that data on many animals exposed only once can be readily obtained. This eliminates the need for reexposure following "recovery," which is an understandable part of the protocol of experi- ments that use complicated physiologic preparations of larger animals. Unless repeated exposure is the factor being tested, reuse of an experimental animal is not sound toxicologic practice. Studies on larger animals permit the use of more elaborate methods of respi- ratory physiology that shed light on the mechanisms that underlie the changes.

Exposure of guinea pigs to sulfur dioxide produces a dose-related increase in flow resistance. The lower curve in Figure shows a dose response curve relating increase in resistance to sulfur dioxide concentration. Because the figure is designed primarily to demonstrate comparative irritant potency of sulfuric acid and sulfur dioxide, the concentrations are plotted as mg of sulfur per m^3 (1 ppm sulfur dioxide = 1.31 mg S/m^3). In exposures to 100 ppm or less of sulfur dioxide alone the resistance returns to control values by the end of a· one-hour postexposure period.

The increase in flow resistance is accompanied by a decrease in compliance of the lungs. The units of resistance are cm H_2O/ml/sec. The units of compliance are ml/cm H_2O. The product of resistance times compliance has the unit of time and it referred to as the time constant of the lungs. An increase in the time constant is associated with a decrease in frequency of breathing. In sulfur dioxide exposures, the increase in resistance is proportionately greater than the decrease in compliance, resulting in an elevated time constant. A decrease in respiratory frequency is observed that becomes statis- tically significant at concentrations above 25 ppm. Tidal volume increases, but not enough to compensate for the decrease in frequency. The end result is a decrease in the minute volume.

This pattern of respiratory response, characterized by increased flow resistance and decreased frequency of breathing, is typical of a variety of irritants. Among them are sulfur dioxide, sulfuric acid, formaldehyde, formic acid, acetic acid, acrolein, zinc ammonium sulfate, zinc sulfate, ammonium sulfate, ferric sulfate and iodine. For this group of irritants, increase in flow resistance is the most sensitive criterion of response. For another group of irritants the pattern of respiratory reponse is different. There is a major decrease in compliance with minimal change in resistance and an increase in the frequency of breathing. This second pattern of response is typical of deep lung irritants.

Dogs also respond to inhalation of sulfur dioxide with an increase in resistance. The gas was given to anesthetized dogs by nose, by tracheal cannula, and by exposing an isolated segment of the trachea to sulfur dioxide while the lungs were ventilated with air. The response

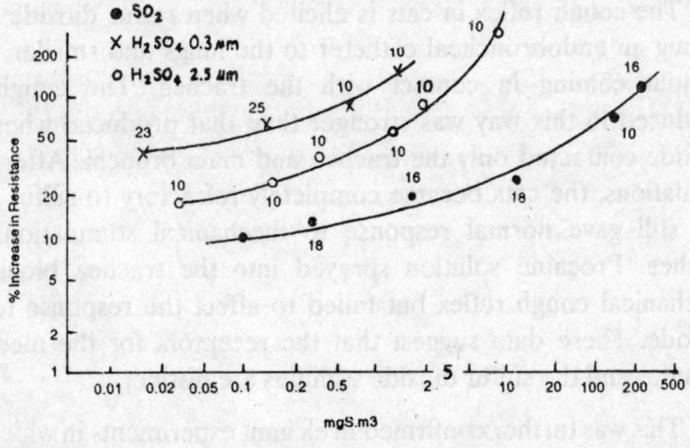

Fig. 11. 1. Dose-response curves of increase in pulmonary flow resistance produced in guinea pigs exposed for one hour to sulfur dioxide or sulfuric acid. The number beside each point represents the number of animals exposed.

was greatest when the gas was introduced directly into the lungs via a tracheal cannula and least when only a segment of trachea was exposed. The fact that resistance in creased by exposure of only the tracheal segment suggests a referred reflex constriction of the bronchi. Nasal resistance also increased in a manner roughly proportional to sulfur dioxide concen- tration over a range of 7 to 61 ppm for 15 minutes. At concentrations above 23 ppm the nasal resistance became progressively greater throughout the exposure period. These changes probably reflect mucosal swelling and/or increased secretion of mucus.

The mechanism of bronchoconstriction produced by sulfur dioxide has been studied in cats. Anesthitized cats were ventilated with a pump via a tracheal cannula. Sulfur dioxide gas was delivered either to the lungs or to the upper airways. Total pulmonary resistance increassed during the first breath when sulfur dioxide was delivered to the lower airways and lungs during a single inflation cycle. It had returned to control values within one minute. Exposing only the upper airways also produced an increase in resis- tance. An IV injection of atropine or coolingof the cervical vagosympathetic nerves abolished these effects; rewarming of the nerve reestablished the reponse. The rapidity of the response and its reversal suggests that changes in smooth muscle tone are the cause of the bronchoconstriction. The response depends on intact parasympathetic pathways.

The cough reflex in cats is elicited when sulfur dioxide is given throug an endobroncheal catheter to the lungs and smaller bronchi without coming in contact with the trachea. The cough reflex produced in this way was stronger than that produced when sulfur dioxide contacted only the trachea and main bronchi. After several inhalations, the cats became completely refractory to sulfur dioxide but still gave normal response to mechanical stimulation of the trachea. Procaine solution sprayed into the trachea blocked the mechanical cough reflex but failed to affect the response to sulfur dioxide. These data suggest that the receptors for the mechanical stimulus and the sulfur dioxide stimulus are distinct.

This was further confirmed in elegant experiments in which single vagal fibers that were excited by inflation of the lungs were dissected for study. One group of these fibers, distributed throughout the lower trachea and main bronchi, was senstitized and then inhibited by sulfur dioxide. These fibers were not inhibited by procaine. Another group of fibers was sensitive to mechanical stimulation but not to sulfur dioxide. Vagal temperatures of 7° to 10° C blocked the fibers sensitive to sulfur dioxide. Sulfur dioxide acts via the sympathetic nervous system as well as via the vagus nerve. The group of fibers sensitive to sulfur dioxide sends fibers to the sympathetic trunk.

Human subjects exposed for brief periods to sulfur dioxide also show alterations in pulmonary mechanics. Sensitive physiologic methods that yield essentially the same respiratory data as discussed above can be made without interrupting exposure to sulfur dioxide. Frank *et al.* (1962) exposed 11 subjects to 1,5 and 13 ppm sulfur dioxide for periods of ten minutes. Al 1 ppm only one individual showed an increase flow resistance. It is of interest to note that this was the individual with the highest control resistance of the group. At a concentration of 5 ppm the average increase in resistance for the group was 39 per cent above control values. Nine out of the group showed a statistically significant increase in resistance. At 13 ppm the resistance of all subjects increased, with an average increase for the group of 72 per cent. It is thus possible by meticulous attention to experimental protocol and by the use of sensitive physiologic methods to demonstrate a doseresponse relationship in human subjects. Extension of some of the exposures to 30 minutes showed that the resistance remained elevated but decreased slightly compared with the ten-minute response. Extension of the exposures to 1 ppm to 30 minutes did not produce an increase in resistance. These data suggest that the response is related to concentration of sulfur dioxide and not to a *C X T* relationship. There was a slight decrease in functional residual capacity at 13 ppm, but the two lower concentrations produced no effect.

Speizer and Frank (1966) compared the effects of breathing sulfur dioxide by nose and by mouth in ten-minute exposures to 15 or 28 ppm. In mouth breathing the pulmonary flow resistance increased in 9 out of 12 experiments. When the same concentrations were breathed through the nose, pulmonary flow resistance increased in only 3 out of 12 experiments and decreased in one experiment. When the gas was breathed by nose, the pulmonary flow resistance was often higher during the postexposure period than during the exposure itself. In 8 of the 12 experiments nasal resistance increased, in three it decreased, and in one it decreased and then increased during the same exposure.

Lawther *et al.* (1975) found that airway resistance increased when subjects inhaled 1 to 3 ppm deeply by mouth. There was a dose-response relationship between the number of breaths taken (up to 32) and the observed increase in resistance. Airway resistance was increased when concentrations of 5 to 30 ppm were inhaled normally. Although concentrations as high as 3 ppm are unlikely as daily or hourly averages, they could be encountered briefly as pockets of pollution from local sources. The fact that as few as eight deep breaths will produce in creased airway resistance could be of significance to indivviduals with diseased lungs. The relationship between ventilation and perfusion could be rurther disturbed with undesirable clinical consequences. Although the subjects used in these studies were normal, healty individuals, 3 out of 25 showed a greater change that persisted for a longer period after exposure was discontinued.

Almost all of the studies of human subjects mention some individual who was more sensitive than the rest of the subjects. Upon occasion, exposures to 5 to 10 ppm have been reported to cause severe bronchospasm. These individuals who were more sensitive have been "normal, healthy subjects," not individuals with respiratory disease. Unfortunately, there are no studies of how these individuals would react to even lower concentrations of sulfur dioxide, nor do we have data that would indicate in any way what percentage of an urban population might be expected to show a greater-than-average sensitivity to sulfur dioxide.

The changes we have been discussing have been observed when animals or human subjects have been exposed for short periods of an hour or less to sulfur dioxide. Alarie *et al.* (1970) exposed guinea pigs to 0.13, 1.01, and 5.72 ppm sulfur dioxide continuously for a year. When the guinea pigs were compared with a comparable control group breathing clean air, no evidence was found of adverse effects on the mechanical propeties of the lug. Measurements included tidal volume, respiratory rate, minute volume, flow resistance, and work of breathing. Monkeys (groups of nine animals) were exposed continuously for 78 weeks to 0.14, o. 64, and 1,28 ppm sulfur dioxide. No detrimental alterations in pulmonary function were detected. A fourth group had been started at a level of about 5 ppm. After 30 weeks on experiment, there was an accidental high exposure for an hour to somewhere beween 200 and 1,000 ppm. These animals were continued for the functional measurements but shifted to room air instead of continuing the exposure. Following the over- exposure, these animals showed a definite deterioration in pulmonary function, which was shown clearly by the change in the distribution of ventilation.

Lewis *et al.* (1969) exposed dogs to levels of 5 ppm sulfur dioxide dabout 21 hours a day for 225 days. One group of dogs was normal and the other had had lung impairment produced by prior exposure to notrogen dioxide (191 days at 26 ppm). The exposure to sulfur dioxide produced about a 50 per cent increase in resistance and about a 16 per cent decrease in compliance. In general, the adverse effects were less in the dogs with impaired function from previous nitrogen dioxide exposure, which might suggest that a lung previously remodeled by a toxicant may be more difficult to alter physiologically than one that had never been exposed to toxic concentrations of irritant. Vaughan *et al.*, (1969) exposed dogs for 16 hours a day for 18 months to a combination of 0.5 ppm sulfur dioxide and 0.1 mg/m^3 sulfuric acid and concluded there was no impairment in pulmonary function.

Well-planned and carefully executed studies of the effect of sulfur dioxide on human subjects exposed continuously for periods of 120 hours were reported by Weir *et al.* (1972). Their subjects lived throughout the exposure period in a 30-m^3 dynamic flow chamber. Daily pulmonary function measurements included airway resistance at varying lung volumes, functional residual capacity, dynamic lung compliance, and total lung volume. In the initial series, 0, 3, 6 and 8 ppm sulfur dioxide were used, Dose-related changes were observed in subjective complaints, clinical evaluations, blood gases, airway resistance, and dynamic lung compliance. The second group of experiments used concentrations of 0.3,1 and 3 ppm sulfur dioxide. No dose-related changes were observed in subjective complaints, clinical evaluations or most pulmonary function measurements at these lower concentrations. Significant but minimal reversible increases were noted in airway resistance accompanied by a decrease in compliance at high frequencies of breathing at the 3-ppm concentration. Unfortunately a detailed report of these studies has never been published.

Biochemical Studies

Information on biochemical aspects of the toxicology of sulfur dioxide is very limited. Some early studies indicated that radioactive sulfur persisted in the lung incorporated into protein, but no evidence was presented on the nature of this complex. Work by Gunnison and Palmes (1974) has indicated the presence in plasma of S-sulfonate formed by the reaction of sulfite with the disulfide bond inproteins. This has been found in rabbits and in human subjects. They examined the plasma from subjects exposed in Weir's experiments and found a positive correlation of plasma S-sulfonate levels with exposure concentrations of sulfur dioxide. Most recently they have found S-sulfonte present in the plasma and aorta of rabbits infused with sulfite. While the biologic significance is not at present understood, this finding represents the first biologic significance is not at present understood, this finding represents the first biochemical alteration observed in target organs.

Toxicology of Sulfuric Acid and Particulate Sulfates

Sulfuric Acid

The guinea pig is the small mammal most sensitive to sulfuric acid inhalation. Rabbits, rats, and mice are much less sensitive. Two early studies

used guinea pigs to examine the lethal effects of eight-hour exposures to sulfuric acid. The concentration required to kill the animals ranged from 18 to 60 mg/m^3. Animals were more sensitive at one to two months of age than at 18 months Particles of 2.7 μm were more toxic than 0.8 μm particles when mortality was the criterion of response. This is probably because the larger particles deposited in areas of airways would be rapidly lethal. Dropping the temperature to 0° C significantly increased the toxicity of the sulfuric acid, probably due to the cold stress on a tropical animal. Sufficient ammonlum carbonate in the chamber to provide an excess of ammonia protected against the effects of sulfuric acid.

The pathologic findings in the lugs were similar in the two investigations. The cause of death in animals that succumbed rapidly appeared to be bronchoconstriction and laryngeal spasm. Animals dying after a longer exposure showed gross pulmonary pathology, including hemorrhage, capillary engorgement, and some edema. The animals surviving the exposure showed spotty areas of old hemorrhage and areas of consolidation, especially around the hilar regions. Such damage is repaired only slowly.

Extending the exposure time to 72 hours did not increase the mortality beyond that observed for eigth-hour exposures, but did increase the lung pathology. The mortality was related to concentration of sulfuric acid, and the degree of lung pathology was related to the total does (concentration X time). This suggests that sulfuric acid has two actions. First, it promotes laryngeal spasm and bronchospasm, which are the causes of death. These actions are related primarily to concentration and individual senstitivity. Sulfuric acid can also cause parenchymal lung damage and this action is related to the total dosage.

Sulfuric acid produces an increase in flow resistance in guinea pigs, the magnitude of which is related to both concentration and particle size. In the following discussion, particle size is expressed as mass median diameter (MMD). Particles of 7 μm produced only a slight increase in resistance even at very high concentration (30 mg/m^3). Since these particles would not penetrate beyond the upper respiratory tract, this response was probably a referred reflex or an increase in nasal resistance. Particles of 2.5 γm gave response that was slow in onset and accompanied by a major decrease in compliance. These mechanical changes where suggestive of closure of large areas of the lung due to constriction or obstruction with mucous secretions. Particles of 1 μm or below produced a swift response similar to that observed with irritant gases. Irritant potency increased with decreasing particle size. At concentrations below 1 mg/m^3 the response was greater for 0.3-μm than for 1- μm particles. Elevated flow resistance produced by sulfuric acid is much slower to return to control values than is the increase produced by sulfur dioxide. The action of irritant particles deposited on the lung surface is more prolonged than that of a gas, which is rapidly cleared from the lungs when exposure ceases.

Dose-response curves indicate that at a given concentrartion of sulfur, sulfuric acid produces a greater response than sulfur dioxide. Even the less irritant 2.5- μm particles produce a two-to threefold greater response than the equivalent amount of sulfur dioxide.

There is evidence that on a chronic basis as well, sulfuric acid is a more potent irritant than sulfur dioxide. These studies involved the exposure of monkeys to either sulfur dioxide of sulfuric acid for periods of X weeks. Table 11.1 indicates some of the changes observed. When compared on the basis of mg S/m^3, it is evident that far higher levels of sulfur administered as sulfur dioxide had no effect on the animals.

The effect of sulfuric acid (0.3 to 0.6 μm) on tracheobronchial mucociliary clearance of radioactively tagged ferric oxide particles has been studies in the donkey (Schlesinger *et al.*, 1978). Previous studies have shown that the donkey is a good experimental model, as clearance changes are consistent with and quantitatively comparable to similar data obtained in human subjects. Three out of four animals demonstrated short term slowing of clearance as result of single one-hour exposures to 194 to 1,364 μm/m^3. Two of the four animals showed a more persistent slowing of control clearance values as the series of weekly exposures progressed. Pulmonary flow resistance, compliance, and regional deposition of aerosol were unchanged. Concentrations of ammonium sulfate up to 2 mg/m^3 had no effect on clearance or on pulmonary mechanics. Sulfur dioxide affected clearance only at concentrations of 300 ppm or greater. This is another piece of evidence indicating the greater toxicity of sulfuric acid as compared to sulfur dioxide.

Particulate Sulfates

The irritant potency of zinc ammonium sulfate, which was reported as a constituent of the Donora fog, was studied at four particle sizes, from 0.29 to 1.4 μm, at concentrations from 0.25 to 3.6 mg/m^3. The lowest concentration is the order of magnitude of what might have occurred during the fog incident. All concentrations tested produced an increase in flow resistance in guinea pigs. The irritant potency increased as the particle size decreased. The slope of the dose-response curves also steepened as the particle size decreased, so that a smaller increment in concentration produced a larger increment in response. This paper points out the utter inadequacy of relying on information on mass concentration alone when attempting to assess the irritant potency of particulate matter. Information on particle size can be equally critical.

TABLE. 11.1

Comparative Toxicity of SO_2 and H_2SO_4; Pulmonary Function of Monkeys Exposed two Years

Compound	*Concentration*	*mgS/M^3*	*Alterations*		
SO_2	*1ppm*	*1.31*		*None*	
			Histopathology	*Distribution of Ventilation*	*Arterial O_2*
H_2SO_4(2.15γm)	0.38 mg m^3	0.12	Slight	0	0
H_2SO_4 (3.60γm)	2.43 mg m^3	0.80	Moderate	Moderate	Slight
H_2SO_4 (0.73 γm)	4.79 mg m^3	1.57	—	Slight	0
H_2SO_4 (0.54 γm)	0.48 mg m^3	0.16	Moderate to severe	Moderate	Moderate

Nadel *et al.* (1967) found that zinc ammonium sulfate produced a response similar to, but lesser in degree than, that produced by histamine. The experiments were made on anesthetized cats artificially ventilated with a pump. The aerosol was submicron in size and the concentration was 40 to 50 mg/m^3. The response to a three-minute inhalation included increased pulmonary resistance, decreased compliance, and increased end-expiratory transpulmonary pressure. Isporoteronol prevented the changes, suggesting that they were due to smooth muscle contraction. Cardiac arrest during histamine inhalation did not prevent the changes. This suggests that the action was directly on the airway smooth muscle and was not dependent on airculoation. To correlate with the physiologic responses, anatomic studies were made after rapid deep freezing of the lungs in the open thorax. The principal sites of constriction were the alveolar ducts and terminal bronchioles. Bronchi and bronchioles larger than 400 μm were not constricted. Physiologic response and anatomic data combine to indicate the locus of action of the submicron particles.

The sulfate ion per se is not irritant; therefore, sulfates vary widly in irriant potency. For comparison, the increase in flow resistance can be calculated per μg of sulfate. Data were obtained with 0.3-μm aerosols (Amdur *et al.*, 1978b). Table 11.2. expresses the response to the sulfate salts as a percentage of the corresponding response to sulfuric acid. Pharmacologic studies also indicate differences among sulfates. Ammonium sulfate causes histamine release by guinea pig lungs, whereas sodium sulfate does not. The removal of intratracheally injected sulfate from the lung is markedly influenced by the cation present. These data suggest that "suspended sulfate" without further characterization is unlikely to be a definitive atmospheric measurement for the assessment of health effects. Certainly it is toxicologically meaningless.

TABLE 11.2
Relative irritant Potencey of sulfates

Sulfuric acid	100
Zinc ammounium sulfate	33
Ferric sulfate†	26
Zince sulfate	19
Ammonium sulfate	10
Ammonium bisulfate	3
Cupric sulfate	2
Ferrous sulfate	0.7
Sodium sulfate ‡	0.7
Manganous sulfate †	-0.9§

‡ Particle size : 0.1 γm.
§ Resistance decreased ; change N.S.

Effect of Aerosols on the Response to Sulfur Dioxide

Measurement of flow resistance in the guinea pog is a bioassay of sufficient sensitivity to detect quantitatively the potentiation of irritant gases by aerosols. The aerosols are "inert," in that they do not affect resistance when used alone. The dose- response curve for sulfur dioxide plus the aerosol can be compared with the dose response curve for sulfur dioxide alone.

Early studies with sodium chloride delineate several factors affecting the potentiation. Particle size of the aerosol is important. Submicron particles potentiate but 2.5- μm particles do not. Concentration of the aerosol is important. At a relative humidity of 50 per cent, 10 mg/m^3 potentiates but 4 mg/m^3 does not. Post- exposure resistance values remain elevated (typical of irritant aerosols) and can be related directly to total aerosol does. The potentiation is relatively slow to develop; at ten minutes the response is the same as that to sulfur dioxide alone. Relative humidity is important

At 80 per cent RH 1 mg/m^3 is as effective as 10 mg/m^3 at 50 per cent RH. At high humidities the potentiation occurs rapidly.

The aerosols that have produced a manjor potentiation of the response to sulfur dioxide are soluble sulfur of such metals as manganese, ferrous iron, and vanadium. These materials form droplets that can dissolve sulfur dioxide, and the metal ions promote oxidation to sulfuric acid. These aerosols potentiate the response about threefold when present at a concentration of 1 mg/m^3 at 50 per cent RH. This is far greater than the potentiation produced by sodium chloride except at a high relative humidity. At a sulfur dioxide concentration of 0.2 ppm, about 10 per cent of the sulfur was converted to sulfuric acid. When this amount of sulfuric acid was added to 0.2 ppm of sulfur dioxide, the response to the gas plus metal aerosol could be duplicated.

The response to sulfur dioxide was not altered by the presence of solid aerosols of carbon, iron oxide, open hearth dust, fly ash, or manganese dioxide. The response was attenuated by aerosols of triphenyl phosphate or motor oil.

Results of studies of flow resistance in human subjects exposed to sulfur dioxide alone and in combination with sodium chloride aerosol are conflicting. Frank *et al.* (1964) exposed human subjects to 1,5 and 15 ppm of sulfur dioxide alone and with the addition of sodium chloride aerosol. In an initial series of experiments the exposures were given in sequence with a 15-to-20- minute recovery period between them. Whether the gas alone or the combination was given first was randomized. The only consistent finding was that the resistance increase in response to the second exposure was less than that evoked by the first exposure. The series was repeated, allowing the lapse of a month between exposures. No difference could be detected between the response to the sulfur dioxide alone and the response to sulfur dioxide plus the aerosol. Although the paper does not so indicate in the discussion, the variability of the response of the same individual on different occasions to the same concentration of sulfur dioxide alone is greater than the potentiating effect of sodium chloride observed in guinea pigs. Burton *et al.* (1969) tested the effect of about 2 mg/m^3 sodium chloride aerosol on the response of ten subjects to 1 to 3 ppm sulfur dioxide. As would have been predicted on the basis of experiments with guinea pigs, at this gas concentration and 4 mg/m^3 sodium chloride, no poten- tiation was observed.

Toyama (1962) reported that sodium chloride did potentiate the response of human subjects to sulfur dioxide. The dose-response curves for sulfur dioxide alone and plus aerosol resemble strikingly data on guinea pigs. The major toxicologic problem in accepting these results is the fact that the exposures were given in sequence and the second exposure was always to the sulfur dioxide plus aerosol. If data were given on responses of a similar number of subjects to sequential exposures to sulfur dioxide alone, the results could be properly interpreted. As was indicated above, Frank *et al.* (1964) found a lesser response to a second exposure given in sequence, which is the reverse of these findings. Sequential exposures are a toxicologic trap into which unwary physiologists fall.

Toyama and Nakamura (1964) observed potentiation by an aerosol of hydrogen peroxide, which would be a liquid droplet and also oxidize sulfur dioxide to sulfuric acid. Again, there is the problem of sequential exposures. No published data exist on the effect of the soluble metal salts on the reponse of human subjects to sulfur dioxide.

The techniques of the experimental toxicologist who can use many animals, exposed only once, may be more suitable for the elucidation of potentiation responses than are the techniques of the physiologist who can expose only a few human subjects.

Components and Formation of Photochemical Air Pollution

Photochemical air pollution arises from a series of atmospheric reactions. The main components are ozone, oxides of nitrogen, aldehydes, peroxyacetyl nitrates and hydrocarbons. From the point of view of a discussion of the toxicology of air pollutants, the hydro-carbons as such do not concern us. The concentrations of hydro-carbons in ambient air do not reach levels high enough to produce any toxic effect. They are important because they enter into the chemical reactions that lead to the formation of photochemical smog. The chemical reactions that lead to the formation of this particular complex of pollutants in the atmosphere are extremely complex.

The oxidant found in the largest amount in polluted atmospheres is ozone (O_3). Several miles above the earth's surface there is sufficient shortwave ultraviolet light to convert O_2 to O_3 by direct absorption, but these wavelength do not reach the earth's surface. Of the major atmospheric pollutants, nitrogen dioxide is the most efficient absorber of the UV light that does reach the earth's surface. This absorption of UV light by NO_2 leads to a complex series of reactions, which may be simplified as follows:

$$NO_2 \text{ UV} \qquad NO + O \tag{1}$$

$$O + O_2 \rightarrow O_3 \tag{2}$$

$$O_3 + NO \rightarrow NO_2 + O_2 \tag{3}$$

Since NO_2 is regenerated by the reaction of the NO and O_3 formed, the overall result is a cyclic reaction that can be perpetuated.

This NO_2 photolytic cycle serves to explain the initial formation of O_3 in polluted atmospheres, but cannot explain the development of concentrations of O_3 as great as those that have been measured. If no additional mechanisms were involved, most of the O_3 would be broken down by reaction with the NO formed, and in steadystate conditions O_3 and NO would be formed and destroyed in equal quantities. The hydrocarbons, expecially olefins and substituted aromatics, become important by providing the necessary added reactants. Oxygen atoms attack the hydrocarbons. The resulting oxidized compounds and free radicals react with NO to produce more NO_2. Thus the balance of the reactions shown in equations 1 to 3 is upset so that NO_2 and O_3 levels build up while NO levels are depleted. These reactions are very complex and involve the formation of intermediate free radicals that are very reactive and undergo a series of changes.

Aldehydes are major products in the photo oxidation of hydrocarbons, and in the reactions of hydrocarbons with ozone, oxygen atoms, or free radicals. Formadehyde and acrolein have been specifically identified in urban atmospheres. About 50 per cent of the total aldehyde is present as formaldehyde and about 5 per cent as acrolein. Peroxyacetyl nitrate, often referred to as PAN, is most likely formed in the atmosphere from the reaction of the peroxyacetyl redical with NO_2. Its chemical formula is CH_3COONO_2. Higher homologs are probably also present, but as present in urban atmospheres.

Toxicology of Ozone

Studies of Mortality and Lung Pathology

Ozone is a deep lung irritant capable of causing death from pulmonary edema. The early studies, many of them from his own research group, have been well reviewed by Stokinger (1965). The LC_{50} following three-hour exposures varies from about 50 ppm for guinea pigs to about 20 ppm for mice. The lethality of ozone is influenced by factors other than concentration and length of exposure. Young mice are more susceptible. Elevated ambient temperature or exercise increases the toxicity. Intermittent exposures are much less toxic than the same total dose administered continuously. Early studies reported gross pumonary edema in mice exposed four hours to 3.2 ppm. Exposures to 1 ppm produced engorged blood vessels and excess leukocytes in lung capillaries. Rats exposed to 2 ppm for three hours showed an increased water content of the lung. More sensitive techniques. which utilize recovery of radiolabeled blood albumin in lung lavage fluid, indicate that the threshold for edema formation in rats is 0.25 to 0.5 ppm for six hours.

Boatman *et al.* (1974) reported desquamation of the cilliated epithelium throughout the ciliated airways of cats exposed four to six hours to 0.25, 0.5 and 1.0 ppm. The degree of damage was dose related. Alveolar damage included swelling and denudation of the cytoplasm of type 1 cells, swelling or rupture of the capillary endothelium, and lysis of erythrocytes. In rats, Stephens *et al.* (1974) observed degenerative changes in type 1 alveolar cells after three-hour exposure to concentrations as low as 0.2 ppm. These are replaced by type II cells beginning a day after exposure.

Rats and monkeys exposed for eight hours a day on seven consecutive days to -0.2,0.5, or 0.8 ppm showed mild but significant morphologic lesions at the lowest concentrations. When an exposure of rats was extended contiuously, lesions reached a peak in three to five days and then diminished. After 90 days at 0.8 ppm there was obvious damage but this was less severe than at seven days. The biochemical and morphologic changes associated with this adaptation are currently under active investigation.

It is important to note that in the case of ozone, pulmonary pathology has been observed in experimental animals following relatively short exposures to concentrations that can occasionally be attained for short periods in polluted urban areas. Early studies suggested that long-term exposure to ozone results in effects on morphology and function of the lung and acceleration of lungtumor formation and of aging. Chronic bronchitis, bronchiolitis, fibrosis, and emphysematous changes were observed in a variety of species exposed to ozone concentrations slightly above 1 ppm.

Electron microscopy has been used to document morphologic changes in the lungs of dogs exposed to 1 to 3 ppm ozone for up to 18 months. Daily exposure times were 8, 16, or 24 hours. The effects increases in severity in a manner related more to concentration than to duration of exposure. A thickening of the terminal and respiratory bronchioles was accompanied at the highest concentration by an infiltration of cells that reduced the caliber of the small airways. Young rats exposed for up to three weeks to 0.5 to 0.9 ppm showed morphologic lesions in respiratory bronchioles, distal portions of the terminal bronchiolar epithelium, the alveolar duct, and alveoli. It is thus evident that damage from ozone occurs in peripheral areas of the lung.

Toerance Development

Early studies indicated that a single brief (on hour) exposure to low concentrations on ozone (0.3 ppm) would protect against subsequent exposure to otherwise lethal concen- trations. Tolerance lasted four to six weeks in rats and up to 14 weeks in mice. Water content of the lungs remained normal and there was less histopathologic damage. Lung alkaline phosphatase and gluta-thione levels remained normal in tolerant animals. Cross-tolerance was found between ozone and other edema- producing irritants such as nitrogen dioxide and phosgene.

The tolerance protects against pulmonary edema but not against alterations in pulmonary function. More recent evidence indicates that the antibacterial defense mechanisms of the lung are not protected. Other investigations showed that repeated exposures to low levels of nitrogen dioxide conferred less protection against subsequent exposures than did a single exposure. It is thus unlikely that tolerance development plays a significant role in protecting human populations exposed to low concentrations of oxidant pollutants.

Studies of Pulmonary Function

Exposure to ozone produces alterations in respiration, the most characteristic of which are shallow, rapid breathing and a decrease in pulmonary compliance. Alterations in resistance occur only at high concentrations. This pattern is typical of materials that have their site of action in the smaller airways and peripheral portions of the lung. In two-hour exposures of guinea pigs to 0.34 to 1.35 ppm), the earliest effects detected were an increase in frequency and a decrease in tidal volume that were observed at all concentrations. Higher concentrations produced a greater change that was present earlier in the exposure period. Higher concentrations also produced an increase in flow resistance. These changes were reversible during a 90 minute recovery period. More recent work from Amdur's laboratory has shown that concentrations of 1.3 to 1.8 ppm will reduce pulmonary compliance by about 50 per cent. A lesser change in compliance was detected as low as 0.4 ppm, but not at 0.2 ppm.

Ozone exposure increases the sensitivity of the lung to broncho-constrictive agents such as histamine, acetylcholine and allergens. Easton and Murphy (1967) found an increased respiratory response to histamine injected s.c. in guinea pigs two hours after a two-hour exposure to 1 to 5 ppm oxone. Increased mortality for higher levels of histamine could be produced by as little as 0.5 to 1 ppm oxone. The increased susceptibility persisted up to 12 hours after a two-hour exposure to 5 ppm oxone. More recent studies on dogs have shown an increased respiratory response to an aerosol of histamine following exposure to 1 ppm ozone. This sensitivity can persist for 7 to 28 days. Compared to air controls, guinea pigs exposed one hour to 0.1 ppm O_3 showed a more marked decrease in compliance in response to S.C. injection of histamine two hours after exposure. These studies raise important questions in regard to the possibility of ozone sensitizing to sub- sequent exposure to other pollutants.

Alterations in pulmonary function have also been observed in human subjects, especially when intermittent light exercise is superimposed on exposure. This entire area, including emphasis of points that are critical to appropriate experimental design, is discussed at length in the 1977 National Academy of Sciences monograph referenced earlier. Hazucha *et al.*, (1973) exposed human subjects for two hours to 0.37 and 0.75 ppm with alternate periods of rest and exercise. These exposures produced alterations in several pulmonary function tests, including a decrease in forced vital capacity, forced expiratory volume at one second (FEV_1) maximal flow rate at 50 per cent of vital capacity, and maximal midexpiratory flow rate. Closing volume and residual volume increased. The subjects in these studies were Canadians. A group of southern Californians studied by Hackney *et al.* (1975) did not respond to a four-to-five-hour exposure to 0.5 ppm ozone alone or combined with 0.3 ppm nitrogen dioxide or 0.3 ppm nitrogen dioxide and 30 ppm carbon monoxide. Another group of subjects, each with either a history of developing symptoms during light activity in smog or a history of asthma, developed clinical discomfort and were unable to complete a similar protocol. These experiments indicate the wide variation in sensitivity and indicate that sensitive people develop significant symptoms under exposure conditions similar to those occuring during pollution episodes. In further exploration of the difference between Canadians and californians, it was found that Canadians also showed a larger increase in recell fragility after exposure. These differences, if real, might suggest an adaptation by the Californians. Studies done by Derr *et al.* (1975) on 20 healthy adults showed alterations in pulmonary function from exposure for six hours to 0.5 ppm oxone. Ten were smokers and ten were nonsmoders. When the smokers were considered separately as a group, no significant alterations in pulmonary function were seen. Cough and chest discomfort were more frequent in nonsmokers.

Experiments in the chamber in Montreal suggested that if 0.37 ppm sulfur dioxide, which had a negligible effect on pulmonary function, was combined with 0.37 ppm ozone, observed effects were much greater than with ozone alone. This effect was not observed in the California chamber at Rancho Los Amigos. A major joint endeavor has finally resolved this conflict conditions in the Montreal chamber led to the formation of about 200 μg cf sulfate, most likely sulfuric acid, which was not present in the Rancho chamber. There is evidence from air pollution incidents in Japan and in Holland that when both ozone and sulfur dioxide are present at about 0.3 ppm, symptoms of respiratory distress are greater than would be expected especially among persons exercising. Although there is no direct evidence, experimental data would suggest that perhaps this increased response was the result of the formation of sulfuric acide aerosol.

Increased Susceptibility to Bacterial Infection

Exposure to ozone prior to challenge with aerosols of infectious agents produces a higher incidence of infection than seen in control animals. It is assumed that this results from in hibition of clearance mechanisms, either mucociliary streaming or phagocytosis. Exposure of mice to concentrations as low as 0.08 ppm ozone for three hours enhanced the mortality from subsequent exposure to a bacterial aerosol of *Streptococcus* (group C). The susceptibility of mice and hamsters to *Klebsiella pneumoniae* aerosol was increased by prior exposure to ozone, as indicated by a higher mortality, shorter survival time, and lower LD_{50} for *K. Pneumoniae* in ozone-exposed animals as compared with controls. It has been shown that exposure to ozone reduced the number as well as the *in vitro* phagocytic ability of pulmonary macrophages in rabbits. This could help to explain the increased survival time of bacteria observed in the lungs of animals preexposed to ozone.

Biochemical Studies

The similarity of some of the effects of ozone to those of radiation suggested that ozone toxicity might result from the formation of reactive free-radical intermediates. Stokinger (1965) and Menzel (1970) have reviewed the evidence on this subject. Ozone and X-irradiation are nearly additive in producing chromosomal aberrations in plants and animals. Antioxidant and radical trapping agents such as quinones, ascorbic acid, and α-tocopherol protect against ozone toxicity. Ozone- induced free radicals may be derived from interaction with sulfhydrylsand/or from oxidative decomposition of unsaturated fatty acids. It is likely that more than one radical is formed either directly from ozone or from its interaction with normal cellular constituents. Further research is needed to determine the exact radicals produced by ozone *in vivo*.

Various investigators have noted protection from oxone by sulfhydryl compounds. Decreases in lung glutathione and succinic dehydrogenase occur following ozone exposure. Exposure of rats to 2 ppm for four to eight hours produced a decrease in both protein and nonprotein lung sulfhydryl groups. The bulk of the glutathione oxidized in rat lung was in the form of mixed disulfides with lung protein sulfhydryl groups. Peak oxidation of nonprotein sulfhydryl groups did not occur until 24 hours after exposure and recovery was evident by 48 hours. Chow and Tappel (1973) have examined the effect of ozone on enzymes active in the defense against intracellular oxidation and oxidant stress. Rats were exposed to ozone at 0.2, 0.5, or 0.8 ppm continuously for eitht days. Increases in glucose-6-phosphate dehydrogenase, glutathione peroxidase, and glutathione reductase were linearly related to dose. Histochemical studies have demonstrated an alteration in NADPH- and NADH-diaphorase activity in the lungs of rats exposed to 0.8 ppm ozone for seven days.

Several lines of evidence indicate that one of the biologic actions of ozone is reaction with unsaturated fatty acids. The biologic implications and resulting products of ozonization of these fatty acids are essentially equivalent to lipid peroxidation.

Incubation of human red cells *in vitro* with ozone or ozonides provides evidence for a role of lipid peroxidation in cellular damage from ozone. Incubation with oxonized liposomes (Tiege *et al.*, 1974) or with ozonized serum was more damanging than incubation with ozoneltself. Lung toxicity similar to that seen after ozone exposure was produced in rats by IV injection of fatty acid hydroperoxidases or ozonides. Conjugated diene bonds were found in extracts of lungs of mice exposed to 0.4 to 0.7 ppm ozone for four hours. One piece of indirect evidence for a role of lipid peroxidation is the finding that animals deficient in vitamin E were more susceptible to ozone. Data on the protective effect of added supplements of vitamin E are conflicting. Menzel *et al.* (1975) found that oral administration of vitamin E did protect against Heinz body formation in cells incubated *in vitro* with ozonides. Alterations in the protein fraction of lung tissue have been observed in rabbits following one-hour exposure to 1 ppm ozone. A variety of aldehydes and ketones were indenti- fied. the aldehydes formed may bring about an intra-or inter-molecular cross-linking of proteins that would alter normal lung structure. There was also evidence of oxidative degradation of hyaluronic acid, which would suggest alterations in the ground substance of the lungs. Since the ground substance serves as a lubricant and matrix for the fibers of the lung, a decreased slippage and flexibility of elastic protein molecules could result. This could be enhanced by croos-linking of aldehydes with fibers of elastic protein molecules. The authors suggest these alterations may help explain the increased sensitivity of ozone-exposed animals to bacterial infection because a breakdown of ground substance by ozone exposure would facilitate penetration of bacteria into the lung tissue.

Extrapulmonary Effects

Various studies have reported effects other than those directly related to the lung in laboratory animals exposed to ozone at a concentration as low as 0.2 ppm. Include are reduction of voluntary activity in mice, chromosomal aberratins in circulating lymphocytes of hamsters, and increased neonatal mortality as well as incidence of jaw abnormalities in offspring of mice exposed to ozone.

Toxicology of Nitrogen Dioxide

Studies of Mortality and Lung Pathology

Nitrogen dioxide, like ozone, is a deep lung irritant capable of producing pulmonary edema if inhaled in sufficient concentrations. This is a practical problem to farmers, as sufficient amounts can be liberated from ensilage to produce the symptoms of pulmonary damage known as silo-filler's disease.

Even over short time intervals, Haber's law does not hold for nitrogen dioxide. This low states that $C X T = K$, where C is the concentration of the toxic material, T is time, and K is a constant. For example, for a 15-minute exposure, 420 ppm nitrogen dioxide kills 50 per cent of the animals. Applying Haber's law, one would predict 105 ppm for one hour or 26 ppm for four hours would kill 50 per cent of the animals so exposed. Experimental results indicate, however, that the rats exposed for longer times can withstand higher concentrations than predicted. The LC_{50} values found for one-hour and four-hour exposures were 166 and 88 ppm, respectively. Even short exposure to the higher concentration was sufficient to produce overwhelming pulmonary edema and death.

Some earlier work in which animals were allowed to remain in an atmosphere of nitrogen dioxide until they died suggested a low toxicity. Because death from pulmonary edema is delayed, not immediate, the animals unquestionably received multilethal doses of nitrogen dioxide. In the case of irritant agents that produce delayed edema, the experimental design that measures survival time under continuous exposure will underestimate the lethal dose. Damage to mast cells in the lungs of rats was produced by exposure to as little as 0.5 ppm for four hours or 1 ppm for one hour. The demage, which was repaired within 24 hours, was interpreted as the potential onset of an acute inflammatory reaction. More prolonged alterations in lung collagen were produced in rabbits by exposure to 0.25 ppm four hours a day for six days. Damage was still evident seven days after the final exposure.

In rats exposed to 17 or 2 ppm nitrogen dioxide, changes in cell proliferation in the lung were estimated by determining the proportion of cells that could be labeled with tritiated thymidine. A dose-related increase was found that peaked on the second day of exposure and had returned to control levels by the fifth day. This occurred in terminal bronchioles and alveoli and represented an increased turnover rate of type II alveolar cell. Type I cells are damaged, slough off, and are replaced by cuboidal cells that have some of the characteristics of type II cells.

Exposure of squirrel monkeys for two hours to 10 to 50 ppm nitrogen dioxide produced primary lesions in the alveoli. The degree of damage was related to the concentration of nitrogen dioxide. At 10 ppm, there were many septal breaks and the alveoli were markedly expanded. In some areas there were large air vesicles with very thin septal walls. At 15 ppm, the alveolar tissue was expanded with minimal wall thinning and patchy interstitial infil- tration with lymphocytes. The bronchioles were normal. At 35 ppm, some lung areas were collapsed and alveolar septa had become very basophilic. In other areas the alveoli were expanded and had thin septal walls. The bronchi were moderately inflamed with some showing epithelial proliferation. Frank edema was produced by 50 ppm. The lungs of these monkeys showed extreme vesicular dilations or total collapse of alveoli with lymphocyte infiltration. The bronchi showed epithelial surface erosion and the absence of cilia.

Over the past 15 years the pathologic effects of chronic exposure of rats to nitrogen dioxide have received extensive and meticulous study with both light and electron microscopy. The emphysematous nature of the changes observed makes these studies extremely pertinent ot cigarette smoking as well as to air pollution. Cigarette smoke contains nitrogen dioxide and increases the incidence of emphysema. Rats grew grew normally and survived their natural lifespans in atmospheres containing 0.8 or 2 ppm nitrogen dioxide. They showed moderate tachypnea without apparent distress. The lungs were grossly normal except for slight bloating. They were able to contract on exposure to the atmosphere, althouth lungs from the rats exposed to 2 ppm retained some air and weighed about 20 per cent more than control lungs. The lungs of the rats exposed to 2 ppm showed histologic alterations in the bronchiolar epithelium. The cells were broader and more uniform than controls. The cilia were either reduced or absent. The possibility exists that such a deficiency in ciliary cleansing mechanisms could lead to increased residence time of materials such as carcinogenic hydrocarbons. The group exposed to 0.8 ppm occasionally showed similar celllular alterations that were not seen in controls of the same age. A 16-week exposure to 4 ppm produced no gross changes in lung volume. The terminal bronchiolar epithelium was tall and hypertrophied. Exposure to 10 and 25 ppm produced large, air-filled heavy lungs without edema or increased blood volume. The animals died of respiratory failure. The development of emphysema-like lungs was accompanied by enlargement of the thoracic cage with dorsal kyphosis. The terminal bronchiolar lumen of the "emphysematous" lungs was narrowed due to hypertropy of the epithelium and to amorphous proteinaceous material, fibrin strands, and alveolar macrophages accumulated at the junctions with alveolar ducts. Such obstructive lesions may have led to the distention of the lungs and to the broken alveolar septa. Lungs from rats exposed for 20 weeks to 18 ppm nitrogen dioxide were stained appropriately for detection of collagen and elastic tissue. In the exposed lungs both materials stained more prominently and were thicker and more prominently and were thicker and more prevalent, especially at the level of the alveolar ducts. The lengthwise elastic tissue fibers were often replicated and fractured into straighter and shorter pieces than could be found in control lungs. Such changes very likely contributed to the increased lung weight and to the reduction observed in pulmonary compliance.

Rats exposed for a lifetime to 15 ppm had voluminous, dry lungs with a large functional residual capacity. At maximal standard pressure the lung volume was greater than in control animals. Terminal bronchioles and alveolar ducts showed loss of cilia, epithelial hypertrophy, and narrowing. Alveoli were distended and exposed animals had approximately one-third the number of alveoli of control animals with a resultant reduction in ventilatory surface. Arterial oxygen tension was reduced and secondary polycythemia was evident. The disease state resembled human emphysema.

Emphysema-like lesions were also produced in the lungs of beagle dogs exposed for six months to 25 ppm nitrogen dioxide. The lungs of one of six exposed dogs showed bullous emphysema. The lungs of the other five exposed dogs showed increased firmness with scattered small bullae. The lungs of all dogs exposed showed a diffuse increase in collagen.

Effects on Pulmonary Funnction

Experimental exposure of animals or human subjects to nitrogen dioxide can cause measurable alterations in pulmonary function. The pattern of changes produced is in general similar to the pattern produced by ozone. This is an increase in respiratory frequency, with a decrease in compliance as the predominant alteration in mechanical behaviour of the lungs. Pulmonary flow resistance is minimally altered. Exposure of guinea pigs for periods of two to four hours at nitrogen dioxide concentrations of 5 to 13 ppm caused an increase in respiratory frequency and a decrease in tidal volume. The overall minute volume remained the same. The magnitude of the changes in frequency and tidal volume was related to the concentration of nitrogen dioxide. In addition, the time at which the maximum response was reached was shorter at higher concentrations. At 5 ppm the respiratory frequency continued to increase and the tidal volume continued to decrease for three hours. After this no further change was produced by an additional hour of exposure. At a concentration of 13 ppm, the maximum change was reached in one hour and no further alteration was noted during a second hour of exposure. The values for respiratory frequency and tidal volume had returned to pre-exposure values by four hours after the end of exposure. The method used for studying pulmonary mechanics in this investigation did not measure compliance. There is, however, little doubt that there was an accompanying decrease in compliance.

In the long-term studies of rates by Freeman's group, an increase in frequency and a decrease in compliance were observed as chronic effects. In the studies of squirrel monkeys discussed earlier, exposure for two hours to 10 to 15 ppm caused a decrease in tidal volume and minute volume, but no increase in respiratory frequency. Recovery was complete within 24 to 48 hours after exposure. Concentrations of 35 to 50 ppm for two hours caused an increase in respiratory frequency and a decrease in tidal volume that was completely reversible in 72 hours. Similar changes were produced by two-months' exposure to 5 ppm.

Abe (1967) exposed five human subjects for ten minutes to 5 ppm nitrogen dioxide. He observed a mean increase in airway resistance of 92 per cent. This response was delayed and did not become apparent until 30 minutes after the exposure ended. The mean value of pulmonary compliance was 40 per cent below control levels by 30 minutes after the exposure. There was a marked reduction in three subjects and no change in the other two. No indication is given of the time interval over which these delayed effects would be reversed. In the late 1970a number of investigators examined the response of human subects to nitrogen dioxide. The results on the whole are more conflicting and less definitive than those obtained in similar studies of sulfur dioxide or ozone. There seems little doubt that alterations of pulmonary function can be produced in normal, healthy subjects by exposure to 2 to 3 ppm and above. There is also the suggestion in these data that a concentration of 5 ppm may render normal subjects more sensitive to bronchoconstrictive agents and that similar effects may occur at far lawer (0.1-pmm) concentrations in some asthatic subjects.

Effects on Suseptibility to Respiratory Infection

Data obtained on a variety of species of experimental animals (mice, hamsters, rabbits, squirrel monkeys) suggest that either short-term or long term exposures to nitrogen dioxide can increase susceptibility to respiratory infection by bacterial pneumonia or influenza virus. The evidence for this effect falls into three categories: (1) increased mortality rates; (2) reduced survival time; and (3) reduced ability to clear pathogenic organisms from the lung, as indicated by the number of viable organisms that can be cultured. Coffin *et al.* (1976) tested the effect of varying both concentration and time of exposure on the mortality of mice exposed to *Strepto- coccus pyrogenes*. A *C X T* value of 7 ppm-hours was used with exposures from 14 ppm for 0.5 hour to 1 ppm for seven hours. The concentration was more critical than time in increasing the mortality from infection. In other experiments, 1.5 ppm for 18 hours increased mortality by 25 per cent but a two-hour exposure to 14.5 ppm caused a 65 per cent increase.

Exposure of two hours to 5 ppm nitrogen dioxide reduced the rate of clearance of *K. pneumoniae* from the lungs of mice and hamsters. The bacterial challenge took place one hour after the irritant exposure. The number of bacteria in the lungs was measured one, three, five, six, seven, and eight hours after infection. In control animals the bacterial population was markedly reduced during the six hours postinfection. The bbacterial count then increased, reaching its initial concentration in about eight hours. In the exposed animals the initial period of clearance was reduced to 4.5 (mice) or 5 hours (hamsters) and the original concentration was reestablished in less than seven hours. In rabbits there is evidence that short-term exposure (three hours) to nitrogen dioxide will inhibit phagocytic activity.

Both increased mortality and reduced clearance rates of *K. pneumoniae* From the lungs were observed in mice exposed chronically to 0.5 ppm nitrogen dioxide. Statistically significant increases in mortality following infection were observed in mice exposed continuously for three months and after six months of daily 6-or 18-hour exposures. The clearance rate of bacteria from the lung was reduced by exposure for 6 or 18 hours a day for nine months. These eifects were more pronounced after 12 months of exposure. When exposure was continuous, a reduced capacity to clear bacteria from the lung was observed after six months of exposure as well as at 9 and 12 months.

Increased mortality was seen in squirrel monkeys exposed for short periods of two hours to 50 ppm nitrogen dioxide or for periods of one or two monts to 10 or 5 ppm when they were infected with *K. pneumoniae*. Monkeys exposed to 10 ppm nitrogen dioxide for two hours and then infected had viable bacteria present in their lungs up to 50 days after challenge. Squirrel monkeys were infected with nonlethal levels of A/PR-8 influenza virus and then exposed continuously to 5 or 10 ppm nitrogen dioxide. All six monkeys exposed to 10 ppm died within three days and one out of three exposed to 5 ppm died. Other experiments suggested that exposure of squirrel monkeys for five months to 5 ppm nitrogen dioxide depressed the formation of protective antibody against this influenza virus.

Biochemical Effects

Both acute and chronic exposure to nitrogen dioxide can cause biochemical alterations as well as histologically demonstrable lung damage. Short exposures of one hour to as little as 1 ppm nitrogen dioxide caused alterations in the configuration of collagen and elastin in the lungs of rabbits Differential ultraviolet spectrophotometry showed molecular alterations in both collagen and elastin in animals killed immediately after exposure. The changes were not observed in an animal killed 24 hours after exposure, which demonstrates the reversibility of the change. It may be remembered that the structural changes in collagen produced by repeated exposure were not reversible up to at least seven days after the final exposure. One might speculate that repeated denaturation of the sort observed here can lead on to irreversible structural change.

It is interesting to note that hydroxylysine glycosides, probably originating from the degradation of collagen, were found in the urine of the Apollo crew of the Apollo-Soyuz mission. They had been exposed to an average of 250 ppm nitrogen dioxide for four to five minutes. Pulmonary damage was evident on roetgenograms. Nitrogen dioxide, like ozone, can cause lipid peroxidation. Four-hour exposures to 1 ppm produced evidence of lipid peroxidation in extracts of lipids from rat lung. Peroxidation wass maximum 24 hours after exposure and lasted for at least 24 hours longer. When six daily exposures were given, lipid peroxidation increased. This suggests a cumulative effect. Rats fed a diet deficient in vitamin E showed more peroxidation in surfactant and tissue lipids than rats fed a diet normal in vitamin E content.

Aldehydes

Various aldehydes in polluted air are formed as reaction products in the photo-oxidation of hydrocarbons. The two aldehydes of major interest are formaldehyde and acrolein. The irritant nature of these materials probably contributes to the odour and eye irritation of photochemical smog. Formaldehyde accounts for about 50 per cent of the estimated total aldehydes in polluted air. Acrolein, the more irritant of the two, may account for about 5 percent of the total aldehydes. A recent study by Kane and Alarie (1978) indicates that these aldehydes act as competitive agonists. Irritation would not be related to "total aldehyde" but to specific concentrations of acrolein and formaldehyde.

Formaldehyde

Formaldehyde is a primary irritant. Because of its high water solubility, it produces signs of irritation of mucous membranes of the nose, upper respiratory tract, and eyes. Concentrations of 0.5 to 1 ppm are detectable by odor, 2 to 3 ppm produce mild irritation, and 4 to 5 ppm are intolerable to most people.

The effect of low concentrations of formaldehyde on the respiration of guinea pigs has been studied. A one-hour exposure to concentrations of 0.3 ppm and above produced an increase in pulmonary flow resistance accompanied by a lesser decrease in compliance. The respiratory frequency and minute volume decreased, but changes in these factors did not become statistically significant until concentrations of 10 ppm and above were use. The over all pattern of respiratory response produce by formaldehyde is similar to that produced by sulfur dioxide. A concentration of 0.05 ppm produced no alterations in any of the respiratory criteria used. Below concentrations of 50 ppm the alterations were reversible within an hour after the exposure. The response to a given concentration of formaldehyde was greater when the gas was inhaled through a tracheal cannula that bypassed the scrubbing effect of the upper respiratory tract and permitted a greater concentration of the irritant to teach the lungs. The response in these animals was also readily reversible and the flow resistance values had returned to preexposure levels by one hour after the end of exposure.

The response to formaldehyde was potentiated by the simultaneous administration of an aerosol of submicron particles of sodium chloride. The values for pulmonary resistance remained above pre-exposure levels for one hour after the end of exposure when the gas-aerosol combination was used. This prolonged response, which is typical of the response of irritant aerosols, suggests that the potentiation is brought about by the attachment of formaldehyde to the particles to form an irritant aerosol. This hypothesis is further supported by the fact that when 3,10 and 30 mg/m^3 concentrations of sodium chloride were used, the potentiation increased with the increasing concentration of particles. The response to a given concentration of formaldehyde plus aerosol breathed by nose was greater than the response to the gas alone breathed through a tracheal cannula. This indicates that the increment added by the aerosol is not due to the transfer of an additional amount of formaldehyde gas as such to the lungs, since it was greater than could be accounted for by the transfer of the full concentration of formaldehyde to the lungs. The particles of Los Angeles type smog are capable of carrying a considerable amount of formaldehyde. This would suggest that the biologic data obtained may have some practical significance.

Acrolein

The fact that it is an unsaturated aldehyde makes acrolein much more irritant than formaldehyde. Concentrations of below 1 ppm cause irritation of the eyes and mucous membrances of the respiratory tract.

The effect of acrolein on the respiratory function of guinea pigs has been studied. Exposure to 0.6 ppm and above increased pulmonary flow resistance, incresed tidal volume, and decreased respiratory frequency. The effects were reversible when the animals were returned to clean air. In the case of irritants or this type, flow resistance is increased by concentrations below those which cause a decrease in frequency. This would suggest that flow resistance increases would be produced by far lower concentrations of acrolein than were tested. Atropine, aminophylline, isoproterenol, and epinephrine partially or completely reversed the changes. Pyrilamine and tripelennamine were without effect. The mechanism of increased resistance appears to be bronchoconstriction mediated through reflex cholinergic stimulation.

Inhalation of acrolein also has effects on various rat liver enzymes. Elevated alkaline phosphatase activity occurred in the liver following 40 hours' continuous exposure to acrolein at concentrations as low as 2.1 ppm. Exposure to higher concentrations for shorter times also increased the enzyme activity, but the effect was not constant with a constant CT. Continuous exposure to 4 ppm acrolein for 4, 8 and 20 hours, resulted in liver AP values that were 135, 222 and 253 per cent of their respective controls. When a constant CT of 80 ppm-hours was given as 4 ppm for 20 hours, 2 ppm for 40 hours, and 1 pp. for 80 hours, the increases in alkaline phos- phatase were 233, 146 and 103 per cent of respective control values. This suggests that a threshold level of irritation must be reached to initiate processes that result in elevated enzyme activity. Liver alkaline phosphatase and tyrosine-keto-glutarate transaminase activities were markedly increased in rats 5 to 12 hours after injection or inhalation of acrolein. These effects could be prevented or substantially reduced by prior adrenalectomy or hypophysectomy or by pretreatment of the animals with chemicals that inhibit protein synthesis. The data suggest that the irritant action of acrolein stimulates the pituitary-adrenal system, leading to hypersecretion of glucocorticoids, which act to induce or stimulated the synthesis of increased amounts of the enzyme proteins by the liver.

Carbon Monoxide

Carbon monoxide would be classed toxicologically as a chemical asphyxiant and its toxic action stems from its formation of carboxy-hemoglobin (COHb). The fundamental factors of the toxicology of carbon monoxide and the physiologic factors that determine the level of carboxyhemoglobin reached in the blood at various atmospheric concentrations of carbon monoxide. The normal concentration of COHb in the blood of nonsmokers is about 0.5 per cent. This is attributed to endogenous production of CO from such sources as heme catabolism. Uptake of exogenous CO increases blood COHb in proportion to the concentration in the air as well as the length of exposure and the ventilation rate of the person. Continuous exposure of human subjects to 30 ppm CO leads to an equilibrium value of 5 per cent COHb. About 80 per cent of this value is approached in four hours and the remaining 20 per cent is approached slowly over the next eight hours. It can be calculated that contiuous exposure to 20 ppm CO gives and equilibrium COHb value of about 3.7 per cent and 10 ppm CO gives an equilibrium value of 2 per cent COHb. The equilibrium values are generally reached after eight or more houres of exposure. The time required to reach equilibrium can be shortened by physical activity.

Analysis of data from air-monitoring programmes in California indicates that eight-hour average values, which may be exceeded for 0.1 per cent of the time, ranged from 10 to 40 ppm CO. Depending on location within a community, CO concentrations can vary widely. Concentrations predicted inside the passenger compartments of motor vehicles in downtown traffic were almost three times those for central urban areas and five times those expected in residential areas. Occupants of vehicles travelling on expressways had CO exposures somewhere between those in central urban areas and in downtown traffic. Concentrations above 87 ppm have been measured in underground garages, in tunnels, and in buildings over highways.

No human health effects have been demonstrated for COHb levels below 2 per cent. Above 2 per cent COHb in nonsmokers (the median value for smokers is of the order of 5 per cent COHb) it has been possible to demonstrate effects on the central neryous system. At COHb levels of 2.5 per cent resulting from about 90 minutes exposure to about 50 ppm CO, there is an impairment of time-interval discrimination; at approximately 5 percent COHb, there is an impairment of other psychomotor faculties. Cardiovasular changes may be produced by exposure sufficient to produce over 5 per cent COHb. These include increased cardiac output, AV oxygen difference, and coronary blood flow in patients without coronary diseas. Decreased cornoary sinus blood P_{o2} occurs in patients with coronary heart disease. Impaired oxidative metabolism of the myocardium may occur. These changes could produce an added burden on patients with heart disease. Some adaptation to chronic low levels of CO may occur through such mechanisms as increased hematocrit, increased hemoglobin, and increased blood volume.

Auto Exhaust and Synthetic Smog

Many investigators have studied the effect of atmospheres designed to simulate photochemical smog. These have included irradiated and nonirradiated auto exhaust or ozonized gasoline. In such experiments the observed effects are those of a mixture of components, some of which are known, others of which are unknown. Such experiments have the advantage of being a step closer to actual urban air pollution than studies of individual specific chemicals and the disadvantage that it is difficult to determine which of the many substances present are critical for the observed effects.

Short-Term Exposures

Exposure of two to three hours to heavy Los Angeles smog containing 0.4 ppm total oxidant or synthetic smog containing 0.5 ppm total oxidant was capable of producing ultrastructural changes in alveolar tissue of mice ranging in age from 5 to 21 months. The severity of the damage increased with increasing age. No change was detectable in the youngest animals. there were definite alterations, but by 14 to 18 hours after exposure there was no detectable difference from control animals. In animals 15 months of age alterations were still present 24 hours after exposure. The endothelial cells were seriously affected but the lining epithelium and basement membrane were intact. In the oldest animals changes that could be interpreted as edema-like occurred in the lining epithelium. The alterations were still present 18 hours after exposure. the implication is clear that loss of regenerative capacity of alveolar tissure following damage is one result of aging.

Exposure of A-strain and C_{57} black mice to 1 to 3.8 ppm ozonized gasoline produced an increased incidence of lung tumors. True carcinomas have been produced in C_{57} black mice by combined exposure to ozonized gasoline and influenza virus.

Mice were exposed for four hours to autoexhaust containing 0.08 to 0.67 ppm oxidant and 12 to 100 ppm CO for four hours. Immediately following this exposure the mice were exposed to a bacterial aerosol of *Streptococcus* (group C) at the rate of 100,000 organisms per mouse. When the exhaust contained 0.35 to 0.67 ppm oxidant and 100 ppm CO, there was enhanced mortality from streptococcal pneumonia: 53 per cent among the exposed and 11 per cent among the controls. The mortality was not enhanced by exhaust containing 0.12 ppm oxidant and 25 ppm CO. The increased mortality was probably related to the oxidant content. The levels involved are well below peak concentrations reported for heavy pollution.

The respiratory function of guinea pigs exposed to irradiated and nonirradiated auto exhaust has been measured. Increases in flow resistance were produced by 150:1 dilutions of irradiated exhaust but not by similar dilution of nonirradiated exhaust. This was attributed to the formation of aldehydes, nitrogen dioxide, and total oxidant by irradiation. The nature of the response would suggest aldehyde as the most likely component responsible for the observed change.

Chronic Exposures

In an extensive experiment designed to assess the long-term effects of auto exhaust, beagle dogs were exposed daily for 16 hours for a total of 68 months. A variety of pulmonary function studies were made at intervals throughout the exposure years. At the end of the exposure period the dogs were moved from the EPA laboratory in Cincinnati to the College of Veterinary Medicine at Davis, California. There a series of physiologic measurements were made both on arrival and two years after the exposure had terminated. The dogs were then sacriviced and extensive morphologic examination by light and electron microoscopy was made of the lungs. These experiments thus provided an opportunity to correlate physiologic and morphologic observations. One hundred and four dogs were divided into eight groups. One group included 20 dogs that served as controls and were exposed in similar chambers to clean air. The seven experimental groups each contained 12 dogs. The exposures were to (1) nonirradiated auto exhaust; (2) irradiated auto exhaust; (3) sulfur dioxide plus sulfuric acid; (4) and (5) the two types of exhaust plus the sulfur mixture; and (6) and (7) a high and a low level of nitrogen oxides. The irradiated exhaust contained oxidant (measured as ozone) at about 0.2 ppm and nitrogen dioxide at about 0.9 ppm. The raw exhaust contained minimal concentrations of these materials and about 1.5 ppm nitric oxide. Both forms of exhaust contained close to 100 ppm carbon monoxide. The values for physilolgic tests done on the control dogs showed no change between the end of exposure and two years after. All other exposure groups had pulmonary function values different from controls and had more functional abnormalities at the end of the two-year postexposure period. Pulmonary function tests suggested that auto exhaust exposure injured the airways and parenchyma while oxides of sulfur or nitrogen injured the parenchyma.

Two important exposure-related pulmonary lesions were observed. Enlargement of air spaces and loss of interalveolar septa in proximal acinar regions were most severe in dogs exposed to oxides of nitrogen, oxides of sulfur, or the latter with irradiated exhaust. Hyperplasia of nonciliated bronchiolar cells was most severe in dogs exposed to raw auto exhaust alone or with oxides of sulfur. These studies indicate that alterations in function that are reflected by morphololigic injury are persistent in nature following exposure to quite relistic levels of mixed pollution.

12

Water and Soil Pollutants

The ultimate sinks for most chemicals produced and used by man are water and soil. Three quarters of the earth's surface is covered by wate, and the remainder that is not covered by water, and remainder that is not covered by asphalt or concrete is covered by soil. Although water and soil are usually considered as separate ecologic systems, one needs to realize that suspended soil particles in water represent an interface between the two systems and serve as a mechanism for contamination of the one by the other. In reality it is impossible to consider any component of the real world in isolation from any other. For our purposes, however, we shall consider the presence, fate, and effects of chemicals in water and in soil as separate systems, as far as that is possible.

Systems of water may be compartmentalized based on their natural occurrence and the use made of them. One may consider separately the naturally occurring bodies of water: marine systems, Fresh water systems, and the interface between them, the estuarine systems. One may also consider these systems on the basis of the use made of water removed from them for drinking purposes or other domestic consumption. Water systems are also the recipients of the products of domestic and industrial sewage systems. Bodies of water, including rivers, lakes, ponds, and the ground water, are also the recipients of runoff from agricultural and urban areas, which greatly modifies their capability to support life and their usefulness for other purposes.

Although water can be ultimately purified tc a specific, difinite chemical entity, soil has no commonly accepted compositional definition. Soils are composed of inorganic and organic constituents.

The inorganic are silt, sand, and clay in varying ratios. These inorganic particles are coated and admixed with organic contituents, living and dead. The behaviour of soil to a major degree is determined by the size and shape of the particles of which it is composed. Soil particles range in size from less than 0.002 mm to about 2.0 mm in diameter. Soils are classified according to particle size ranges as follows: clay, 22.002 mm; silt, 0.002 to 0.2 mm; fine sand, 0.02 to 0.20 mm; and coarse sand, 0.2 to 2.0 mm. The most important use of soil is for agriculture. Soil is the ultimate support of man's sources of most food and much fiber. In addition, the soil has been the final disposal site for much of the industrial and urban waste generated by man's societies.

The interface between soil and water is an intimate one. Virtually all water systems contain suspended soil particles, and virtually all soil contains at least a small amount of water. The sediment that is the end product of soil erosion is by volume the greatest single pollutant of surface waters and is the principal carrier of most pollutants found in water. In a joint study the United States Department of Agriculture and the Environmental Protection Agency have estimated that potential annual water erosion losses range from negligible to more than 100 tons of soil per acre. About 20 per cent of the 438 million acres of crop land in the United States averages more than 8 tons of soil loss per acre per year; 30 per cent averages less than 3 and 8 tons; and the other 50 per cent between 3 and 8 tons. In fact, the sedimentary materials in water resulting from soil erosion accumulate more than 700 times more than those derived from sewage discharges. Thus, any treatment of the environmental toxicology of soil and water must consider each as a two-phase system, each containing the other and interacting through the water-sediment system.

Sources of Chemicals in the Environment

Chemicals in the environment may be classified in variety of ways. In this chaper, we have chosen to consider chemicals primarily according to their use; secondarily, by chemical properties. Thus, we will consider chemicals by their source as follows: (1) industrial, (2) agricultural, (3) domestic and urban, and (4) naturally occurring. No matter what use is made of chemicals, contamination of the environment may be either from point sources or from nonpoint sources. The results of point source pollution are generally easy to identify and the remedies are frequently more attainable. No point

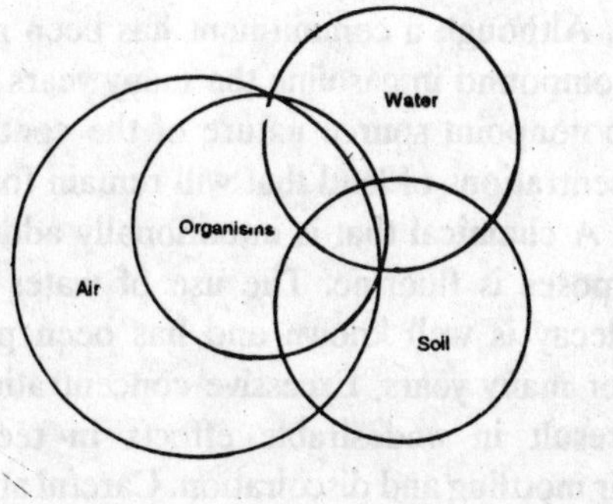

Fig. 11.1. Overlapping relationships of environmental compartments.

source pollution, on the other hand, is generally less dramatic in its initial effects, but is more difficult to contain or correct.

The production, use and disposal of industrial chemicals all lead to contamination of soil and water. Production activities lead to soil and water contamination when by-product chemicals are not properly conserved during manufacturing processes. For example, in various smelting operations toxic chemicals present in ores may not be properly controlled. Naturally occurring arsenic in copper ores, for example, frequently finds its way into soil and water. Accidental spillage of industrial chemicals may also result in contamination, sometimes dramatically, of soil and water. Careless manufacturing practices in a small chemica firm in Hopewell, Virginia, led to serious contamination of the James River and Chesapeake Bay by the pesticide Kepone. Even though these practices have now ceased, the contamination of the estuarine system will be present for many decades. Another example is the contamination of the Ohio River with carbon tetrachloride resulting from an accidental dumping of the material from a chemical plant. Such incidents of point source contamination of water can generally be prevented or controlled by the appropriate use of technology. The result, however, of not controlling such point source pollution is frequently a very high cost for decontamination, where that is even possible, and frequently both acute and chronic detrimental effects on organisms.

The use of chemicals for their intended purpose often leads to contamination, sometimes undesirably, of soil and water. Lead contamination of soils and occasionally water near highways results from the use of tetraethyl lead as an antiknock component of gasoline for automobiles. Although a commitment has been made to reduce the use of this compound in gasoline the many years during which it was use and the nonpoint source nature of the contamination have resulted in concentrations of lead that will remain for many years in soils and water. A chemical that is intentionally added to water for therapeutic purposes is fluorine. The use of water fluoridation to prevent tooth decay is well known and has been practised in the United States for many years, Excessive concentrations of fluorine, however, can result in undesirable effects in teeth, manifested primarily by their mottling and discolration. Careful attention must be paid to the use of fluorine to prevent overflouridation with its undesirable side effects.

The disposal of industrial chemicals following their use presents a major problem in several industries. Detergents used in clothes laundering are discharged into sewage systems and ultimately into rivers, lakes and strams. Phosphate detergents then serve as nutrients for algae and other organisms that can cause major difficulties in these bodies of water. The green scum resulting from algal blooms is a familiar sight in some areas. The contamination of rivers resulting from discharge of water containing organic mercurials used in paper manufacturing presents a problem in some local situations. Other chemicals resulting from paper making can also present serious water pollution problems. Asbestos tailings resulting from mining operations have also contaminated water systems in some parts of the United States, and this has resulted in concern over the potential health effects of the material that then finds its way into drinking-water systems.

The use of chemicals in agriculture results in contamination of soil and water through the direct use of pesticides and fertilizers. Pesticides are of course, applied directly to the soil in some cases to control insects, weeds, and plant diseases. Some of these chemicals can persist for many years and thereby cause concern about their potential movement from soil into water systems and from both soil and water into organisms that live in and on water and soil. The effects of persticides in the food chain are now generally familiar. Likewise, fertilzers applied to the soil to promote plant growth and productivity can leach or run off from soil and find their way into natural water systems, causing an upset in the ecologic balance to the degree that organisms living in those systems can be either enhanced or otherwise effected.

The domestic and urban use and disposal of chemicals also result in the contamination of soil and water. Domestic wastes are concentrated in sewage systems and landfill operations. Frequently, large buildups of heavy metals occur as a result. Pesticides and fertilizers used in subrurban and some urban situations for lawn and home garden purposes or pest control in other situations also are serious problems when improperly used. Detergents may also cause difficulties as referred to above. Recently, the discovery that the process of purifying water can result in the chlorination of certain small organic chemicals to produce chlorinated hydrocarbons that are potential carcinogens has generated concern. In some parts of the country concentrations of such chemicals above levels considered to be safe have been found in drinking-water systems.

Finally, metals, minerals, and plant or animal toxins are found in the environment as natural components of water and soil systems. Although they have always been there and always will be, man's activities frequently result in excessive production or movement of such chemicals found naturally occuring in the environment and can result in concentrations detrimental to man or other organisms. Furthermore, the possibility of interaction of synthetic chemicals and pollutants with naturally occurring metals, minerals, and toxins must be considered.

Transport, Mobility and Disposition

The fate and distribution of chemicals in the environment are determined by several variables that can interact in numerous ways. An analogy between pharmacodynamics and chemodynamics can be drawn to illustrate some basic similarities in each approach. First, one must appreciate the physicochemical properties of a chemical, such as water solubility, lipid solubility, partitioning behaviour, vapour pressure, pKa for ionic species, chemical stability, etc., if one is to predict the behavior of a chemical in a system – be it man or an ecosystem. Second, the processes that act within the system must be considered. Transport, via serum proteins versus suspended sediments; circulation, via the circulatory system versus the hydrologic cycle; degradation, in liver versus soil microorganisms; and excretion,

Table 12.1 Classes of Materials Related to the Effect of pH on Adsorption

Molecular form *Class*	*Example*	*pK*$_a$	*Low pH*	*High pH*	*pH Effect*
Storng acid	Linear alkylsulfonates	Anion	Anion	Small	
Weak acid	Picloram	3.7	Free acid	Anion	Large adsorption; pH a
Strong base	Diquat	Cation	Cation	Decrease at very low pH (18 N H_2SO_4	
Weak base	Ametryne	Cation	Free base	Increasing adsorption to pH approx, p	
then decrease					
Polar mokecul:s	Diuron	Nonionized	Un-ionized		Small
Neutral molecules		DDT	Nil	Nonionized	Probably none

via urine, feces, and expired air versus dilution in water and air to nondetectable levels or deposition in ultimate sinks such as deep ocean sediments, are all processes that act on chemicals to determine the mobility and final disposition of a chemical in a system. The analogy can be carried one step further to include target organs or tissues affected by the chemical in comparison with the susceptible species of an ecosystem. The fundamental difference in this consideration is one of scale, in both time and dimension, which then requires models of varying scale. Mathematical transport models for predicting the fate of chemicals remain primitive and approximate at best.

Water solubility. The water solubility and latent heat of solution are critical properties of a chemical that affect its environmental fate. Many environmental toxicants are hydrophobic, having solubilities in the parts-per-million (ppm, mg/1) to parts-per-billion (ppb, μm/1) range. Reported solubility values vary with the method used for determination. Water solubilities are affected by pH (for ionizable chemicals), presence of dissolved salts and organics and temperature.

Soil adsorption. Adsorption to particulate matter is a major mechanism by which chemicals are removed from solution. Adsorbent materials in soils and seciments can be divided into clay minerals and soil organic matter. Clay minerals include various hydrous silicates, oxides, and layer silicates. The clay minerals have been extensively studied and are characterized by physical structure or layering type, either 1 : 1 or 2 : 1, swelling ability, cation exachage capacity, and specific surface (m^2/g). These parameters are important considerations in the behaviour of organic cations, polar organic molecules, and metal ions in soils. High specific surface is associated with small particle size; therefore, the colloidal fraction of the soil is a dominant factor in chemical-soil interactions. Cation exchange capacity of the inorganic fraction is a function of the magnitude and distribution of the structural charge. Exchangeability is dependent on the adsorbed actions, usually sodium, potassium, or calcium, and the nature of the replacing cations.

The water associated with clay plays an important role in defining its characteristics. Adsorbed water on clay surfaces is more ordered than free water. Water on the clay surface may also be more ionized than otherwise. Thus, the hydrogen ion concentration of the clay surface is high.

Adsorption data for chemicals in soils is usually expressed by the Freundlich isotherm, $x/m = KCn$; x/m is the amount of chemical sorbed per weight of the adsorbent, C is the equilibrium concentration of the chemical, and K and n are constants. The constant K represents the extent of adsorption while the value n sheds light on the nature of the adsorption mechanism and the role of the solvent, water.

Soil organic matter usually ranges from 0.1 to 7.0 per cent and serves as the most important sorptive surface for nonionic chemicals. Above a few per cent organic matter, all the soil mineral surfaces are effectively blocked and thus no longer function as adsorbents. Soil organic matter can be divided into two main groups: (1) nonhumic substances, which are fresh or incompletely decomposed plant and animal material, and (2) humic substances, which are more or less completely altered or resynthesized materials. The former serve as a source for the latter. Nonhumic materials include well-known organic chemical groups with definite characteristics: proteins, carbohydrates, organic acids, sugars, fats, waxes, resins, lignins, pigments, and lowmolecular-weight compounds. These materials comprise 10 to 15 per cent of the soil organic matter. Their com- position and residence times are quite variable. Humic substances account for 85 to 90 per cent of soil organic matter and their nature is not well understood. Humic substances are fractionated to give fulvic acid, which is soluble in both alkali and acid; which is soluble alkali but not in acid; and the humin fraction, which cannot be readily extracted with cold alkali. Humic acid and fulvic acid are aromatic polymers with molecular weights that range from 5,000 to 100,000 and from 2,000 to 9,000, respectively. Functional groups that have been identified on humic substances are carboxyl, phenolic hydroxyl, alcoholic hydroxyl, carbonyl and methoxy. Heterocyclic rings with oxygen and nitrogen atoms are also present. A hypothetic structure for humic acid has been proposed by Kononova (1966) as follows:

Vaporization

Vaporization from soil, water, or plant surfaces is a major transport process for many chemicals. The volatility of a chemical is a function of its vapor pressure, but the rate of vapourization also

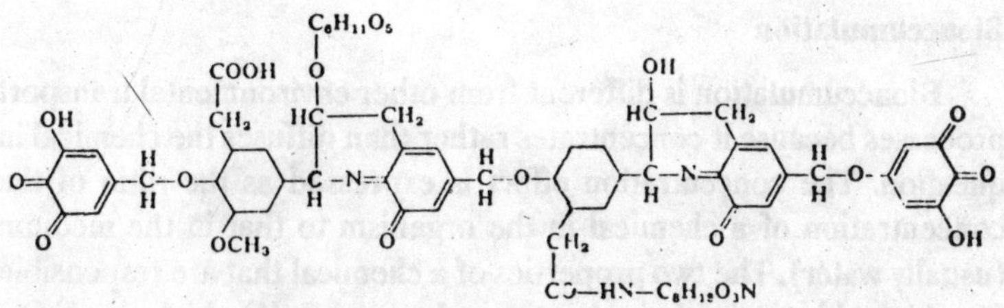

depends on environmental conditions such as temperature, degree of adsorption, soil properties, and soil water content. Airflow over the evaporating surface affects vaporization rate since air movement continuously replaces and mixes air around the evaporating surface. Many chemicals evaporate simultaneously with water, which leads some resarchers to believe that chemicals such as DDT "condistill" with water. This phenomenon can be demonstrated in laboratory distillations at 100° C, but does not occur at normal environmental temperatures. Instead, water evaporation and DDT volatilization occur independently. Higher vapour loss from most soil surfaces correlates with pesticide volatilization, but is due to desorption of chemicals from soil adsorption sites by water molecules and the mass flow of chemical to the soil surface by the "wick effect." This phenomenon has been noted with chemicals such as 2,4.D esters, thiolcarbamates, triazines, organochlorine insecticides, and *N*-methyl-carbamates. Volatility of organic chemicals from water increases with decreasing water solubility. As a result, a chemical with both low vapor pressure in the solid phase and very low water solubility would be much more volatile from aqueous solution than might be expected. DDT is again the example.

Partitioning

Many chemicals have low water solubility but high solubility in nonpolar solvents (lipophilicity). The partition coefficient as a measure of these properties is often determined in a system of *n*-octano/water. The importance of partitioning behaviour of chemicals lies primarily in the phenomenon of bioaccumulation (biomagnification or bioconcentration). Orga- nisms in contact with a solution of chemical that has a high partition coefficient will act as the nonpolar phase of the binary system and accumulate it through a partitioning process.

Bioaccumulation

Bioaccumulation is different from other environmental transport processes because it concentrates rather than diffuses the chemical in question. The concentration effect is expressed as the ratio of the concentration of a chemical in the organism to that in the medium (usually water). The two properties of a chemical that are responsible for high bioaccumulation ratio values are (1) high partition coefficient, i.e., lipophilic, and (2) recalcitrance toward all types of degradation. Bioaccumulation ratios have been determined for a variety of envirnmental chemicals in laboratory model ecosystems and correlate well with the *n*-octanol/water partition coefficients

Degradation

Transformations of chemicals in soil and water occur by chemical, photochemical, and biochemical reactions. Degra- dation results in the true "disappearance" of a chemical's molecular form as opposed to transport processes, which merely move chemicals from one environmental compartment to another. However, it must be recognized that transport processes that move chemicals to ultimate sinks, such as deep ocean sediments, for all practical purposes do remove chemicals from the environment. Chemical transformation are classified as hydrolyses, oxidations, reductions, nucleophilic substitutions involving water, and free radical reactions. These re- actions may be catalyzed by the presence of metal ions, metal oxides, clay sufaces, organic compounds, and organic surfaces. The pH of solutions and the effective pH of clay surfaces, which may be quite different from the surrounding aqueous environ- ment, can significantly influence rates of degration. Other obvious conditions that affect degradation rates are tempe- rature, moisture content in soils, and other environmental processes that alter chemical concentrations. The kinetics of degradation rates are dependent on the mechanism of degradation. Some degradative processes follow first-order kinetics. while others are best described by a "hyperbolic rate model".

Photochemical reactions of chemicals occur in air and water but are probably of little or no significance in soils. For a chemical to undergo a photochemical reaction, it must absorb light energy from an appropreate portion of the spectrum or have the light energy

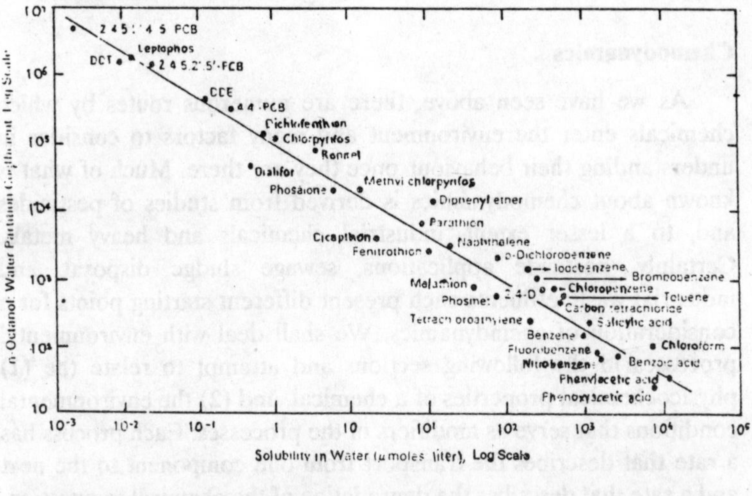

Fig. 12.2. Relationship between water solubility and *n*- octanol/water partition coefficient.

transferred through an intermediate substance know as a sensitizer. Ultaviolet light (4 to 400-nm wavelengths) has sufficient energy to break existing chemical bonds, but light above 450 nm, which represents an energy of 65 kcal/mole, is usually not sufficient to initiate reactions. Light of wavelengths shorter than 295 nm does not reach the earth's surface in appreciable amounts. The principal reactions are photo-oxidations and photo reductions which proceed through light-formed free radicals and which then react with molecular oxygen or abstract hydrogen from organic compounds, respectively.

Biologic reactions of chemicals in soil and water are mediated primarily by microorganisms. Microorganisms are quite versatile when confronted with foreign chemicals. The major reactions involved are dehalogenation, hydrolysis, oxidations, reductions, conjugations and methylations. They are also very important in the natural cycles of many elements, such as mercury and arsenic. These natural cycles can be disturbed by introduction of various forms of metals and can increase formation of toxic species, e.g., methyl-mercury. The types and rates of microbiologic reactions are determined by the microbial ecology of any given system. Thus, pol, temperature, redox potential, nutrient availability and microbial interactions will affect the microbial degradation of a chemical.

Chemodynamics

As we have seen above, there are numerous routes by which chemicals enter the environment and many factors to consider in understanding their behaviour once they are there. Much of what is known about chemodynamics is derived from studies of pesticides and, to a lesser extent, industrial chemicals and heavy metals. Certainly, pesticide applications, sewage sludge disposal, and industrial waste effluents each present different starting points for a consideration of chemdynamics. We shall deal with environmental processes in the following sections and attempt to relate the (1) physicochemical properties of a chemical, and (2) the environmental conditions that serve as modifiers of the processes. Each process has a rate that describes the transport from one component to the next and a rate that describes the degradation of the chemical in question. A complete analysis of all of the rates for entry, transport, and degradation of a chemical will describe its ultimate fate in the water and soil.

PESTICIDES

The major classes of pesticides have been grouped as "nonpersistent" or "slightly residual," "moderately persistent" or

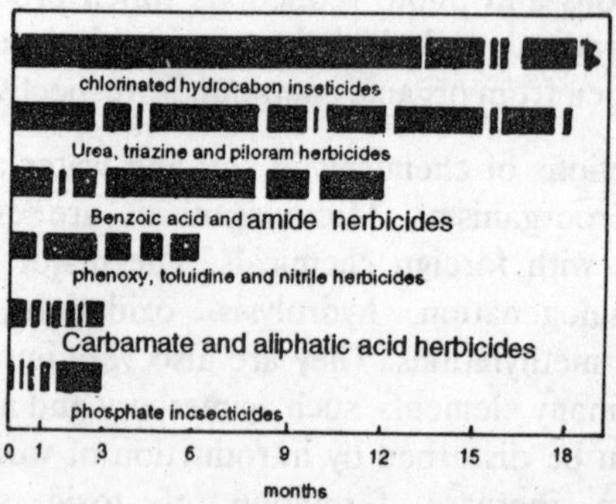

Fig. 12.3. Persistenc in soils of several classes of insecticides and herbicides.

"moderately residual," or "persistent" or "highly residual". Persistence times reflect the time required for 75 to 100 per cent disappearance of pesticide residues from the site of application. Non-persistent pestides have persistence values of 1 to 12 weeks; moderately persistent pesticides, 1 to 18 months; and persistent pesticides, two to five years. Persistence times vary with environmental conditions and the generalizations about the classes are subject to several exceptions by individual pesticides within the class.

Persistent Pesticides

Chlorinated hydrocarbon insecticides. This group of chemicals includes DDT, DTE (a major metabolite of DDT), methoxychlor, and related chemicals; the cyclodiene insecticides, aldrin, dieldrin, endrin, heptachlor, chlordane, mirex, and Kepone; the hexachlorocy clohexanes (commonly refereed to as BHC) and the purified gamma isomer, lindane; and the polychloroterpenes, toxaphene and Strobane. The technology, application, and biologic and environmental aspects of this class of insecticides have been authoritatively reviewed in a two-volume treatise by Brooks (1974). The persistence of pesticides in the environment has been the subject of numerous books and reviews.

DDT, its major metabolites, DDD (TDE) and DDE (which are collectively referred to as DDT-R), and dieldrin, which is both an insecticide and the major metabolite of the insecticide aldrin, are ubiquitous residues and the prime examples of persistent pesticides. Studies with DDT and dieldrin have elucidated several important concepts in environmental toxicology. First, persistence is not a desirable attribute as originally believed. Second, the transport and disposition of persistent pesticides is affected by physical and biologic processes, which occur from the micro to the global scale. Third, high lipid solubility combined with chemical and biologic stability can lead to bioligic magnification of pesticide residues.

Persistence is primarily a function of physico chemical properties of substances. In additions, the sorption/desoption process is one of the important factors controlling the fate of pesticides in soils. Sorption of pesticides in soils has been reviewed by numerous authors. Chlorinated hydrocarbon insecticides are highly soluble in lipids and most organbic solvents, but have low water solubilities and relatively low vapor pressures. Studies of the adsorption of chlorinated hydro-carbons on various soils are difficult because of their low water solubility. However, studies with PCBs, which have many similar physicochemical properties to the chlorinated hydrocarbons, show that adsorption of the hydrophobic chlorinated hydrocarbons depends strongly on the presence of soil organic matter. Once adsorbed, these chemicals do not readily desorb. Two implications are readily apparent: (1) such compounds will not leach or diffuse in soils, and (2) transport into the hydrosphere from contaminated soils will be through erosion of soil particles or sediment, not by desorption and dissolution. When chlorinated hydrocarbons are poorly adsorbed, as in sandy soils, the vaporization loss will be singnificant as compared to that in soils with higher organic matter.

Volatilization of pesticides into the atmosphere from water and soils is also a transport route. The volatility of a chemical from soil or water is a function of its vapour pressure, but the actual vaporization rate depends on several environmental parameters. Temperature, Soil properties, soil water content, and other physicochemical proper-ties such as water solubility and degree of adsorption affect the vaporization rate of pesticides from soil. Soil properties such as high organic matter that cause the pesticide to be strongly adsorbed reduce volatility greathly. The importance of soil moisture in volatilization of chlorinated hydrocarbons led to the use of the term "codistillation." DDT condistills with water in the laboratory. However, the effect observed in soils is more adequately described as displacement of the sorbed pesticide by water molecules, plus a carrier action by water to the soil-air interfact. The distribution of a pesticide between water solubility and vapor pressure. As a result, compounds like DDT with very low water and air is dependent on both water solubilities are quite volatile form water.

Degradation of the chlorinated hydrocarbons is quite slow as compared to the other classes of insecticides, and in soil and water is due mainly to the action of microorganism. To a lesser extent, chemical reactions and photochemical reactions degrade the chlori- nated hydrocarbons under certain conditions. A summary of the pathways for DDT emphasizes the importance of dechlorination and dehydrochlorination rections. Oxidative reactions are only moderately important in the scheme. The formation of DDCN in sewage sludge and lake sediment is an example of the unique reactions carried out by

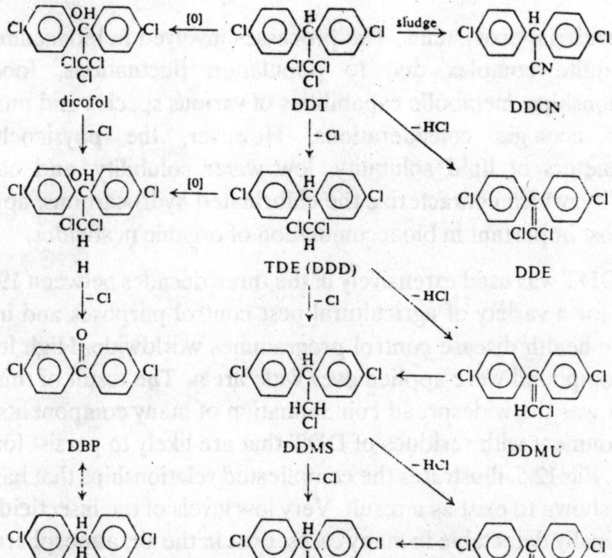

Fig. 12.4. Microbiologic and environmental degradation of DDT, illustrating the importance of dechlorination and dehydrochlorination reaction.

microorganisms. Epoxidations and rearrange- ment reactions are common among the cyclodiene insecticides. The most thoroughly studied of these reactions is the epoxidation of aldrin to dieldrin and heptachlor to heptachlor epoxide. The rearrangement products are mainly compli- cated "caged" structures that are still toxic. The caged compounds mirex and Kepone undergo very little detectable degradation. Toxaphene and BHC are degraded initially by dechlorinations and dehydrochlorinations.

The bioaccumulation of the chlorinated hydrocarbons DDT and dieldrin is well documented by environmental residue data (Edwards, 1970). Bioaccumulation rations relate organism residues to environ- mental residue levels and are higher in aquatic ecosystems as opposed to terrestrial ecosystems. The processes involved in bioaccumulation are quite complex due to population fluctuations, food web relationships, metabolic capabilities of various species and numerous other ecologic considerations. However, the physicochemical parameters of lipid solubility, low water solubility, and chemical stability, which characterize the chlorinated hydrocarbons, appear to be most important in bioaccumulation of organic pesticides.

DDT was used extensively in the three decades between 1942 and 1972 for a variety of agricultural pest control purposes and in many public health disease control programmes worldwide. High levels of the compound were applied over wide areas. The result of this heavy usage was the widespread contamination of many components of the environment with residues of DDT that are likely to persist for many years. Fig.12.5. illustrates the complicated relationships that have now been shown to exist as a result. Very low levels of the insecticide, only marginally detectable in many cases, exist in the air, atmospheric dust, and rainwater. From there the compound is transported to soil and water ecosystems where it becomes available to a variety of organisms. These residues are then bioaccumulated, with man, of course, sitting atop the food chain as the ultimate consumer. Although direct toxicity to man has not been documented as a result of the accumulation of DDT, the thin eggshells and reproductive failures of birds and chronic toxic effects on fish served as the warning signs that led to restriction of the use of this chemical.

It may be observed that the effects of bioaccumulation of DDT-R and dieldrin are manifest primarily at the tops of food chains. Predatory fish and birds suffer from acute toxicity, chronic toxicity, and reproductive failures. Behavioural changes in DDT-treated fish have also been demonstrated. Thus, the effects can range from obvious toxicity to subtle behavioural changes, the ecologic consequences of which are unknown.

Cationic Herbicides. The two chemicals of importance in this group are diquat and paraquat, which are used in "no-till" methods of farming. These compounds readily dissolve and dissociated in aqueous solution. As cations, they are strongly adsorbed to soil particles by cation exchange reactions.

$$\text{Paraquat}^{2} + 2\ \text{Na-clay} \qquad \text{Paraquat-clay} + 2\ \text{Na}^{+}$$

TABLE 12.2
Bioconcentration of DDT-R Residues in Plants or Animals form its Environment

Environment	Plant or Animal Organism	*(DDT-R Residue in Organism Divied ny Residue in Environment)* Maximum Value Observed	Minimum Value Observed
Soil	Earthworm	73	0.67
	Beetles	2.81	0.31
	Slugs	3.70	2.33
	Crop roots	0.13	0.04
	Crop foliage	0.08	–
Water	Sea squirt	1,000,000†	200†
	Sea hare	178,000†	–
	Eastern oyster, clam	70,000†	60
	Shrimp	2,800†	280
	Crabs	144	–
	Crayfish	97	17
	Snails	1,480†	–
	Plankton	16,666†	250
	Fish	829,300 †	5-(1,450) †
	Fish (DDD)	9,214 †	417
	Algae	33	0.34
	Aquatic plants	100,000 †	0.45
Diet	Pheasant	2.91	–
	Woodcock	4.5	2.6
	Bald eagle		
	Brain	0.1	–
	Liver	1.9	–
	Fat	35.7	–

+ DDT may be present in excess of solubility in water.

X-ray studies show that these planer molecules interlayer between the parallel silicate sheets of various clays. The adsorption behavior is also related to surface charge densities on various clays. Adsorption isotherms for these cationic compounds indicate a high affinity of the solute for the adsorbent until the cation exchange capacity is reached.

Paraquat and diquat as soil-bound residues are rest ant to microbial degradation and photodecomposition. Tightly adsorbed residues are not biologically available and therefore persist indefinitely. Both diquat and paraquat are nonvolatile and are not transported in the vapour phase. Environmental transport is thus tied to sediment transport processes.

Moderately Persistent Pesticides

Triazine Herbicides. The triazines behave as weak bases in aqueous solution with pK_a values that range from 1.1 to 4.3. Water solubilities are therefore determined by the pH level, with the triazines being more soluble at low pH levels. The behaviour and fate of triazines in soils has been extensively studied and reviewed. Adsorption of triazines through an exchange process to organic matter and clay minerals is dependent on the pH of the solution and the acidity of the adsorbent surface. Hydrogen bonding and hydro-phobic bonding are other mechanisms by which soil organic matter adsorbs triazine herbicides, especially at higher pH levels.

Hydrolysis and oxidation are the general routes of soil metabolism for triazine herbicides. Photodecomposition appears to be minimal on soils. Vapour transport losses of triazines are dependent on vapor pressure and pH of the evaporating surface, since ionized compounds are less volatile. Transport from soil to water occurs in solution and in sediments.

Phenylurea Herbicides. Between 20 and 25 different substituted pheylurea compounds are presently commercially available as herbicides for the control of annual and perennial grasses in alfalfa, cotton, sugar cane, pineapple, grapes, apples and citrus. The phenylureas can be divided into three categories based on their water solubilities, which in turn seem to be related to a number of other properties that each group holds in common. Fenuron, the most water soluble of the phenylureas at approximately 2,900 to 3,850 ppm, is in a category by itself. This compound is also the most mobile in soil systems of all the phenylureas. Fenuron has been shown to move substantially in a lateral direction over the soil surface and in a vertical direction into the subsoil. Its movement is also related to soil texture and organic matter content. Movement was greater in coarse-textured soils and was decreased at higher organic contents.

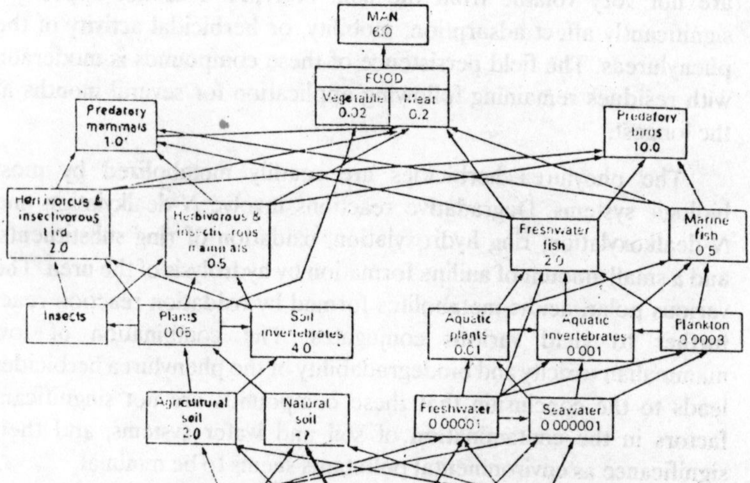

*Estimate based on very few samples
? Insufficient data available

Fig. 12.5. Typical levels of DDT (ppm) in environmental compartment, media, terrestrial and aquatic organisms, and man and the relationship between them.

The middle group of phenylureas in terms of water solubility includes monuron, diuron, linuron, monolinuron, fluometuron, metobromuron, norea, and siduron, in which water solubilities range from 18 to 580 ppm. Compounds in this category are moderately mobile in the soil; their relative movement decreases as the water solubilities decrease. Movement of the compounds also decreases as organic matter content of the soil increases, and in turn the herbicidal activity of the phenylureas decreases as the compounds are bound to the soil organic matter. The least soluble of the phenylureas are neburon and chloro-xuron, where water solubilities range from 2,0 to 4.8 ppm. These compounds are rather immobile in the soil. Most of the phenylurea compounds have relatively low vapour pressurea and are not very volatile from the soil. Soil pH does not appear to significantly affect adsorption, mobility, or herbicidal activity of the phenylureas. The field persistence of these compounds is moderate, with residues remaining following application for several months at the longest.

The pheylurea herbicides are readily metabolized by most biologic systems. Degradative reactions involve *N*-dealkylation and *N*-dealkoxylation, ring hydroxylation, oxidation of ring substituents, and a small amount of aniline formation by hydrolysis of the urea. The various polar, acidic metabolites formed by oxidation reaction react further to yield various conjugates. The combination of low mammalian toxicity and biodegradability of the phenylurea herbicides leads to the conclusion that these compounds are not singnificant factors in the contamination of soil and water systems, and their significance as environmental pollutants seems to be minimal.

Substituted Denitronilines. The substituted dinitroanilines are an important group of herbicides. Included are trifluralin, benefin, nitralin, and related materials. These compounds are only slightly soluble in water, have generally low vapor pressures, and are relatively immobile in soil systems, remaining esssentially where they are applied. They have been classed among the least mobile of the herbicides. The dinitroanilines are considered to be moderately persistent herbicides in the soil. They are generally considered to have a very low degree of toxicity to mammals and are degraded in the environment to products without singnificant adverse effects on organisms.

Nonpersistent Pesticides

Phenoxy and Related Acidic Herbicides. A large group of compounds may be designated as acidic herbicides, since they are chemicals that possess carboxyl or phenolic functional groups and which ionize in aqueous systems yielding pK_a's of The behaviour of these chemicals is closely correlated with their acid character. The most significant factor with respect to mobility of these compounds is the organic matter content of the soil. Various adsorption studies have shown that the compounds are readily adsorbed by soil organic matter. Furthermore, in acidic systems these compounds are also adsorbed by clay particles. A number of these compounds are commercially available in either the acid form or as esters. The behaviour of the esters might be expected to be considerably different from that of the acid forms.

Included within this group of herbicides are several important herbicide classes. The phenoxyacetic acids, including 2,4-D and its esters, were introduced following World War II for their high activity against many broad-leaved weeds. The chlorinated aliphatic acids include dalapon and trichloroacentic acid, compounds used against perennial weeds. The benzoic acid herbicides include chloramben, dicamba, diclobenil, and several other compounds. This is a very heterogenous group of herbicides used for a variety of puposes. The dinitrophenols are group of broad-spectrum herbicides, the most common if which is dinitro- *o*-cresol (DNOC).

Phenylcarbamate and carbanilate herbicides. The phenylcarbamate herbicides are much more water soluble than the substituted anilines. In spite of this, however, they are very immobile in soil systems. Again, these compounds have been shown to be inactivated by adsorption to soil organic matter. Compounds in this group onclude propham, chlorpropham, barban, terbutol and diclormate. The mechanism of adsorption to soil organic matter is thought to involve hydrogen bonding between the carboxyl groups of the organic matter and the nitrogen and carbonyl oxygen of the carbamate.

Ethylenebisdithiocarbamate fungicides. The metal derivatives of the ethylenebisdithiocarbamates are one of the most important groups of fungicides currently used in agriculture. The principal compounds in this group are the manganese and zinc derivatives, maneb and zineb, and the disodium derivative nabam. A closely related group is dithiocarbamates, represented by ferbam, ferric dimethyldithiocarbamate, and diram, zinc dimethyldithiocarbamate. From a toxicologic standpoint these compounds have caused concern because of their degradation to ethylene thiourea, a known carcinogen. This group is thoroughly considered in a two-volume treatise on antifungal compounds.

These compounds may be used as seed protectants and foliar fungicides. Only small quanitites find their way into the soil , and once there, are rapidly degraded. In soil ethylene thiourea, ethylenethiuram monosulfide, CS_2, and H_2S result from treatment with nabam, zineb and maneb. Ethylene thiourea is further degraded to ethyleneurea. These fungicides appear to have some systemic activity, and plants growing on treated soil will take up residues of the compounds, as well as ethylene thiourea. Because of the carcinogenic activity of ethylene thiourea, the significance of residues of this compound found in soils, plants, and food is of concern. Although the residues of ethylene thiourea itself may be small, it has been shown that cooking vegetables containing residues of ethylenebisdithiocarbamates releases ethylene thiourea. Hence, the toxicologic significance of residues of these chemicals in soil, once thought to be of little consequence, is now a matter of considerable attention.

Organophosphorus and carbamate insecticides. Incontrast to the persistent insecticides particularly the chlorinated hydrocarbons, the organophosphorus and carbamate insecticides are relatively nonpersistent in the environment. They are typically applied to crops, sometimes directly to the soil as systemic insecticides, for the control of phytophagous pests. These chemicals generally persist form only a few hours through several weeks to months. Only in rare instances are organophosphorus or carbamate residues found in crops beyond the growing season during which they were applied. These chemicals are rapidly replacing the chlorinated hydrocarbon insecticides as the principal weapons in the arsenal of American agriculture against the invasion of the pests that compete with man for food and fiber. More than half (57 per cent in 1971) of the total volume of insecticides applied in the United States in made up of members of this group of chemicals. These compounds have been the subject of two recent monorgraphs.

The organophoshates and carbamates used as insecticides are neutral esters of phosphoric and carbamic acids. A large number of these compounds representing a variety of chemical, physical, and biologic properties are presently in commercial use, allowing their specific application to be tailored to particular needs. They act as anticholinesterase agents by phosphorylating or carbamylating acetylcholinesterase, freeing in the process a leaving group that is generally easily further degraded. This reaction may take place as well in the environment by chemical or potochmical mechanisms. Thus, these compounds do not represent a serious problem as contaminants of soil and water. Their breakdown products are generally nontoxic,

Table 12.3. Effect of Temperature and pH on the Hydrolysis in Water of some Organophosphorus Insecticides

Half-Life
Temp. (^0C)† *Parathion Paraoxon Methyl Parathion*
Disulfoton Trichlorofon *Dichlorvos* *Azinphos- methyl*

Days

103,0001,2007604,8302,4002401,070

206903201751,11052661.5240

30180934529014017.361.5

40502912.578415.818

50159.64.02410.71.665.46

604.753.21.347.83.20.581.9

701.651.20.470.471.130.1640.61

Hours

70

pH 13418.515.462322.324

3212311.262333.49

519.524.410.76015.32.88.9

77.811.56.927.60.70.454.8

92.72.11.51.0

being composed of lowmolecular- weight, volatile molecules that are easily degraded and utilized by organisms.

The specific rate of chemica and biochemical transformation of the organophosphorus and carbamate insecticides depends on the specific properties of the individual compounds. Some of these compounds are relatively soluble in water. Being esters, they are also susceptible to hydrolysis. The half-lives of a number of common organophosphorus insecticides at various temperatures and pH values are given in Table 12.3. Most organophosphorus and carbamate compounds are stable at acid pH values. However, under alkaline conditions hydrolysis is rapid, the breakdown rate increasing approximately tenfold for each pH unit above 7. An increase of 10° of temperature will increase the hydrolysis rate approximately fourfold.

Organophosphorus and carbamate insecticides may contaminate soils by either direct application or through runoff from applications to crops. When these compounds are present in the soil, their disappearance is influenced by their interaction with the physical characteristics of the soil, the water content of the soil, and the micro-plora present. They may be tightly bound in certain types of soils, even to the point where they are unvailable for biologic decomposition. Under such conditions very little movement takes place even though water may be running through the soil or over its surface. The combination of interaction with soil compoments and rapid chemical and biochemical degradation in the soil results in minimal contamination of water supplies and soil to which compounds have not been applied.

The detection of organophosphorus and carbamate insecticides in soil can be difficult. Their interaction with soil components renders them unavailable to exert their toxic action on organisms and makes them difficult to remove by conventional solvent extraction techniques. Hence, methods that depend on remmoval of residues from soil particles often underestimate the presence of these compounds. The analysis of organophosphorus and carbamate residues in water, on the other hand, is relatively uncomplicated and easily done. Extraction of large quantities of water with organic solvents and subsequent analysis provides a very sensitive assay for small quantities of organophosphorus and carbamate insecticides in water, although other water pollutants may complicate the analyses when they are coextracted with the insecticide.

When organophosphates or carbamates reach the soil, their subsequent disposition is influenced by interaction with the mineral components of the soil, soil organic matter, soil pH, and soil moisture. In addirion, the flora and fauna present in the soil are responsible for the degradation of these insecticides into innocuous breakdown products a number of workers have found that after treatment of soil in laboratory or controlled field studies, extraction of the material from the soil was increasingly difficult as time progressed. For example, in a comarative study with ^{14}C-dimethoate applied to three different soil types at two different times of the year, recovery of dimethoate-^{14}C-equivalents from the soil with organic solvent extraction decreased from nearly 100 per cent immediately after application to between 60 and 80 per cent by 37 days after application, depending on the soil type and moisture content. In greenhouse experiments where Mocap was applied to the soil in pots, very little of the applied ^{14}C- Mocap was taken up by bean or corn plants, and again the extractability of the compound from soil was increasingly difficult with the passage of time. In these experiments, Mocap was labeled either at the ^{14}C-propyl position or the ^{14}C-ethyl position, and it was noted that more radioactivity was recovered from the ethyl-labeled compound than the prophyl-labelled compound.

The persistence of these comopounds is, at least in part, a function of interactions with the mineral components of soil. Metallic ions in soils interact with organophosphorus insecticides. Malathion, for example, is quickly incorporated into the montmorillonite clay interlayer region where it is adsorbed as a double layer. The mechanism was shown to be hydrogen binding between carbonyl oxygen atoms and the hydration water shells of cations. In this case adsorption was so strong that no degradation of the malathion was observed. In these studies sodium, calcium, copper, iron, and aluminum montmorillonite were used. Similarly, Saltzman and Yaron (1972) have shown a strong affinity of arathion for sodium montmorillonite. On the other hand, diazinon and chloropyrifos were decomposed rapidly upon contact with copper (II)-montmorillonite. Calcium and magnesium montmorillonites were relatively ineffective in degrading these componds.

Binding of organophosphorus and carbamate insecticides correlates well with the organic matter content of soil. It has been shown, for example, that the amount of mevinphos bound by soils increased with increasing organic matter content. Furthermore, the absorption of phorate from the soil by plants appeared to be in competition with the binding of the compound with the organic matter content of the soil. In a kinetic study of the adsorption of carbaryl to soil organic matter surfaces, Leenheer and Ahlrichs (1971) found that carbaryl was more readily adsorbed to acid soils than to neutral or alkaline soils. The may be due to decreased displacement of the carbamate from the active sites by water at lower pH values.

Soil moisture has a major influence on the availability and extractability of residues of organophosphorus and carbamate insecticides, apparently because of competition between the insecticides and water for the adsorption sites on the soil particles. Harris (1964) has shown, for example that, diazinon, parathion, trichlorfon, and mevinphos are 135-fold, 28-fold, 20- fold, and 1,4-fold, respectively, more active in moist soils than in dry soils. Analytic procedures for the recovery of residues from soils generally recommend the addition of water for the desorption of the residues from soil particles before extraction with an organic solvent. However, even though there is a major interaction between these insecticides and water, they do not appear to move freely in soils with water, and loss by leaching does not appear to be a major factor in the disappearance of these compounds from soils. In studies designed to test the behaviour of phorate in soils, Getzin and Chapman (1960) applied the radiolabeled compound on the surface of various soil types and applied 10 to 30 mm of water every two days for a total of 250 mm over 24 days. More than 80 per cent of the radiolabeled phorate equivalents remained near the surface with silt loam and muck soils, although only 50 per cent remained in sandy soil and 76 per cent in quartz sand.

The fact that microorganisms exert a major influence on the behaviour of pesticide residues in soil has been demostrated by observing the effect of soil sterilization on the breakdown of a number of compounds. Getzin (1968) showed that zinophos degraded faster in nonautoclaved soil, although the decomposition of diazinon was unaffected by autoclaving. Getzin and Rosefield (1968) showed that malathion, Ciodrin, dichlorvos, mevinphos, parathion, methyl parathion, Supracide, dimethoate, and chloropyrifos were all degraded faster in nonsterile soils. Lichtenstein *et al.,* (1968)showed that both sodium azide treatment and autoclaving reduced soil bacteria and resulted in an increased persistence of parathion. It was also shown in these studies that sodium azide affected diazinon residues in the soil directly, but the studies were unable to assess the effect of soil sterilization on this compound. Bro-Rasmussen *et al.* (1968) reported that sterilization extensively affected diazinon degradation in loam and sandy loam soils. The effect of sterilization of soil on pesticide degradation may appear to be somewhat ambiguous because of changes other than the destruction of soil microorganisms that would have an effect on the degradation of compounds applied. Autoclaving, for example, is known to change the physicochemical properties of soil and to destroy heat-sensitive soil components other than microorganisms that participate in the degradation of organophosphorus insecticides. As in the case of diazinon, chemical sterilization may result in direct reactions between the insecticide and the soil sterilant. In general, however, there is at least strong evidence that microorganisms exert a major influence on the degradation of pesticides in soils.

Harris has proposed a classification scheme in which all pesticides are ordered into three groups. Group I is defined as "highly residual" compounds; group II, "moderately residual;" and goup III, "slightly residual". Various studies by many workers allow the classification of the organophosphorus and carbamate insecticides in groups II and III. In group II are dazanit, carbofuran, dimethoate, methomyl, Mocap, and Mobam, and in group III, para- thion, diazinon, phorate, chloropyrifos, bromophos, and disulfoton. Group II compounds persist between 4 and 16 weeks, and group III coompounds persist between two and four weeks in soils.

There is little information on the effects of organophosphates and carbamates on organisms living in water and soil. In general, only minute amounts of residues of the insecticides and their toxic degradation products are found in natural water systems. Thus, their biologic effect seems to be minimal. In soil, however, there is greater likelihood of the presence and buildup of toxic residues. Several studies have shown that some compounds can cause reduction in bacterial populations. Garretson and San Clemente (1968) showed that parathion inhibited nitrifying chemolithotrophic bacteria, although malathion did not. Sommer (1971) showed that organo-phosphates had little effect, while carbamates markedly inhibited nitrification. In studies to assess the effect of diazinon on soil microorganisms Gunner and cowarkers have shown that the compound exerts a selective effect on both soil and rhizosphere microflora expressed as selective enrichment of coccidioidal rods. Although the numbers of fungi were unaffected, a large number of the genus *Streptomyces* appeared as a climax population. In another instance the microflora that arose in response to diazinon belonged primarily to one species of *Arthrobacter*. Similar results were obtained by Stojanovic *et al.* (1972) with carbaryl as the test insecticide. The various studies show that these insecticides may cause a variety of effects on the soil flora and fauna not always expressed as directly toxic effects.

NONPESTICIDAL ORGANIC CHEMICALS

Data gathered by the United States Environmental Profection Agency through November, 1975, indicate that 253 different organic chemicals have been detected in drinking water in the United States. There is every reason to believe that an additional nmber, yet unidentified may also be present, and the total number of chemicals could be considerably larger. The list of chemicals could be considerably larger. The list of chemicals includes both aliphatic and aromatic hydrocarbons, pesticides, industrial chemicals, plasticizers, and solvents. Many of these materials are halogenated and some are produced by chlorination of the water during the purification process. Others appear through industrial and municipal discharges, urban and rural runoff, natural sources, and sewage purification practices.

The principal objective of the nationwide survey of organic chemicals in drinking water made by the Environmental Protection Agency was do determine the extent and significance of the occurrence of suspected carcinogens in water.

Low-Molecular-Weight Halogenated Hydrocarbons

Of particular concern is the possibility of the production of low-molecular-weight chlorinated hydrocarbons through the use of chlorination for water purification. This concern focuses principally on the four trihalomethanes, chloroform and bromoform; and carbon tetrachloride and 1,2,-dichloro-ethane. A two-part study was conducted by the Environmental Protection Agency to ascertain the presence of these chemicals in water, and whether they were produced by chlorination. The results are summarized in Table 12.4. It was noted that all of these materials were found in drinking water and most were found in the raw water before chlorination. The second part of the study, an intensive study of the situation in midwestern United States, led to several conclusions:

TABLE 12.4

Organice Detected in Water in the Epa National Organics Reconnaisance Survey

	Raw Water Analysis		*Finished Water Analysis*		
				Concentration	
Compound	*No. of Locations Where Detected*	*Concentration Range (μm/1)*	*No. of Locations Where Detected*	*Range (μg/1)*	*Median*
Chloroform	49	-0.9	80	-311	21
Bromodichloromethane	7	-0.8	78	0.3-116	6
Dibromochloromethane	1	3	72	-110	1.2
Bromoform	0	—	26	-92	5
Carbon tetrachloride	4	-4	10	-3	—
1,2-Dichloroethane	11	-3	26	-6	—

1. Raw water with low turbidity resulted in finished water that was relatively free of chloroform and related halogenated compounds.

2. Chloroform, bromodichloromethane, dibromochloromethane, and bromoform do result from chlorination of precursors in the raw water. On the other hand, carbon tetrachloride, methylene chloride, and 1,2-dichlorethane do not appear to be produced chemically during the treatment process.

3. There appears to be a correlation between chloroform, dibromochioromethane, bromodichloromethane, and bromoform concentrations. The ratio between the four chemicals appears to be relatively constant in all water examined, indicating the probability of a common precursor or group of precursors for these halogenated hydrocarbons.

4. The use of granular activated carbon treatment was not effective in removing volatile organic compunds during the water treatement process.

The four halogenated hydrocarbons produced by chlorination seem to result from naturally occurring humic substances as precursors. The maximum concentrations of these materials found were chloroform, 54 μg/1; bromodichloromethane, 20 μg/1; dibromochloromethane, 13 μg/1; and bromoform, 10 μg/1.

Studies were conducted to compare the rate and extent of chloroform formation when chlorine was added to raw river water, filtered water, and activated-carbon-treated water. These experiments were carried out at constant pH and 25° C. When sufficient chlorine was added to satisfy the chorine demand, chlorination of raw river water yielded approximately seven times as much chloroform as did chlorination of the filtered water and approximately 80 times as much as chlorination of activated-carbon-treated water. Similar results were obtained when the same experiments were conducted with realistic concentrations of humic acid. Concentrations of humic materilas are probably reduced during coagulation, settling, and filtration, thereby reducing the rate and extent of chloroform formation by chlorination. Thus, it may be possible to reduced the quantity of chlorinated hydrocarbons formed during chlorination by altering the water purification process so that chlorination is performed following the removal of humic meterials through filtration and coagulation steps.

As yet, there is no generally accepted substitute for the use of chlorine as disinfectant in water purification. However, the confirmation that chlorination produces a number of halogenated hydrocarbons has stimulated an extensive investigation of other chemicals that could be used for this purpose, such as chloramines, chlorine dioxide, ozone, bromine and iodine.

Aromatic Halogenated Hydrocarbons

In recent years a number of halogenated aromatic com- pounds have engendered increasing concern about their effects as environmental pollutants. The polychlorinated biphenyls (PCB) have appeared as ubiquitous contaminants of soil and water. Chlorophenols used for a variety of purposes have been detected in surface waters and drinking water. The extremely toxic 2,3,7,8-tetrachlorodibenzop-*p*-dioxin (TCDD) has contaminated large areas of both water and soil through industrial accidents and through widescale application of herbicides containing small quantities of the chemical as a contaminant.

Polychlorinated biphenyls. PCBs are very stable materials of low flammability, which contain from 12 to 68 per cent chlorine. They are exceptionally persistent in the environment, even more persistent than the chlorinated hydrocarbon insecticides, with which they often have been confused in anlytic studies of environmental samples. They have been used as insulating materials in electrical capacitors and transformers, plasticizers in waxes, in paper manufacturing, and for a variety of other industrial purposes. The diversity of their use patterns, the large quantities used, and their stability have led to widespread occurrence of these compounds in soil and water. Fish from the upper Hudson River and Lake Ontario have been found to contain PCB concentrations from 5 to 20 ppm. Fish from a number of other rivers throughout the United States have also been found to contain comparable quantities of PCBs. Waterfowl have also accumulated high concentrations of PCBs. These examples of PCB pollution have occurred in spite of efforts to restict and eventually eliminate the release of such compounds into the environment.

Chlorophenols. Pentachlorophenol has been used in significant quantities since 1936 as a wood preservative. As a result of this use surface water and treated drinking water have been found to contain as much as 0.70 and 0.6 ppb, respectively, of pentachlorophenol. Hexachlorophene (2,2,-methylene-*bis*-[3,4,6-tri- chlorophenol]) has been widely used as an antibacterial agent in a number of consumer products, including soaps and deodorants. It has been detected in surface waters as high as 48 ppb and in drinking water at 0.01 ppb. Hexachlorophene is resistant to metabolic attack and tends to persist in environement and bio accumulates in food chains.

Pentachlorophenol has a fairly high acute toxicity and has been shown to cause reproductive failures in rats. However, the full extent of its health effects has not been completely evaluated. The acute toxicity of hexachlorophene is also quite high. The compound has exhibited neurotoxicity in dogs, sheep, and rats. The presence of these chemicals in water needs to be closely monitored because of their high toxicity and the possibility of adverse health effects in man.

2,3,4,8-Tetrachlorodibenzo-p-dioxin. What was probably the most dramatic and catastrophic occurrence of environmental pollution by a toxic chemical occurred in Seveso, Italy, July 10, 1976. On that date a safety disk in a reaction vessel being used to manufacture 2,4,5-trichlorophenol, an intermediate in the production of hexachlorophene and 2,4,5-trichlorophenoxyacetic acid (2,4,5-T), ruptured, releasing a chemical cloud over the region. The cloud contained predominantly 2,4,5,-trichlorophenol. However, an estimated 3 to 16 kg of 2,3,7,8- tetrachlorodibenzo-*p*-dioxin (TCDD), potent teratogen, was also released. The area was thus contaminated with the greatest concentration of TCDD ever found in the environment, up to 51.3 ppm in some samples.

When the safety disk of the reaction vessel ruptured, the chemicals blew up into a plume 30 to 50m high above the factory, then cooled and came down over a cone-shaped area about 2 km long and 700 m wide, all over a matter of minutes. The result was that some people become ill immediately, while others began to show symptoms ranging from skin burns to stomach pains and internal bleeding a few days later. Animals and vegetation within the area began to die. Shortly after the incident the entire area was evacuated and cordoned off. It was expected that a 123-acre area, which became a wasteland of shriveled and dying plants and evacuated homes, would have to remain entirely closed for ten years or longer before the levels of TCDD would degrade sufficiently for the area to be safe for habitation.

TCDD is extremely toxic, as indicated by its acute oral LD_{50}, reported to be between 0.6 and 115 μg/kg for several animal species. The compound causes degenerative changes in liver and thymus, porphyria, altered serum enzyme concentrations, and loss in body weight. Thymic atrophy is a very sensitive index of TCDD exposure. TCDD is also a potent fetotoxic agent in various animal species.

The singnificance of TCDD as an environmental pollutant lies primarily in the fact that it is a trace by-product in the synthesis of 2,4,5-T, an important herbicide that has been widely used in agricultural and other weed control programmes. It was extensively used as a defoliant. Fortunately, water transport of TCDD is limited since its solubility in water is only 0.2 ppb. Its only accumulation, therefore, will likely be in soils to which heavy applications of 2,4,5-T have been made. TCDD is firmly bound to soil where it tends to persist for more than one year. Degradation of the chemical in the soil will be largely by photochemical mechanisms.

Phthalate Ester Plasticizers

The phthalate ester plasticizers are used in virtually every major product category, including construction, automotive, household products, apparel, toys, packaging, and medical products, resulting in the widest possible distribution of these materials. The industry doday comprises 13 major suppliers who produced in 1972 approximately 1 billion pounds of 20 different compounds. The two most abundantly produced phthalate ester plasticizers are di-2-ethylhexylphthalate (DEHP) and di-*n*-butylphthalate (DBP).

The phthalate esters are now know to be ubiquitously distributed in the environment. They have been found complexed with the fulvic acid components of humic substances in soil and in both marine and estuarine waters. Fulvic acid apparently functions as a solubilzer for the rather insoluble phthalate esters and thus serves to mediate the mobilization, transport, and immobilization of these materials in soil and water. Hites (1973) reports the presence of phthalate esters in the Charles and Merrimack Rivers in Massachusetts. In the Charles River it was shown that the phthalate concentration increased as one moved upstram to about river mile 7 where apparently the phthalate esters were added to the river. The concentration decreased as the contaminated water flowed down-stram, diluting the phthalates present. Concentration at river mile 7 of phthalate esters was approximately 1.9 ppb, while at the mouth of the river concentrations had decreased to 0.97 ppb. In another study phthalate ester plasticizers were detected in the open ocean environment of the Gulf of Mexico and the North Atlantic, DEHP and DBP were found in almost all samples analyzed including a deep sea jellyfish, *Atolla,* from 1,000-M depths in the North Atlantic. Concentrations of DEHP in suface water ranged from 4.9 to 130 ng/1. DBP ranged from a nonderectable level to 95 ng 1. Lower levels of both compounds were found in sediment. It has become clear that the phthalate ester plasticizers are general contaminants of virtually all soil and water ecosystems; it has become very difficult to analyze any soil or water sample without detecting the presence of phthalate esters.

Because of the widespread occurrence of these compounds, their toxicity is of concern. In general, the phthalate esters have a low order of acute toxicity. For example, the intraperitoneal LD_{50} dose in mice ranges from 1.5 to 14.2 g/kg (Rubin and Jaeger, 1973). Furthermore, 90-day and two-year feeding studies of DEHP in rats and two-year feeding studies in guinea pigs and dogs indicated a low order of chronic toxicity. On the basis of these results the Food and Drug Administration approved DEHP for use in plastic warpping of food. Although these studies indicated a low order of toxicity, they emphasized evaluation of gross overt effecrts, not taking into account subtle toxicologic effects. Rubin and Jaeger (1973) investigated a number of more subtle effects of DEHP on a variety of biologic systems. They found of particular note a sensitivity of cultured beating chick embryo heart cells to DEHP. As little as 4 μg/m in the culture medium was lethal to 97 to 98 per cent of the cells. This concentration could be reached in human blood stored in vinyl plastic bags for a period of one to two days. These results indicated the need for caution in assuming the safety of low concentrations of phthalate esters.

Data reported by Mayer and Sanders (1973) indicate that DEHP and DBP may also be detrimental to the reproduction of some aquatic organisms at low concentrations. *Daphnia magna* reproduction was decreased by approximately 80 per cent by continuous exposure of 30 μg 1 DEHP for up to 21 days. Reproduction in zebra fish and guppies was also decreased by low concentrations of DEHP. Although the concentrations of phthalate esters in soil and water are quite low and the toxicity of the compound is also quite low, some concern is evident because of the ubiquitous nature of environmental contamination by these compounds. They have been shown to occur in drinking water in the United States, are undoubtedly present in food, and may reach man through other mechanisms as well.

Metals

The toxicology of metals, including their use, occurrence, and effects, In this section we/will deal only with some aspects of the natural cycles of elements and conditions that alter the process involved. It is necessary to limit this treatment to the best-studied examples of environmentally important elements: mercury, cadmium, lead, arsenic, and selenium.

Mercury. Methyl mercury pollution of Minimate Bay and the subsequent human poisoning from consumption of contaminated seafood has stimulated much research on the origin and fate of methyl mercury. An important theme throughout the discussion of metals is the question of chemical species. Mercury, for example, exists in the inorganic form as free mercury, Hg^0, mercury ion in salts and complexes. Hg^{2+} ; or as organic mercury compounds, such as phenylmercuric salts, which have been used as fungicides and herbicides, and the alkyl-mercury compounds including methyl mercury. In natural systems, a dynamic equilibrium that is determined by the physiochemical and biologic conditions of the soil-water system exists between the various chemical species. Some organic forms of mercury are man-made, such as the phenylmercuric compunds, while methyl mercury is produced by man, sediment microorganism, nonbiologically in sediments, and possibly by some fish. As expected, each species of mercury has its own set of physical, chemical, and toxicologic properties.

Mercury is transported to aquatic ecosystems via surface runoff and through the atmosphere. It is complexed or tightly bound to both organic and inorganic particles. Sediments with high sulfur content will strongly bind mercury. Organic acids such as fulvic and humic acid are usually associated with the mercury that is not bound to particles. Methylation of mercury by microorganisms is a detoxication response that allows the organism to dispose of heavy metal ions as small organometallic complexes. Conditions for methylation by sediment microorganisms are strict and occur only within a narrow pH range. The rate of synthesis of methyl mercury also depends on the redox potential, composition of the microbial population, availability of Hg^{2+}, and temperature. Vitamin B_{12} derivatives are believed to be the methylating agents, since mechanistically they are the only methyl carbanion and methyl radical donation coenzymes know. An understanding of the biomethylation reaction mechanisms together with oxidation reduction chemistry of elements allows predictions of the environmental conditions necessary for the biomethylation of mercury and several other metals. However, the best conversion rate for inorganic mercury to methyl mercury under ideal conditions is less than 1.5 per cent per month.

Little or no methyl mercury is found in sediments. Conversion of inorganic mercury to methyl mercury results in its desorption from sediment particles at a relatively fast rate. Demethylation by sediment microorganisms also occurs at a rapid rate when compared to methylation. Methyl mercury released in surface waters can undergo photodecomposition to inorganic mercury. However, methyl mercury can also be bioaccumulated by plankton algae and fish. In fish, the rate of absorption of methyl mercury is faster than that for inorganic mercury, and the clearance rate is slower with a net result of high methyl mercury concentrations in the muscle tissue. Selenium, which is present in seawater and seafood, readily complexes with methyl mercury and is believed have an important protective action against the toxic effects of methyl mercury. In summary, the danger of methyl mercury poisoning, as occurred in Minimata, arises from direct methyl mercury contamination rather than methylation of environmental sources of inorganic mercury.

Cadmium. Cadmium has long been recognized as toxic element. Its importance as an environmental contaminant was demonstrated in the outbreak of *itai-itai* disease caused by smelter wastes that contaminate rice paddies. Cadmium deposits are found as sulfides with zinc, copper, and lead deposits, and cadmium is recovered as a by-product of smelting processes for those metals. A major environmental source of cadmium is vapor emissions that contaminate surrounding soil and water through fallout during smelting. Natural soil concentrations of cadmium are less than 1 ppm and average about 0.4 ppm. Sewage sludge is often contaminated with cadmium, which can the concentrate in plants grown on contaminated soils. The problem of heavy metal contamination, especially cadmium, has been one of the most serious concerns impeding the use and disposal of domestic sewage sludge on agricultural lands. Cadmium also enters agricultural soils as a contaminant of phosphate fertilizers. There is some evidence for the leaching of cadmium in soils.

Cadmium concentration in fresh waters are usually less than 1 ppb while sea water ranges from 0.05 to 0.2 ppb and averages about 0.15 ppb. Higher concentrations of cadmium in surface water are usually due to metallurgic plants, plating operations, cadmium pigments, batteries, plastic manufacture, or from sewage effuent. Mine drainage and mineralized areas also contribute significantly to cadmium fluxes. Drinking water in soft water areas can serve as a source of cadmium through corrosion of plumbing. However, this source is estimated to be small in relation to food intake. As in the association of selenium and mercury, there appears to be a protective effect with zinc and calcium against cadmium toxicity.

Lead. The use of lead, its mining, and its processing date back several centuries. Changing usage patterns rather than increased consumption determine present environmental inputs from man's use of lead. Betteries, gasoline additives, and paint pigments are major uses, but combustion of gasoline additives is the major source of environmental pollution by lead. Thus, lead is primarily an atmospheric pollutant that enters soil and water as fallout, a process determined by physical from and particle size. The net result is a buildup of lead near heavily traveled roads.

Lead enters aquatic systems from runoff or as fallout of insoluble precipitates and is found in sediments. Typical fresh water concentrations lie between 1 and 10 μg/1 while natural lead concentration in soil range from 2 to 200 ppm and average 10 to 15 ppm. Deep ocean waters, below 1,000 m, contain lead at 0.02-to 0.04-μg/kg concentrations, but surface waters of the Mediterranean Sea and Pacific Ocean contain 0.20-and 0.35-μg/kg levels. Drinking water concentrations of lead may be greatly increased in soft water areas through corrosion of lead-lined piping and connections. However, average drinking water intake is considerably less than food sources.

The biologic methylation of inorganic lead to tetramethyl lead by lake sediment microorganisms has been demonstrated. However, the fate of this volatile, water-insoluble form of lead is unknown.

Arsenic. Arsenic is widely distributed in the environment. Man's input of arsenic into the global cycle occurs through smelting, coal burning, and the use of arsenical pesticides. Speciation of arsenic is an important consideration in the fate, movement, and action of this element. The chemical and biochemical transformations of arsenic include oxidation, reduction, and methylation, which affect the volatilization, adsorption, dissolution and biologic disposition of the arsenic species and involved.

Arsenic contamination of soils from point sources such as copper smelters or coal-burning power plants is easier to control than the dispersive use of arsenical pesticides, resulting in nonpoint source pollution. Various forms of arsenic are used as pesticides. Arsenic acid (H_3AsO_2) is a leaf desiccant used in cotton production, lead and calcium arsenates are insecticides, and organic arsenicals, which include methnearsonic acid and its sodium salts as well as dimethylarsinc acid (cacodylic acid), are used as postemergence herbicides. The transport of arsenic in the environment is largely controlled by adsorption/desorption processes in soil and sediments. Therefore, sediment movement is responsible for transfer or arsenic soil residues to their ultimante sinks in deep ocean sediments. The clay fraction, plus ferrous and aluminum oxides that coat clay

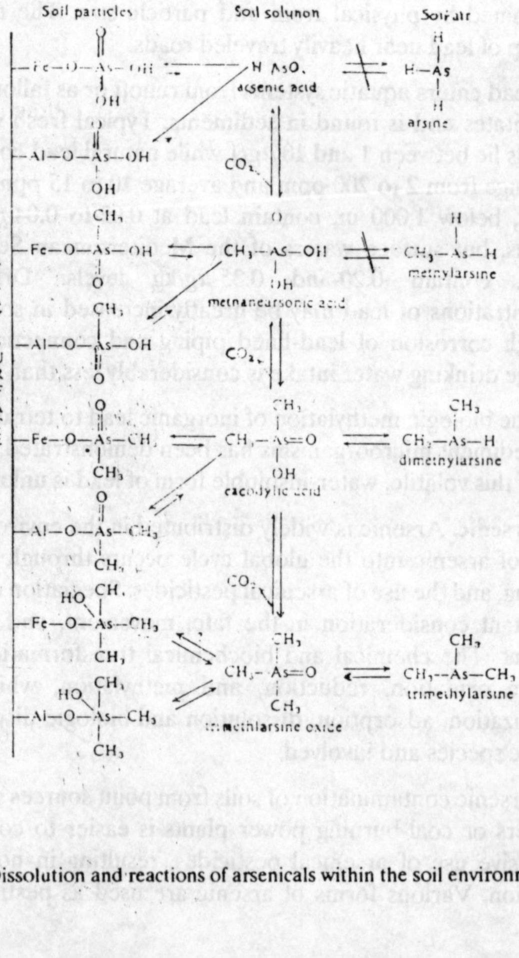

Fig. 12.6. Dissolution and reactions of arsenicals within the soil environment.

particles, adsorbs arsenicals. The reactions of arsenicals in soil include oxidation, reduction, methylation, and demethylation. Conversion of arsenic to volatily alkylarsines leads to air transport loss from soils. The transformation processes of arsenic and its transport processes are intimately linked.

Arsenic concentrations in water are generally much lower than in sediments. In Lake Michigan, the concentrations in water range 0.5 to 2.3 μg/1 while sediment concentrations range from 7.2 to 28.8 mg/kg. Inorganic arsenic exists in water in different oxidation states, depending on the pH and E_n of the water. Arsenate is apparently reduced by bacteria to arsenite in marine environments since the ratio of arsenate to total arsenic is much lower than is predicted thermodynamically. Methylation of arsenic occurs in both freshwater and marine systems and where arsenic is detected as arsenate, arsenite, methanearsonic acid, and dimethyl-arsinic acid.

Bioaccumulation of arsenic species occurs readily in some aquatic organisms. Some seaweeds fresh water algae, and crustaceans accumulate significant amounts of arsenic. Some arsenic in *Daphnia magna* and algae occurs as arseno analogs of phospholipids, indicating the mistaken accumulation and utilization of arsenate in place of phosphate. Carbs, lobsters, and other marine organisms accumulate organoarsenicals along the food chain.

Selenium. Selenium concentrations in natural waters depend largely on the occurrence of seleniferous soils. Average concentrations for selenium in natural waters are less than 10 μg/1, but can reach several hundred micrograms per liter in certain areas of some Western states. Dietary sources of selenium are usually more important than drinking-water sources. Environmental redistribution of selenium through man's activities is due to copper smelting; lead, zinc, phosphate, and uranium mining and processing; manufacturing of glass ceramics and pigments ; and burning of fuels. Selenium can be methylated as also demonstrated for metcury, arsenic, lead and tin. Sediment microorganisms are responsible for the production of dimethyl selenide and dimethyl diselenide from both inorganic and organic selenium compounds. The importance of volatile methyl selenide compounds in the biogeo- chemical cycling of this element is still uncertain.

Inorganic Ions

Nitrate, phosphate, and fluoride are inorganic ions that have caused considerable concern over their environmental effects. With nitrates and fluorides the concern is principally human health, but nitrases and particularly phospates also cause eutrophication of lakes and ponds, a process that is considered environmentally undesirable. Midsummer algal blooms are familiar sights in some parts of the United States.

Nitrates. Man has altered the nitrogen cycle through his agricultural and technologic practices; changing patterns in agriculture, food processing, urbanization, and industrialization have had an impact on the accumulation of nitrate in the environment. Intensive agricultural production has consumed an increasing amount of nitrogen-based fertilizers, particularly with corn, vegetables, other row crops and forages. Nitrogenous wastes from livestock and poultry production as well as urban sewage treatment have contributed nitrogenous wastes to the soil and water environments. Nitrate and nitrite are used extensively for colour enhancement and preservation of processed meat products. These practices inevitability lead to increased exposure of man and animals to significant nitrate levels in food, feed, and water.

The nitrate form of nitrogen is of concern because of the high water solubility of this ion and consequent teaching, diffusion, and environmental mobility in soil and water. Nitrate can contaminate ground water to unacceptable levels. The current recommended limit for nitrate introgen in drinking water is 10 mg/1 (45 ppm nitrate). In the Community Water Supply survey of the Bureau of Water Hygiene in 1969, the range of nitrate concentrations found was 0 to 127 mg/1. Nineteen systems, about 3 percent of those examined for nitrate, had concentrations in excess of the recommended limit. Nitrate is formed from nitrate or ammonium ion by certain microorganisms in soil, water, sewage, and the alimentary tract. Thus, the concern with nitrate in the environment relates to its conversion by biologic systems to nitrite.

Methemoglobinemia is caused by high levels of nitrite or indirectly form nitrate in humans. It results in difficulties in the oxygen transport system of the blood. Poisoning of infants from nitrate in well has been reported. Cases numbering in the thousands have now been reported, mostly in rural areas, mostly involving poisonings in infants.

Of more recent concern is the production of nitrosamines in food by the reaction of nitrite with secondary amines. Other nitroso compounds can result from the analogous reactions of nitrites with amides, ureas, carbamates, and other nitrogenous compounds. Various dialkyl and related nitrosamines have been shown to produce liver damage, hemorrhagic lung lesions, and convulsions and coma in rats. *N*-Nitroso compounds represent a major class of important chemical carcinogens and mutagens. The various forms of cancer for which the environmental *N*-nitroso compounds are suspected to play a causative role occur after long latency periods and at relatively low absolute frequency in the general population. These factors make it difficult to establish cause-and- effect relationships between specific carcinogens and disease incidence.

Phosphates. Although the principal problem of phosphates in the environment is not directly related to human health, there is considerable concern about the effects of phosphorus from various sources on water quality. There remains considerable disagreement and controversy about the principal source of phosphate found in water ecosystems. Some maintain that phospate fertilizers are a major contribution to the levels of phosphates found in water while others claim that phosphate detergents are the major contribution.

Phosphorus applied to the soil as fertilizer moves primarily by erosion because phosphate adsorbs strongly on soil particles. However, some soluble phosphorus compounds do move in runoff water. The total phosphorus content of soils ranges from 0.01 to 0.13 per cent. The phosphorus fertilizers applied as soluble orthophosphate soon revert to insoluble forms in soil. This conversion limits leaching and leads to a higher phosphorus concentration in sediments than in the original soil since phosphorus seems to be associated with finer particles. Control of phosphate pollution from agriculture will result form efforts to reduce erosion and sediment loss by modified agricultural practices. The contribution of phosphorus to water from detergents is likely to be associated with the degree of urbanization. Efforts to control phosphates in water have concentrated on the detergent problem. Some states and local areas have resticted or banned the use of phosphate detergents completely.

In some areas secondary treatment of sewage waster results in the precipitation and removal of phosphates from the effluent before discharge.

Phosphate is a major contributor to the eutrophication process in lakes and ponds. The observer of a lake undergoing eutrophication notices first an extraordinarily rapid growth of algae in the surface water. Planktonic algae cause turbidity and flotation films. Shore algae cause ugly muddying, films, and damage to reeds. Decay of these algae causes oxygen depletion in the deep water and in shallow water near the shore. This rapid growth of large gives rise to a number of undesirable effects on treatment of the water for consumption, on fisheries, and on the use of lakes for recreational purposes.

Fluorides. The beneficial effect of low levels of fluorides in preventing dental caries has led to the extensive use of fluoride in drinking water. The Safe Drinking Water Committee of the National Academy of Sciences has evaluated this practice as follows:

Fluoride is found widly in water supplies, but the concentration is usually not great enough to be undesirable. The maximum concentration found for the 969 supplies studied in the 1969 Community Water Supply Survery was 4.4 mg/liter. Most supplies that were not intentionally fluoridated had fluoride concentrations less than 0.3 mg/liter.

Small amounts of fluoride, on the order of 1 mg/ liter, depending on the environmental temperature, in ingessted water and beverages. are generally conceded to have a beneficial effect on the rate of occurrence of dental caries, particularly among children. Two forms of chronic toxic effects recognized generally as being caused by excess in intake of fluoride over ling periods of time. These are mottling of teeth enamel or dental fluorosis, and skeletal fluorosis. In both cases, it is necessary to consider the severity since the very mild forms are considered beneficial by some. The most sensitive of these effects is the mottling of tooth enamel, which, depending on the temperature, may occur to an objectionable degree with fluoride concentrations in drinking water of only 0.8-1.6 mg/liter. (These observations were made a number of years ago and there have been no recent studies to determine if these levels still cause mottling.) Apparently there has been little systematic investi- gaion of the degree to which consumers of drinking water with several mg/liter of fluoride regard the resultant mottling as an adverse health effect.

Skeletal fluorosis has been observed with use of water containing more than 3 mg/liter. It now appears that there is some probability that objectionable dental mottling and increased bone density may occur in those with ling-standing renal disease or polydipsia who consume water containing more than 1 mg/liter of fluoride for long periods of time. Increased bone density, however, has often been regarded as a beneficial rather than an adverse effect. (This therefore makes the implications of such changes unclear.) Intake of fluoride for long periods in amounts greater than 20 – 40 mg/day may result in crippling skeletal fluorosis. Epidemiological studies where water is naturally high in fluoride have shown no adverse effects other than dental mottling except in rare cases. Controlled studies with fluoridation at the 1 mg/liter level have reported no instances of adverse effects. Available evidence does not suggest that fluoridation has increased or decreased cancer mortality rates.

Asbestos

Asbestos is a general term applied to a family of silicate minerals that have a number of properties in common that render them useful for several commercial purposes. These minerals are fibrous in structure and have electrical and thermal insulating properties as well as being sufficiently flexible that they can be woven into fabrics. The production and use of such materials has been described by Rosato (1959). Approximately 88 per cent of asbestos use is in the construction industry. including cement products, floor tile, paper products, and paint and caulking, with the remainder being used in transportation and plastics industries

The definition of asbestos listed in the Glossary of Gelsclogy is as follows:

(a) The commercial term applied to a group of fibrous silicate minerals that readily long, thin strong fibers of sufficient be woven, are heat resistant and possess a high electric insulation and therefore are suitable for uses as in cloth, paper, paint, brake linings, tiles, cement, fillers, and filters), where imocob nonceonducting, or chemically marerial is required.

(b) Mineral of the asbestos group, principally (best adapted for spinning) and certain various of amphibole (example: tremolite, aunmulate and crocidolite).

The mimeral fibers that comprise the asbestos group are the serpentine: chrysotile; and the amphiboles actinolite, amosite (a cunningtonitegrunente mineral), anthophyllite, crocidolite, and tremolite. Asbestos minerals are mined in Canada and the United States, where chrysotill accounts for about 95 per cent of the production and crocidolite make up most of the remainder. The largest chrysotile deposit in the world as found between Danville and Chaudiere, Queb, Canada. Other deposits are found in northern Ontario, northern British Columbia, and in Canada, and in California, Vermos, arizona, and North Carolina in the United States.

Asbestos is made up of fibrils of individual tubes of single crystals that bind together to produse a fiber. The size of the individual fibers varies greatly for the various minerals making up the asbestos group. Minimum fiber widths range between 0.06 μm for crocidolite to 0.25 μm for anthophylite. Fiber lengths in general range between 0.2 and 2.0 μm. Occasional longer fibers up to 100 m are found, although these are much rarer in the general environment than in occupational situations.

Solubility is an important consideration in assessing the presence and impact of chemicals in soil and water. Asbestos minerals are soluble in acid solution to varying degree. The isoelectric point of the various minerals differs widely; chrysotile has an isoelecticpomt of 11.8 while the amosite isoelectric point falls between 5.2 and 6.0. As the pH of an aqueous medium falls below the isoelectric point, the charge of suspended astbestos particles will become more positive, thereby attracting other dissolved minerals that can interact with them. Therefore, the mobility, transport, disposition, and biologic properties of asbestos will vary widely depending on the mineral involved, the pH of the medium, and the presence of other materials with which the asbestons may interact.

A major difficulty in assessing the environmental impact of asbestos is the difficulty in detecting and analyzing it. Since asbestos is a very heterogeneous material, its detection is also difficult. A number of methods have been proposed for the identification and quantitation of asbestos in air, water, and biologic materials. Optical and electron microscopy, x-ray diffraction, and differential thermal analysis have all been proposed. Analytic problems are complicated by the difficulty of distinguishing between asbestos fibers and other fibers and particles of minerals that may be present in the same sample with them. The quantities present in environmental samples, furthermore, are generally quite small, and the particles present may exist in a wide range of sizes, making identification difficult and greatly complicating the quantitation of the mineral present. It is generally felt that transmission electron microscopy is the most satisfactory method for the detection of asbestos.

Asbestos is found ubiquitously in the environment. Chrysotile asbestos is a common air pollutant in most large urban areas. In fact, because of the industrial use of asbestos, the highest concentrations found in air and water are generally in metropolitan areas. Asbestos fibers have been detected as contaminants of domestic water supplies derived from Lake Superior, generally thought to result from mine waste discharges.

An examle of the contamination of a domestic water supply by asbestos minerals in illustrated by the case of Duluth, Minnesota. An iron ore mining company located at Silver Bay, Minnesota, discharges tailings into Lake Superior to the extent of approximately 70,000 tons per day. These tailings are the residue from the processing of taconite ore into pellets and are predominantly of the amosite type of asbestos. Bottom currents in Lake Superior carry some of this discharged tailing material to Duluth, approximately 70 miles southwest. Duluth draws its water directly from Lake Superior and distributes and uses it unfiltered. The water in the Duluth domestic supply had been shown to contain numerous amphibole fibers and pieces as well as other crystalline material. The concentration of verified asbestos mineral fibers in the Duluth water supply ranges from approximately 20×10^6 to 75×10^6 fibers per liter of water. These concentrations correspond to approximately 5 to 30 μg of asbestos fibers per liter of water Nicholson, 1974). An analysis of the water system of Superior, Wisconsin, the neight borcity of Duluth, which draws only a portion of its water from Lake Superior, shows considerably fewer fibers. An analysis of the water from Grand Marais, Minnesota, which is upstream from Silver Bay, detected no asbestos fibers.

The health effects of asbestos in water have so far been incompletely ascertained. Asbestos is known to lead to asbestosis characterized primarily by pulmonary fibrosis, the formation of plural plaques, a greatly increased risk of bronchogenic carcinoma, plural mesothelioma, and peritoneal mesothelima after occupational exposure to inhaled asbestos dust, as discussed elsewhere in this book. It is not clear, however, whether the ingestion of asbestos-contaminated water will lead to the same or similar diseases in man. Epidmiologic studies of cancer death rates in Duluth, Minnesota, have not yet revealed any increase in such conditions as compared with other areas in relation to the contamination of the water with mineral fibers. However, the contamination of the Duluth water supply began about 20 years ago, and it is not certain whether such conditions might show an increase within the next 5 to 15 years because of charactersitic long latency periods. The assessment of the carcinogenicity of inhaled asbestos is complicated by the fact that synergism has been demonstrated by cigaratte smoking. Whether some sort of synergistic effect may also occur with asbestos in drinking water is entirely unknown.

Impact of Chemicals on Soil and Water Systems

The traditional view of the environment embodied in the phrase "balance of nature" represents an outmoded conceptualization of the forces that control environmental processes. There is, in fact, no simple balance of nature. The environment is composed of many systems and subsystems, each internally balanced in a dynamic way and influenced by many external processes that tend to interact and influence the structure and function of the whole system. The thrust of nature's "balance" is an evolutionary movement toward greater diversity, greater speciation, and more complex structure.

Man has been altering the course of evolution through technologic advances in agricultural and industrial practices. A side effect of a number of these advances is the introduction of chemicals resulting from agricultural and industrial practices to the soil and water ecosystems and the resulting impact of these chemicals on organisms residing there. The effects of chemical pollution are three-fold: (1) a tendency toward simpli- fication of communities through the elimination of more sensitive species and their replacement by larger populations of tolerant species, (2) the change in species relationship within communities, whereby the species that earlier might have enjoyed only a minor niche dominated by other species are allowed to expand into a dominant role in the ecosystem by the disappearance of the control species, and (3) alterations in nutrient cycles, which may have a long- lasting effect on the basic composition of the ecosystem. Alteration of nutrient cycles may lead in turn to permanent changes in an ecosystem through erosion and leaching, which in turn change the basic physical stucture.

Effects of pollutants are seem promarily at the tops of food chains and are observed by man usually as changes in population levels of predator species. The chlorinated hydrocarbon pesticides and industrial chemicals, for example, may cause reproductive difficulties in birds, such as the peregrine falcon. Mink are highly sensitive to methyl mercury, while apparently other mammals are not so sensitive. Contamination with methyl mercury can thus alter the diversity and dominance characteristics of the ecosystem. Disturbances in the ecosystem can be detected in nutrient cycling even though no effects are measured in the diversity or population of the community. Several studies have now shown that changes in nutrients, such as nitrates, are more sensitive than biologic paramenters to chemical stress. This results from the fact that changes in nutrient pools must eventually directly affect the productivity of the entire ecosystem, even though the effects may not be measurable in biologic terms until a number of years later. The net effect of decreased diversity in an ecosystem is a more unstable system. Such communities are subject to wide fluctuations in populations of organisms and are more easily influenced by outside pressure such as chemical pollutants. This leads in turn to the necessity for man's further intervention in an attempt to stabilize the system, a process that historically has sometimes been self-defeating.

In terms of human ecology we are just beginning our efforts to understand the impact of chemicals in the environment. The full range of effects of the loss of species diversity in the ecosystem on man is yet to be understood. Changes in the dominance characteristics of eco-systems will have a major effect on man's activities as they cause him to change strategies of pest control, alter his use of water systems for consumption and recreational purposes, and change his perceptions of the aesthetic quality of the environment. Changes in nutrient cycling lead to the expenditure of energy and resources to correct resulting imbalances and to manage resource commitments and waste effluents. As man changes the agricultural ecosystem to his advantage, it is necessary to add nutrients in the form of phosphate and nitrate fertilizers frequently leading to additional imbalances and problems in managing the contamination of water systems by these nutrients that results. The effect of chemicals on man is known in many instances only indirectly through laboratory experimentaion with test organisms, such as laboratory animals, at high doses. The same chemicals in the environment will not necessarily affect man in the same direct ways since they are always found in the presence of other chemicals with which they may interact. Interactions in the environment may alter mobility, transport , availability, disposition, and toxicologic effect, thus completely altering the properties of a chemical in its interactions with biologic systems.